THERMODYNAMICS AND CHEMISTRY

A Non-Mathematical Treatise for Chemists and Students of Chemistry

MAXWELL PRESS

THERMODYNAMICS AND CHEMISTRY

A Non-Mathematical Treatise for Chemists and Students of Chemistry

P. Duhem

Correspondant de l'Institut de France; Professor of Theoretical Physics at the University of Bordeaux.

MAXWELL PRESS

Chennai Trichy New Delhi

This edition has been published in india by arrangement with Carson Books, UK

ISBN 978-93-5528-077-0 **Maxwell Press**

No. 44, Nallathambi Street,
Triplicane, Chennai 600 005

MJP 1302

Publisher : C. Janarthanan

Publisher's Note

The legacy of a country is in its varied cultural heritage, historical literature, developments in the field of economy and science. The top nations in the world are competing in the field of science, economy and literature. This vast legacy has to be conserved and documented so that it can be bestowed to the future generation. The knowledge of this legacy is slowly getting perished in the present generation due to lack of documentation.

Keeping this in mind, the concern with retrospective acquiring of rare books has been accented recently by the burgeoning reprint industry. Maxwell Press is gratified to retrieve the rare collections with a view to bring back those books that were landmarks in their time.

In this effort, a series of rare books would be republished under the banner, "Maxwell Press". The books in the reprint series have been carefully selected for their contemporary usefulness as well as their historical importance within the intellectual. We reconstruct the book with slight enhancements made for better presentation, without affecting the contents of the original edition.

Most of the works selected for republishing covers a huge range of subjects, from history to anthropology. We believe this reprint edition will be a service to the numerous researchers and practitioners active in this fascinating field. We allow readers to experience the wonder of peering into a scholarly work of the highest order and seminal significance.

MAXWELL PRESS

AUTHOR'S INTRODUCTION TO AMERICAN EDITION.

MY DEAR COLLEAGUE:

You wished that my little book *Thermodynamique et Chimie* appear in English, and you have been willing to undertake its translation. Both of these determinations give me pleasure.

I am glad, in the first place, to have my treatise rendered easily accessible to American students; one of the objects which I had in mind when writing it was to make the work of J. Willard Gibbs known and admired; I like to think it will contribute to enhance, within your active universities, the glory of your illustrious countryman.

Furthermore, this glory is more and more resplendent every day; more and more clearly the author of the *phase law* appears as the initiator of a chemical revolution; and many do not hesitate to compare the Yale College professor to our Lavoisier.

Chemists had fixed upon a certain number of properties by which they recognized a substance to be a definite compound; these characteristics are effaced by the *phase rule;* many substances, to which formulæ had been attributed, are erased from the number of combinations; chemical science as a whole needs a revision at which the laboratories of America and Europe are working most diligently.

Nevertheless, whatever be the outcome of this revolution, it seems to me there is injustice in making the glory of Gibbs consist in this alone, by seeing in him merely the author of the phase rule. In his immortal work, *On the Equilibrium of Heterogeneous Substances,* this rule is not all; it is but one theorem, and is accompanied by other propositions whose importance is not less; the theorems on indifferent points, the laws of dissociation of perfect gases, the properties of dilute solutions, the conditions of osmotic equilibrium, the theory of the voltaic cell, bear, not less than the phase rule, the mark of the genius of their author.

The phase rule is not, therefore, by a great deal, the whole of Gibbs's work; *a fortiori* it is not the whole of chemical thermodynamics; other ideas, other principles, play an essential rôle in the development of this science.

When a chemical system is studied, it is assuredly very important to determine its *variance,* whose value fixes the form of the equilibrium law for the system; but before even calculating the variance, it is expedient to answer this question: Is the equilibrium of the system stable, indifferent, or unstable?

Thermodynamics teaches us that no chemical equilibrium is unstable. The equilibrium conditions which the chemist meets may therefore be classed in two categories: those which are stable and those which are indifferent. This classification seems the most natural, this division the most radical which may be conceived. The systems in stable equilibrium possess a whole ensemble of properties which systems in indifferent equilbirium do not possess; the atter, in their turn, all have certain properties which are not met when systems in stable equilibrium are studied; this opposition has been exposed with very great clearness by your countryman, Dr. Paul Saurel, in the thesis he presented some time since at Bordeaux.

The equilibria of which I speak at present are always found at the common limit of two reactions opposed to each other; among the equilibrium states studied by the chemist they are only particular cases or, better, limiting cases. The states of equilibrium with which the chemist has actually to deal are those which I have studied under the name of *false equilibria.* At the beginning of his immortal work Gibbs has shown how important is the distinction between these states and those of which thermodynamics treats; to these last alone does the phase rule apply.

The classification, borrowed from the phase rule, of systems into *monovariant, bivariant, trivariant,* etc., is therefore of extreme utility; it arranges in admirable order a great number of questions in the discussion of chemical equilibria; it is, however, neither the only conceivable way of classifying these states of equilibrium nor the most essential. This is why I have not thought it wise to follow the example given by several recent treatises on chemical mechanics and base the plan of my book upon the phase rule alone.

The history of Gibbs's work appears to me to contain valuable lessons; the natural continuation of Lagrange's *Mécanique analytique*, it is a powerful algebraic attempt to express in equations the problems of Thermodynamics in the most general and abstract form; but here is this work of a mathematician overturning chemistry!

This example is quite fitting to indicate the close relationship uniting the various sciences, or, better, the innateness of the separations by which we keep apart the various intellectual disciplines. Being methods suitable to discover the varied aspects of the truth, which is single, they cannot be made independent of one another; whosoever by laziness or narrow-mindedness would pretend to use but a single one of these methods would certainly, by isolating it, risk rendering it sterile.

At the start the wholly algebraic doctrine of Gibbs did not find in the country of its creator experimentalists to transform it into chemical theory; it first found them in Holland. From this again, it seems to me, we may learn a lesson. The full discovery of the truth requires the concurrence of all peoples, their diverse intellectual aptitudes, their different ways of conceiving an idea, of developing it and of expressing it. In this respect exclusiveness would again be punished by sterility.

You agree with this, for you have judged it useful to mitigate the vigorous and assertive initiative which the American universities impart to their students by the discipline of equilibrium and of proportion which French teaching imposes; what you have tried in your own intellectual formation you wished to procure for others; you have desired to make known to your countrymen the book in which I have tried to expose, with French neutrality, ideas coming from America; I could not wish an interpreter better prepared to diffuse my thought.

Believe me, my dear colleague, yours most sincerely,

P. DUHEM.

PREFACE.

THE development which Thermodynamics has undergone during the past fifty years merits the attention of men engaged in the most varied studies.

The opinions, formerly admitted without protest, concerning the object and influence of physical theories have been overthrown. Mechanics has ceased to be the ultimate explanation of the inorganic world; it is now but a chapter, the simplest and most perfect, of a general discipline which rules all the transformations of matter; furthermore, it is no longer a question of discovering the nature and essence of these transformations, but merely of coordinating their laws by means of a small number of fundamental postulates. The philosopher follows with keen interest the phases of this, one of the most considerable evolutions which Cosmology has undergone.

Mathematical Physics, at the beginning of the nineteenth century, had furnished geometers with most beautiful and fruitful problems; the efforts made to solve these problems had given birth to more than one branch of modern analysis; but it might be feared that the veins worked by so many geniuses were exhausted. The new doctrine generalizes very greatly the statements of the problems formerly attacked; it posits entirely new ones, and in this way opens vast galleries to the researches of the mathematician.

The different branches of physics seemed isolated from each other; each of them invoked its own principles and depended upon particular methods. To-day the physicist recognizes he is not concerned with a bundle of branches independent of each other, but with a tree whose branches start from the same trunk; all the parts of science which he cultivates appear related as are the members of an organism.

Finally, the laws formulated by Thermodynamics impose a rational order upon the most confused chapters in Chemistry; a few simple, clear rules bring order out of what was a chaos; the circumstances under which the various reactions are produced, the conditions which stop them and assure chemical equilibrium, are fixed by theorems of a geometrical precision.

Thus the philosopher, the mathematician, the physicist, the chemist, are all equally eager to know the Thermodynamics of to-day, to understand clearly its principles, its methods, its results. But each of them is interested in a different aspect of this science; a separate treatise is necessary for each.

It is for the chemist these lessons are intended.

What the chemist expects above all from Thermodynamics are simple, clear rules, easy to use, which shall serve him as conducting thread through the frightful labyrinth of chemical facts already known, which shall guide him in the course of his researches, which shall show him exactly, for every reaction, the variable conditions at his disposal and the essential conditions which he is held to determine.

We have done our best to formulate these rules rigorously and clearly. We have accompanied each of them with numerous examples; in this way we have wished not merely to note their importance and fruitfulness, but also indicate the precautions to be taken in their application.

Is it sufficient, however, for the chemist, to have stated for him the propositions to which Thermodynamics leads, without analyzing the principles from which they arise? Many think so; others say so; we cannot believe it.

Just as it is unworthy of a man who thinks to take certain aphorisms as guides for his scientific activity without seeking to know the titles these aphorisms lay claim to, the sources whence come their authority, so this intellectual laziness would have, in practice, disastrous consequences.

It is often said there are no rules without exception. Concerning the rules which Thermodynamics lays down for Chemical Mechanics, it would be more exact to say that every rule follows after hypotheses, and that no hypothesis is legitimate outside of certain precise and determined conditions. In Physics there is no principle which is true at all times, in all places, under every circumstance. Now the field within which a rule applies with security has for boundaries the limits of exactness of the hypotheses upon which the rule depends. Whoever, therefore, does not know where a rule comes from runs the risk of employing it in cases where its usage is proscribed, and of finding in it, not a safe guide, but a false adviser.

This is why, before developing the laws of Chemical Statics and Dynamics, we have insisted upon examining the foundations upon which these sciences are built.

Our first five chapters are devoted to this examination; we have taken the greatest pains to free the statements of the primary ideas of Thermodynamics from all algebraic complications; the calculus is not used; *in fact we have supposed on the part of the reader no knowledge in mathematics or physics beyond that possessed by the graduate of a good high school.*

It is by means of the fundamental hypotheses of algebra that rules useful to the chemist are found; the mechanism of this deduction cannot, therefore, be separated from the mathematical formulæ by which alone it operates; not wishing to write for the geometer, we have not deemed it necessary to analyze the parts of this mechanism; but this omission is hardly of any importance to the chemist; when the latter has an exact knowledge of the conditions in which it is legitimate to use a principle, when he sees clearly the practical consequences which are related to this principle, he may, with entire confidence, trust in the chain whose two ends he holds in a firm hand; for the intermediate links, which he has not tested, have the rigidity of algebra.

Besides, if any one inclined and prepared for this study desires to fill in this chain and follow, in all its developments, this linking together of Chemical Mechanics, he may easily satisfy his desire to learn; we have elsewhere tried to help him in this.

We have given very considerable space to the most recent applications of Thermodynamics to Chemistry. We have, in particular, developed the applications of that admirable *phase law*, an algebraic theorem conceived by the genius of J. Willard Gibbs and which the chiefs of the Dutch school, Van der Waals, Bakhuis Roozboom, and Van't Hoff, have been able to rear into one of the most precious guiding principles of modern Chemistry.

We have studied with great care those systems with fallacious properties which have for a long time passed as definite compounds: mixed crystals, eutectic conglomerates, indifferent states of double mixtures. We have neglected nothing of what may put the experimenter on his guard against these appearances of chemical analysis.

We have not wished, on the other hand, that the exposition of these chapters of chemical mechanics, so new and so full of promise, detract from the study of the discoveries which have received the sanction of time and which to-day are classic. Disciple of Moutier, Debray, Troost, Hautefeuille, Gernez, we have not wished either to forget or let be forgotten that the union of Thermodynamics and Chemistry was accomplished in France in the laboratory of the immortal Henri Sainte-Claire Deville.[1]

P. DUHEM.

BORDEAUX, January 2, 1902.

---

[1] It was, in fact, the 19th of June, 1871, that J. Moutier communicated to the Academy of Sciences a note entitled *Sur la dissociation au point de vue de la Thermodynamique* (*Comptes Rendus*, v. 72, p. 759). In this note the possibility of applying Clapeyron and Clausius' equation to the dissociation of calcium carbonate studied by Debray was remarked for the first time. The same year Peslin published, in the *Annales de Chimie et de Physique* (4th Series, v. 24, p. 208), an article in which Clapeyron and Clausius' equation, combined with the measurements of dissociation tensions given by Debray, gave the numerical value of the heat of formation of calcium carbonate; this value agreed with Favre's thermochemical determinations.

After these first writings, which inaugurated the union of Thermodynamics and Chemistry, we have to wait until 1873 to find, concerning this question, new investigations, due in part to J. Moutier and in part to Horstmann (Liebig's *Annalen der Chemie und Pharmacie*, v. 170, p. 192).

TABLE OF CONTENTS.

CHAPTER III.

CHAPTER IV.

CHAPTER V.

CHAPTER VI.

CHAPTER VII.

CHAPTER VIII.

CHAPTER IX.

CHAPTER X.

CHAPTER XI.

CHAPTER XII.

CHAPTER XIII.

CHAPTER XIV.

CHAPTER XV.

CHAPTER XVI.

CHAPTER XVII.

CHAPTER XVIII.

CHAPTER XIX.

CHAPTER XX.

THERMODYNAMICS AND CHEMISTRY.

CHAPTER I.

WORK AND ENERGY.

1. Work of a force applied to a movable point.—An elementary definition of work is given when the following conditions are fulfilled:

Under the action of a force F, constant in magnitude and direction (Fig. 1), a material point moves a length $\overline{MM'}=l$ in the direction of the force.

M M′ F

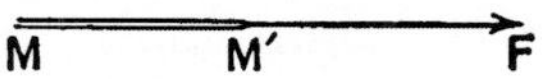

Fig. 1.

We call the *work* done by the force F the product Fl, with the + sign if the material point moves in the direction of the force, with the − sign if the material point moves in the opposite direction.

This definition is sufficient to fix the unit of work. When a material point under the action of a constant force equal to the unit of force moves, in the direction of the force, a unit's distance, the work done is equal to unity.

In the C.G.S. system, in which the unit of length is the centimetre and the unit of force the dyne, the unit of work is the *dyne-centimetre* or *erg*.

The preceding definition may be so generalized that it applies to the case in which under the action of a force F, constant in magnitude and direction (Fig. 2), a material point moves a length $\overline{MM'}=l$ along a straight line which makes a certain angle with the direction of the force. In this case, the force F is projected upon the path of the material point; let f be such projection. The work of the force F is then the product fl, either + or − in sign according as the point moves in the d.rection of the force f or in the opposite sense.

Fig. 2.

Let α be the angle (less than two right angles) that the direction of the force F makes with the direction $\overline{MM'}$ of the displacement of the material point; by the definition of the *cosine* of an angle, the work we have just described will be represented in magnitude and direction by the formula

$$W = lF \cos \alpha. \tag{1}$$

This formula may be interpreted in another way. $l \cos \alpha$ represents the projection of the d splacement $\overline{MM'}$ on the direction of the force F, this projection having its proper sign according to the ordinary convention, that is, counted positive or negative according as it is in the direction of the force F or oppositely directed. We may then say that when a point, under the action of a force constant in magnitude and direction, moves in a straight line, the work of the force is, in magnitude and direction, the product of the magnitude of the force by the projection of the displacement of the material point on the direction of the force.

Suppose that a material point describes a rectilinear segment $\overline{MM'} = l$ (Fig. 3) under the action of the force F which rests constant in magnitude and direction during this displacement; that it then describes a second rectilinear segment $\overline{M'M''} =$ ' under the action of another force F'; then a third segment $\overline{M''M'''} = l''$ under the action of a third force F'', and so on. The work done while the material point describes the broken path $MM'M''M''' \ldots$ is, by definition, the sum of the work done during the separate rectilinear components of the broken path; this work has then the value

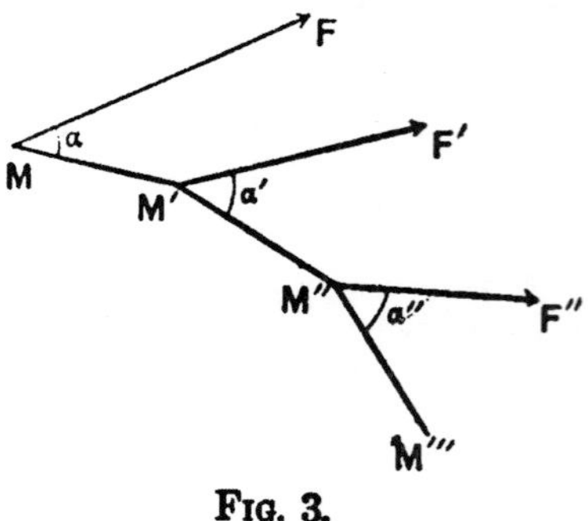

Fig. 3.

$$W = Fl \cos \alpha + F'l' \cos \alpha' + F''l'' \cos \alpha'' + \ldots \tag{2}$$

Consider now a material point which describes a curved path MN (Fig. 4), while the force F acting upon it changes continually in magnitude and direction.

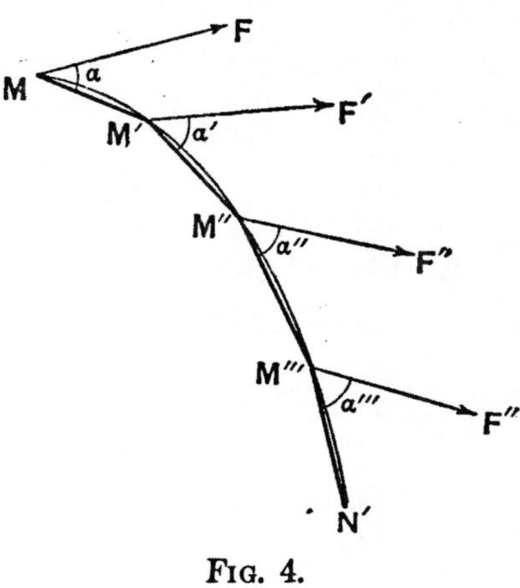

Fig. 4.

Within the curve MN inscribe a rectilinear path $MM'M''M'''$ . . . N. Let F, F', F'', F''' be the values of the variable force at the instants the material point occupies the positions M, M', M'', M'''. Let α be the angle of the two directions F and MM', α' the angle of the two directions F' and $M'M''$, and so on.

If the material point considered describes the rectilinear segment MM' under the action of the constant force F, the segment $M'M''$ under the action of the force F', the segment $M''M'''$ under the action of the force F'', finally the segment $M'''N$ under the action of the force F''', the work done will have, according to formula (2), the value

$$Fl\cos\alpha+F'l'\cos\alpha'+F''l''\cos\alpha''+F'''l'''\cos\alpha'''.$$

Suppose now that we increase indefinitely the number of points of division M', M'', M''', . . . marked on the curve MN, causing each of the segments $\overline{MM'}$, $\overline{M'M''}$, $\overline{M''M'''}$, . . . to approach zero. The broken path $MM'M''M'''$ . . . will approach the curved path MN. At the same time the preceding sum will approach a limit. This limit is, by definition, the work done while the material point describes the curved path MN.

2. Application to the case of weight.—Let us apply these definitions to the very simple case of a movable point under the action of its weight.

Let M be a material point; if we denote its mass by m and the intensity of gravity by g, its weight P is a constant, vertical force, directed downwards and of magnitude $P=mg$.

Suppose that under the action of its weight the point describes the rectilinear path MM' (Fig. 5). Let z be the distance of the point M, and z' the distance of the point M', above an arbitrary horizontal plane H; it is clear that the projection $\overline{MN}$ of the path MM' on the direction of the force P, that is on the vertical directed downwards, is represented in magnitude and direction by the difference $(z-z')$. The work done during the rectilinear displacement MM' of the material point has then for value

$$P(z-z')=mg(z-z').$$

It is equal to the product of the weight of the material point and the height of the fall.

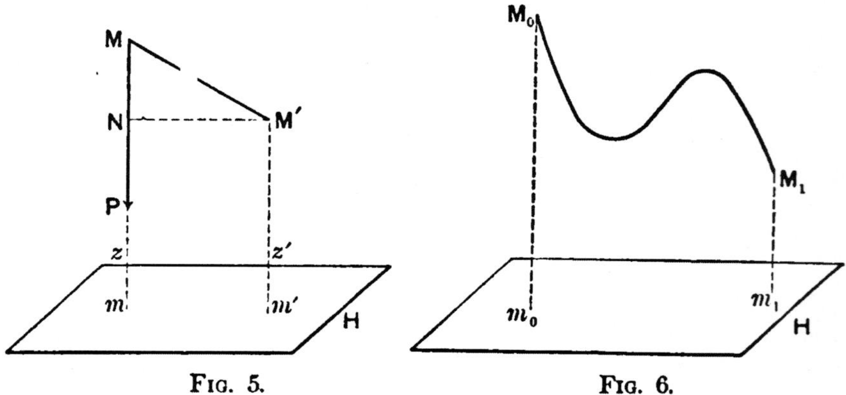

FIG. 5.   FIG. 6.

Consider now a point having mass which moves along a broken line; by definition the work done will be the sum of the products obtained by multiplying the weight of the material point by the height of the partial fall relative to each of the rectilinear segments which compose the broken trajectory; it is then clear that the work done when a material point describes a broken path is obtained by multiplying the weight of the material point by the height of the total fall.

By the method of limits, this result may be immediately extended to the case of a material point which describes any curved path whatever M_0M_1 (Fig. 6). Let z_0 be the initial distance $\overline{M_0m_0}$, and z_1 the final distance $\overline{M_1m_1}$, of this point above an arbitrary horizontal plane H; the work done by gravity during the displacement of the point will have the value

$$W=mg(z_0-z_1). \tag{3}$$

3. Work of forces applied to a system.—When a mechanical system, composed of material points, is displaced and deformed under the action of forces which act on these various points, the work done is, by definition, the sum of the work done in the displacements of the various material points.

4. Case of gravity.—Let us apply this definition to the work done when a system of material points is deformed and displaced under the action of gravity.

Let m, m', m'' . . . be the masses of the various material points composing the system; z_0, z_0', z_0'' . . . the distances of the various points above an arbitrary horizontal plane H, at the beginning of the displacement.

In virtue of the equality (3) and of the preceding definition, the work done will have for value

$$W = mg(z_0 - z_1) + m'g(z_0' - z_1') + m''g(z_0'' - z_1'') + \ . \ . \ .$$

But $mgz_0 + m'gz_0' + m''gz'' + \ . \ . \ .$ is the sum of the moments, with respect to the plane H of the weights of the various material points, taken in their initial positions; it can be shown that this sum is equal to the moment with respect to the plane H of the total weight of the system, this weight being applied at the initial position of the centre of gravity.

If, then, we write $M = m + m' + m'' + \ . \ . \ .$ for the total mass of the system, and Z_0 for the initial distance of the centre of gravity of the system above the horizontal plane H, we have

$$mgz + m'gz_1' + m''gz'' + \ . \ . \ . = MgZ_1.$$

Whence it follows that

$$W = Mg(Z_0 - Z_1). \qquad (4)$$

When a material system having weight is deformed and displaced in any manner whatever due to its weight, the work done is the product of the weight of the system by the height of the fall of the centre of gravity.

5. Case of a system under uniform pressure.—Take another example where it will be easy to calculate the work done in a deformation of the system.

A reservoir R (Fig. 7) contains a gas; a pipe, supposed cylindrical, is soldered to this reservoir; the pipe encloses a liquid which exerts a certain pressure on the gas; S is the level of the liquid; at all points of the surface S the pressure exerted on the gas is vertical, directed upwards, and has the same value P; we will suppose that this value stays constant, while the level of the liquid falls from S to S', and we shall calculate the work done.

Take on the surface S a very small area s; it is acted upon by a vertical force, directed upwards, of value Ps. In the displacement of the surface S the various points of the area s follow paths parallel to the generatrices of the cylinder, so that the surface s is brought to s'; the projection of a point in the surface s' on the direction of the displacement force will be, in magnitude and direction, equal to $-h$, designating by h the height of the cylinder included between S and S'. The force applied to the surface s does work equal to $-Phs$. To obtain the total work it will be necessary to compute the separate work done on the areas s into which the area S may be decomposed and find the sum of these work components. Now, in the various terms of this sum, $-Ph$ will be a common factor; noting that the sum of the areas such as s is equal to the area S, we see that the work sought is $-PhS$. But hS is the volume of the cylinder included between the surfaces S and S'; it is thus the increase undergone, during the modification considered, by the volume of the gas; if V_0 is the initial value of this volume and V_1 the final value, we have evidently $hS = V_1 - V_0$, and the work sought has the value

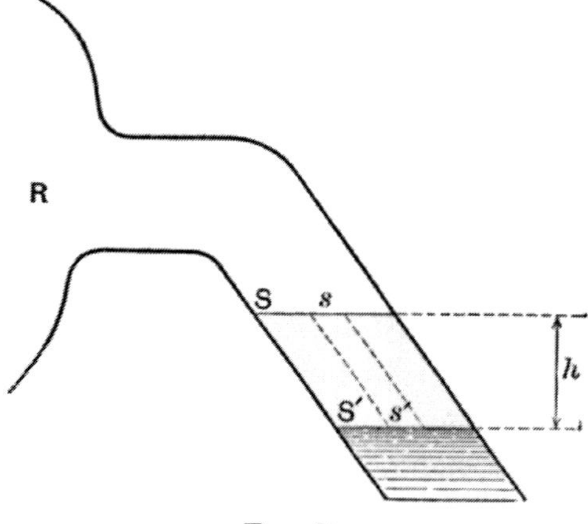

FIG. 7.

$$W = P(V_0 - V_1).$$

This is the product of the pressure on the gas and the decrease (or increase with sign changed) of the volume that it occupies.

The result that we have just found may be generalized in a manner that we shall state without demonstration.

Every time that a body passes from volume V_0 to the volume V_1, while the surface sustains a uniform normal pressure of constant value P, the work done has the value

$$W = P(V_0 - V_1). \tag{5}$$

Suppose now that a body changes its volume while it sustains a pressure which is again normal and uniform, but whose magnitude varies in the following manner:

While the volume of the body passes from the value V to the value V', the pressure keeps the invariable value P; it has the value P' while the volume changes from V' to V''; the value P'' while the volume changes from V'' to V''', and so on. The work done during such a transformation is the sum of the work done during the partial transformations, each of which is performed under constant pressure; it has for value

$$W = P(V - V') + P'(V' - V'') + P''(V'' - V''') + \dots \tag{6}$$

This work is obtained by multiplying each of the diminutions of volume sustained in the system by the constant pressure that it supports during this diminution of volume and taking the sum of the products thus obtained.

By the principle of limits this method may be extended to the case in which the value of the pressure varies in a continuous manner during the change in volume which the system undergoes.

6. Geometrical representation of preceding results.—Take a right angle VOP (Fig. 8) whose sides OV, OP are *coordinate axes* and their intersection O the *origin of coordinates.* One of the axes OV is considered as having the direction from left to right; the other axis OP is directed vertically upwards; the first is called the *axis of abscissæ* and the second the *axis of ordinates.*

Consider a system occupying the volume V under the pressure P. Along the axis of abscissæ lay off from the point O a length OV, measured by the member V; along the axis of ordinates lay off from O a length OP, measured by the member P; through the point V draw a parallel to OP and through P a parallel to OV; these two lines meet in a point M which represents the system; this point has for *abscissa* the member V and for *ordinate* the member P; these two members are the *coordinates* of the point.

Suppose that, under a pressure which remains fixed, the volume of a system passes from the value V_0 to the less value V_1. The point representing the system, whose ordinate keeps the invariable value P, will describe (Fig. 9) a straight line parallel to OV

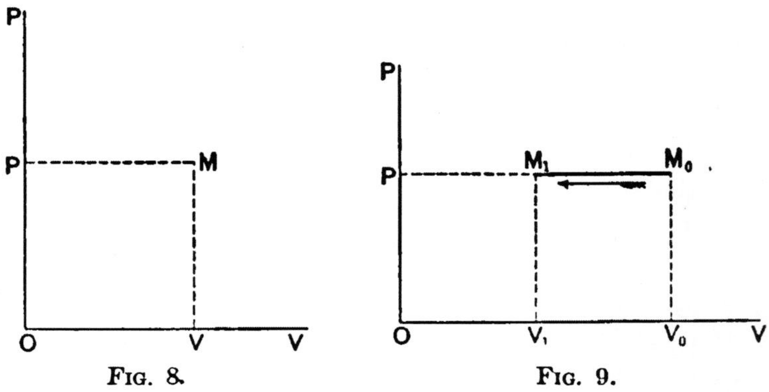

FIG. 8. FIG. 9.

from right to left starting from the point M_0, abscissa V_0, to the point M_1, abscissa V_1. The work done, according to equation (5), will be $W=P(V_0-V_1)$, so that we shall have, in magnitude and direction,

$$W=\text{area } M_0M_1V_1V_0M_0.$$

Suppose, again, that the volume of the system diminishes continuously from V_0 to V_1; but let us not suppose that this diminution takes place under constant pressure; imagine that the volume decreases from V_0 to V' under the constant pressure P, from V' to V'' under the constant pressure P', from V'' to V''' under the constant pressure P'', finally from V''' to V_1 under the constant pressure P'''. The point representing the system (Fig. 10) will describe the broken line $M_0m'M'M''m''M'''m'''M_1$. The work done is, by formula (6),

$$W=P(V_0-V')+P'(V'-V'')+P''(V''-V''')+P'''(V'''-V_1).$$

This equality may also be written

$$W=\text{area } M_0m'M' \ldots M'''m'''M_1V_1V_0M_0.$$

Suppose finally that each of the diminutions of volume undergone by the system between the extreme values V_0, V_1 becomes smaller and smaller; that these diminutions are more and more numerous; that from one of these diminutions of volume to the next the pressure changes less and less: the volume of the system will tend to diminish in a continuous manner; the broken path $M_0m'M' \ldots M_1$, described by the representative point (Fig. 11), will approach the curved path $M_0M' \ldots M_1$, drawn from *right*

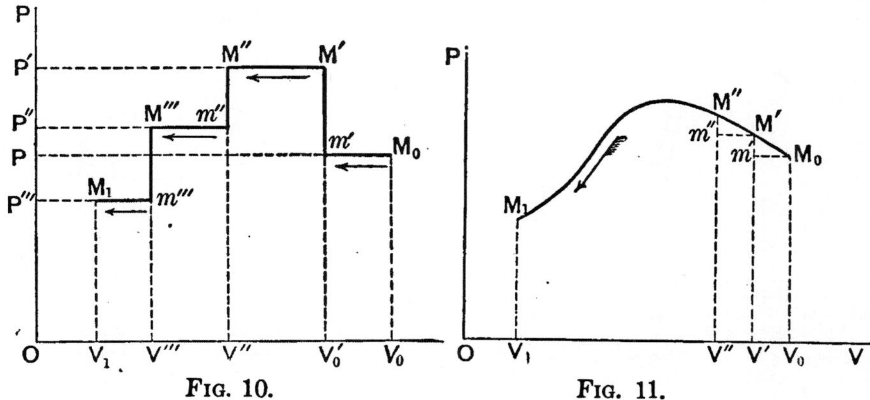

FIG. 10. FIG. 11.

to left. From Art. 5 it follows that the work done during this continuous transformation will be represented by the area bounded in part by the curve traced:

$$W = \text{area } M_0M_1V_1V_0M_0 \tag{7}$$

Instead of decreasing indefinitely, the volume of the system might increase indefinitely from the value V_0 to the greater value V_1, while the pressure varies in a continuous manner; the point representing the system (Fig. 12) would describe from *left to right* the curved line M_0M_1. A series of considerations analogous to the preceding would show that the work done has for value

$$W = -\text{area } M_0M_1V_1V_0M_0. \tag{7'}$$

Thus, when the volume of the system has a single direction of variation between the two extreme values V_0 and V_1, the area $M_0M_1V_1V_0M_0$ gives always the absolute value of the work done; but this absolute value must have the sign + or − according as the representative point describes the curve M_0M_1 from right to left or left to right.

The volume of a system which is being modified may not always change in the same sense; it may, for example, decrease at first from V_0 to V_1, then increase from V_1 to V_2, and decrease finally from V_2 to V_3. The representative point (Fig. 13) will first describe the curve M_0M_1 from right to left, then the curve M_1M_2 from left to right, and finally the curve M_2M_3 from right to left,

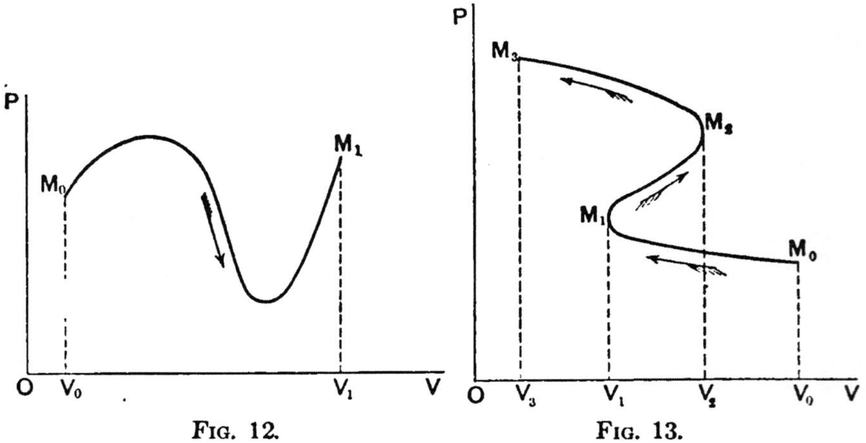

FIG. 12. FIG. 13.

the points M_0, M_1, M_2, M_3 having respectively for abscissæ the numbers V_0, V_1, V_2, V_3.

To each of the three partial transformations where the volume has a single direction of variation, one or the other of the two preceding theorems is applicable; be ides, the work done in the total transformation is the sum of the work done in each of the three partial transformations; this work has then the value

$$\begin{aligned} W = &\text{area } M_0M_1V_1V_0M_0 \\ &-\text{area } M_1M_2V_2V_1M_1 \\ &+\text{area } M_2M_3V_3V_2M_2. \end{aligned}$$

Evidently an evaluation of this kind will always be applicable, even for the most complicated cases.

Let us apply this mode of evaluation to the case in which the system occupying in the first place the volume V_0 under the pressure P undergoes a modification which brings it back to the same volume and pressure.

Suppose, as in Fig. 14, that the curve consists of a single loop described by the representative point *in a direction opposite to that of the motions of the hands of a watch.* We shall have, by equation (8),

$$\begin{aligned} W = & \text{area } M_0M_1V_1V_0M_0 \\ & -\text{area } M_1mM_2V_2V_1M_1 \\ & +\text{area } M_2M_0V_0V_2M_2, \end{aligned}$$

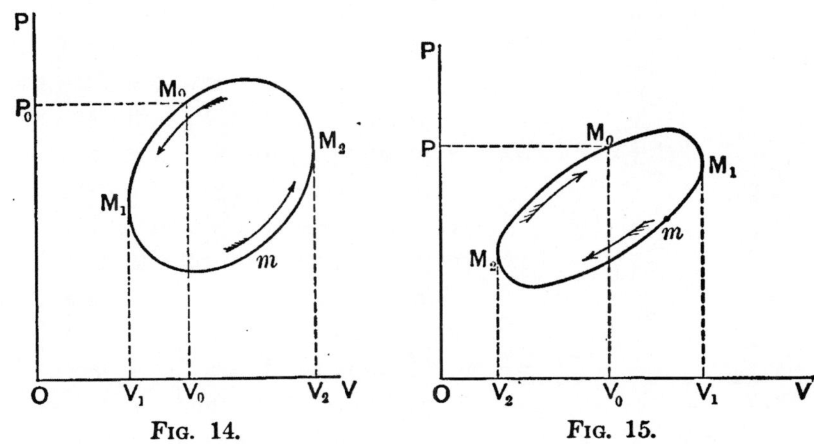

FIG. 14. FIG. 15.

or, in uniting the two areas having the sign +,

$$W = \text{area } M_2M_0M_1V_1V_2M_2 - \text{area } M_1mM_2V_2V_1M_1;$$

or finally

(9) W = area included by the closed curve.

If, as in Fig. 15, the closed path of the representative point is composed of a single loop described *in the direction of the motion of the hands of a watch,* we find in an analogous manner that the work done has the value

(9′) $W = -$area included by the closed curve.

7. Application to a gas that obeys Mariotte's Law.—Suppose we compress from volume V_0 to the less volume V_1 a mass of gas whose temperature is kept constant, and that the gas obeys Mariotte's Law.

Let P_0 be the pressure supported by this gas at the moment it occupies the volume V_0; when it occupies the volume V, included between V_0 and V, it will support a pressure P given by the equation

$$(10) \qquad P = \frac{P_0 V_0}{V}.$$

This equation gives the ordinate of the representative point M when V is the abscissa (Fig. 16); we notice that this ordinate is the greater as V diminishes; it increases beyond limit when V_0 tends towards zero and approaches zero when V increases indefinitely. The locus of the point M is what the geometers call a branch of an *equilateral hyperbola* having for *asymptotes* the two axes OP and OV.

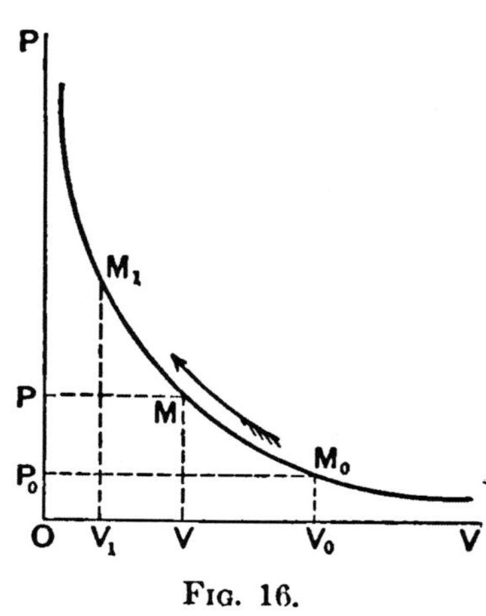

Fig. 16.

From what we have seen in Art. 6, the work done by the pressure when the gas is compressed from the volume V_0 to the volume V_1 is measured by the area bounded by the four following lines:

1°. The arc M_0M_1 of the equilateral hyperbola defined by equation (10);

2° and 3°. The two lines V_0M_0, V_1M_1, parallel to the axis of ordinates;

4°. The part V_0V_1 of the axis of abscissæ.

This area may be found by geometry and has the value

$$(11) \qquad W = 2.3025 P_0 V_0 \log \frac{V_0}{V_1},$$

the symbol log designating a logarithm in Briggs's tables.

8. In some cases, the work of the forces applied to a system depends only on the initial and final states of this system.—The comparison of equation (4), which gives the work due to gravity, with equation (6), which gives the work of a normal and uniform pressure, leads to one of the most important principles in mechanics.

In order to know the work done by gravity acting on a system of given mass, it is sufficient to know the initial and final height of the centre of gravity; it is quite superfluous to know the form of the path the body has taken in its fall and the changes of shape or of dimensions that it may have undergone during this fall. *The work done by gravity depends exclusively on the initial and final states of the system on which it acts.*

9. **In general, the work done by the forces applied to a system depends upon every modification the system undergoes.**— It is not true, as in the case of weight, that the principle announced in Art. 8 holds for the work done by any force.

Imagine that a system, under a uniform, normal pressure, passes from the initial state in which it occupies the volume V_0 at the pressure P_0 to a final state in which it occupies the volume V_1 at the pressure P_1.

From the initial to the final state it may pass in an infinitely great number of ways; let us choose two and calculate for each the work done by the pressure:

1°. The system expands, at constant pressure P_0, from the volume V_0 to the volume V_1; then, without change of volume, change the pressure from the value P_0 to the value P_1; suitable changes of temperature allow accomplishing these two operations; during the first part of the transformation the pressure does work equal to $P_0(V_0-V_1)$, by equation (5); during the second part the representative point describes a straight line parallel to OP, so that no work is done. Work done by the pressure, in the modification considered, has the value

$$W_1=P_0(V_0-V_1).$$

2°. Under the constant volume V_0 the pressure changes from P_0 to P_1; then, at the constant pressure P_1, the volume changes from V_0 to V_1; during the first part of the modification the pressure does no work; the work done during the whole modification reduces to that done during the second part and, from equation (5), has the value

$$W_2=P_1(V_0-V_1).$$

Since P_1 and P_0 are different, the work W_2 done during the second modification is not equal to W_1, the work done during the first.

Suppose that the system is brought, in the initial and final conditions, to the same temperature; we may make it pass from the first to the second without variation of temperature; this third modification will differ, in general, from the two preceding; if the system is a gas that obeys Mariotte's Law, we may find the value of the work done during this last modification ; this value, given by equation (11), will differ again from the two preceding.

Thus, *the value of the work done when a system, supporting a uniform, normal pressure at every point of its surface, passes from a given initial state to a definite final state does not depend upon these two states alone, but also upon all the intermediate states through which the system has passed.*

These two examples we have just given justify the following propositions:

In general, when a material system, acted upon by certain forces, undergoes a certain modification, it is not sufficient, in order to find the work done by these forces during this modification, to know the initial and final states of the system; it is necessary to know also the series of intermediate states the system has passed through and the forces acting upon the system during each of these states.

However, *if the forces acting upon the system belong to certain particular categories, in order to find the work done during a modification it is sufficient to know the initial and final states; any knowledge of the intermediate states is superfluous.*

10. Potential.—Let us consider this particular class of forces, of which gravity is an example

Consider a system acted upon by such forces, and let us choose once for all a definite state of the system which we shall call the state α.

Suppose the system passes into this state α starting from another state x; the forces considered do work which is entirely determined by the knowledge of the initial state x and of the final state α; in order to change the value of this work, it would be necessary to change at least one of the two states α or x; since we suppose the state α chosen in an irrevocable manner, we may say that the value of this work depends merely on the choice of the state x; we shall designate it by Ω_x. We propose to calculate the work W, done by the forces considered, when the system passes from any initial condition 0 to any final condition 1, and with this object let us consider the following modification:

The system passes first from the state 0 to the state 1, then from the state 1 to the state α; by definition, the forces considered do work, during the first part of the modification, equal to W, and during the second part the work Ω_1, or in all the work $(W+\Omega_1)$.

But the modification considered causes the system to pass from the state 0 to the state α; the work done has the value Ω_0.

We have then the equation

$$W+\Omega_1=\Omega_0,$$

or

(12) $$W=\Omega_0-\Omega_1.$$

When the work of the forces applied to a system is entirely determined by a knowledge of the initial and final states, we may represent each state of the system by a quantity Ω, variable from one state to another; the work done in the course of a certain modification is equal to the excess of the initial value of Ω over the final value of this same quantity.

The quantity Ω is called the *potential* of the forces which act upon the system.

Instead of saying that the work of the forces which act on a system depends only on the initial and final states of the system, we say that these forces have a potential.

11. Potential due to gravity.—The comparison of equations (4) and (12) show us that gravity depends upon a potential and that this potential is

(13) $$\Omega=Mgz,$$

M being the total mass of the body and z the distance of the centre of gravity above an arbitrary horizontal plane H.

12. Forces which have a potential in virtue of the restrictions imposed upon the system.—A system of forces which, in general, does not admit a potential may, in certain cases, have one provided that certain restrictions are imposed on the transformations to be studied.

Thus, we have seen that if the surface of a body supported a uniform normal pressure, this pressure, in general, does not admit a potential.

But suppose that we compel the pressure to keep an absolutely invariable value P; the work of the forces which act upon the system will be given by equation (5), which taken together with equation (12) furnishes the following proposition:

When the forces which act on a system reduce to a normal, uniform, constant pressure P, the forces admit the potential

$$\Omega = PV, \tag{14}$$

where V is the variable volume of the system.

Suppose, similarly, that a system supporting a normal, uniform pressure is enclosed in a reservoir of invariable volume. From equation (6), a transformation taking place under these conditions occasions no work due to the pressure. So that, *when a system supporting a normal, uniform pressure keeps an invariable volume, the forces which act upon it have a zero potential.*

13. Energy.—The notions of work and of potential have an extreme importance in all parts of mechanics, in statics as well as in dynamics; we shall have some idea of this when treating of *The Theorem of Energy,* as we shall do now.

Consider a system formed of material points of mass m, m', m'', . . .; suppose this system in motion, and at a given instant let V, V', V'' . . . be the velocities of the various points; multiply the mass of each point by the square of its velocity; find the sum of the products thus obtained; finally, take half of this sum; we form the quantity

$$W = \tfrac{1}{2}(mV^2 + m'V'^2 + m''V''^2 + \ . \ . \ .).$$

This quantity is called the *kinetic energy* of the system at the instant considered.

The mass of a material point is essentially positive; so is the square of the velocity, unless the point is at rest; consequently *the kinetic energy of a system is a quantity essentially positive, unless the system is at rest, in which case the kinetic energy is equal to zero.*

The Theorem of Energy may be stated as follows:

For a system in motion during any lapse of time, the increase in kinetic energy, during this time, is equal to the work done by the forces which act on the system.

Suppose that, during the lapse of time considered, the kinetic energy changes from the value W_0 to the value W_1; let W be the work done by the forces which act upon the system; the preceding theorem is expressed by the equation

$$W = W_1 - W_0. \tag{15}$$

14. Principle of virtual displacements.—From this theorem we deduce immediately an important corollary.

Suppose that the initial velocities of the different points of the system are a zero; W_0 will be equal to 0 and equation (15) reduces to $W = W_1$. If the velocities of the various parts of the system are not all zero in the final state, W_1 is positive, so that the following proposition may be stated:

When a system, starting from an initial state in which the velocities of its various points are equal to zero, is set in motion and attains a state in which the velocities of the various points are not all zero, the work done by the forces which act on it is assuredly positive.

When a system starts from a state 0 where its various points have zero velocities and is put in motion, it may be that this motion brings it into another state 1 where its various points still have zero velocities. Thus, when a pendulum is pulled aside a certain angle to the left from its equilibrium position, and is set swinging without initial velocity, it returns towards the equilibrium position, passes this, and at the instant it attains to the right an angle equal to the previous devia ion to the left, the velocities of its various points all recover the value 0. But it is clear that between these two states in which the various points have their velocities equal to zero, the system passes through a continuous series of states in which certain of the points have velocities different from 0; in other words, a system cannot quit a state in which the velocities of its various points are zero except by first traversing a series of states in which these velocities are not all zero.

From the preceding theorem, taking account of the last remark, the following proposition is evident:

A system placed without initial velocity in a given state cannot be put in motion unless the beginning of this motion corresponds to positive work of the forces applied to the system.

This proposition will furnish us a means of recognizing if a material system, acted upon by given forces, is surely in equilibrium in a given state.

Let us consider the various ways in which the system may be removed from the given state and the different displacements from this state it may be supposed to take; each of these imagined displacements is called a *virtual displacement.*

If the commencement of every virtual displacement imposed on a system in a given state corresponds to negative or zero work of the forces applied to the system, the latter, placed without initial velocity in the state considered, will remain in it in equilibrium.

This proposition has the name of the *Principle of Virtual Displacements.*

15. Conservation of energy. Conservative systems.—The various propositions we have stated assume a remarkable and simple form when the system studied is acted upon by forces which admit a potential Ω. In this case, the work done by these forces is expressed by means of equations (12) and (15), and the theorem of energy becomes

$$W_1 - W_0 = \Omega_0 - \Omega_1. \tag{16}$$

When a system acted upon by forces which depend on a potential is in motion, the increase in kinetic energy during a certain time is equal to the decrease in the potential during the same time.

Equation (16) may also be written

$$\Omega_1 + W_1 = \Omega_0 + W_0, \tag{16'}$$

and the preceding theorem be stated as follows:

When a system is in motion under the action of forces which admit a potential, the sum of the potential and the kinetic energies maintains an invariable value throughout the duration of the motion.

Consider a system having a motion which, after a certain time, brings it back to its initial state, such as a pendulum after a complete oscillation; the potential will retake its initial value; according to the preceding proposition, the same will be true of the kinetic energy; from this, the name *principle of the conservation of energy* is given to the preceding proposition, and the name *conservative systems* is sometimes given to systems acted upon by forces that depend on a potential.

To illustrate, let us apply equation (16′) to a system formed of a single material point of mass m. By equation (13) we shall have $\Omega = mgz$, z being the distance of the point above an arbitrary horizontal plane, while the kinetic energy reduces to $W = \frac{1}{2}mV^2$. Noting that the mass m of the material point is a quantity of invariable magnitude, we see that equation (16′) becomes

$$gz_1 + \frac{V_1^2}{2} = gz_0 + \frac{V_0^2}{2}$$

and includes the following proposition:

Every time that a material particle passes through a given level it has a velocity whose direction may have changed, but whose value is ever the same.

16. Principle of virtual displacements for conservative systems. Stability of equilibrium.—In order that the forces which act upon a conservative system do positive work, it is necessary and sufficient that the potential of the system decreases. From this, starting from a certain state in which the velocities of its various points are all zero, a conservative system cannot attain another state in which some at least of these velocities are different from 0, unless the potential has a less value in the second state than in the first.

In particular, a conservative system cannot quit a state in which its various points have zero velocities, unless the potential decreases, at least at the beginning of the motion. If then all the virtual displacements that may be imposed upon a conservative system in a given state commence by causing the potential to increase or to keep its value constant, the system is assuredly in equilibrium.

All the virtual displacements imposed on a system in a given state will commence necessarily by increasing the potential if, in this state, the potential has a smaller value than in all neighboring states; whence the following theorem:

A conservative system is surely in equilibrium if placed, without initial velocity, in a state in which the potential has a minimum value.

It is shown in mechanics by methods that we cannot expose here that the following proposition is true:

A conservative system is surely in STABLE *equilibrium when it is, without initial velocity, in a state in which the potential has a minimum value.*

Let us apply these last propositions to a system having weight. The potential of such a system is given by the equation

$$\Omega = Mgz, \tag{13}$$

in which M is the mass of the system and z the distance of the centre of gravity above an arbitrary horizontal plane; the weight Mg of the system being a positive quantity of invariable value, the preceding theorems give us the following propositions, which have an important place in the development of mechanics:

A system having weight, placed without initial velocity in a certain state, cannot quit it without occasioning a lowering of its centre of gravity.

A system having weight is surely in stable equilibrium when its centre of gravity is as low as possible.

CHAPTER II.

QUANTITY OF HEAT AND INTERNAL ENERGY.

17. Loss of kinetic energy and generation of heat in a machine left to itself.—Consider what in the older mechanics was called a *machine*, that is, a more or less complicated arrangement of bodies each of which is movable, but keeping invariable its size, form, and state.

Put this machine in motion; its various parts having velocities different from 0, its kinetic energy will also have a value W_0 that is positive.

Abandon this machine to itself or, to speak more precisely, let the system move so that no force acts upon it; will the machine continue indefinitely in motion?

If, to reply to this question, we consult the theorems announced in the preceding chapter, we shall be led to conclude that the motion of our machine will be indefinitely conserved; thus, the forces which act upon the system, being always zero, will do no work; the kinetic energy of the system will keep forever an invariable value; positive at the start, it can never become zero and, set going, our machine will always have some part in motion.

Now, this reply is evidently contrary to the experience that every machine teaches us: left to itself, it stops sooner or later. The kinetic energy of the system, far from remaining constant, decreases incessantly. To keep the machine in motion, to keep the kinetic energy constant, it is necessary to submit this machine to the action of forces which do positive work upon it unceasingly.

We see from these observations that real machines differ notably from ideal machines to which the theorems of the first chapter apply. Other differences may also be made evident; these ideal machines are made of bodies whose state is invariable; on the contrary, the bodies of which a real machine is composed undergo a modification: they heat up, their temperature increases; to keep them in an invariable state, to keep their temperature constant, it is necessary to cool them, to oblige them to cede a certain quantity of heat to foreign bodies.

We are thus led to recognize, for every system in motion which is formed of bodies kept in an invariable state and temperature, the two following properties, the first of which contradicts the theorem of energy, and the second cannot be predicted by the principles set forth in the first chapter:

1°. The work W done by the forces which act upon the system during a definite lapse of time surpasses the increase $(W_1 - W_0)$ which the kinetic energy undergoes during the same time:

$$W-(W_1-W_0)>0.$$

2°. The quantity of heat liberated by the system during the same lapse of time is positive:

$$Q>0.$$

18. Mechanical equivalent of heat.—A natural and simple hypothesis consists in supposing that the two quantities $(W-W_1+W_0)$ and Q, whose sign is always the same, are in a constant ratio, or, in other words, to suppose that *there exists a definite positive number E, such that*

$$\frac{W-W_1+W_0}{Q}=E. \quad (1)$$

This number E is called the *mechanical equivalent of heat.* The correctness of the preceding proposition is subordinate to the following conditions: the bodies which constitute the system are movable, but the size, form, state, and temperature of each remains invariable; W represents the work of all the forces that act upon the system, including the forces that originate in bodies exterior to the system as well as forces by which the various parts of the system act upon each other.

19. Principle of the equivalence of heat and work.—The usefulness of the preceding principle becomes greatly limited from the obligations to consider only collections of bodies each of which keeps an invariable size, form, temperature, and state; so that it is advantageous to replace this principle by another which is analogous but not identical.

To state this new principle, we will make use of an idea that we shall meet often in what follows, the idea of *closed cycle.*

When, at the close of a transformation, the bodies which compose a system reassume the arrangement, size, form, temperature, physical and chemical states that they possessed at the beginning of the transformation, we say that by the effect of this transformation the system has described a closed cycle.

The various parts of the system cannot possess the same velocities at the beginning and at the end of a closed cycle, so that the kinetic energy cannot retake its initial value at the moment the system finishes describing a closed cycle.

Suppose that a system describes a closed cycle during which there is set free a quantity of heat Q while the kinetic energy passes from the value W_0 *to the value* W_1; *let* W_e *denote the* EXTERNAL WORK, *that is, the work done by the forces that bodies foreign to the system exert on this system; we shall then have the equation*

$$\frac{W_e - W_1 + W_0}{Q} = E. \tag{2}$$

This equation may also be written

$$W_e - W_1 + W_0 = EQ. \tag{2'}$$

This fundamental hypothesis is what we shall call the *Principle of the equivalence of heat and mechanical work.*

20. Value of the mechanical equivalent of heat.—Depending on this principle and on the results of various experiments whose description is beyond the scope of this work, Robert Mayer, Joule, and a great number of other physicists have occupied themselves with the determination of the value of the mechanical equivalent of heat E; they found that approximately

$$E = 425,$$

when the *gramme-metre* is taken as unit of work and the *gramme-calorie* or *small calorie* as unit of heat.

If, instead of the gramme-metre, the unit of work taken is the *erg*, which is 98100 times smaller, the numerator of equation (2) is represented by a number 98100 times greater; the value of the denominator does not change if the small calorie is kept as the heat-unit; the new value of the mechanical equivalent is then

$$E=425\times 98100=41692500.$$

21. Extension of the principle of the equivalence of heat and work to an unclosed cycle.—The principle of the equivalence between heat and work, such as we have stated it in Art. **19**, requires that the modification to which it is desired to apply it is a closed cycle; this restriction is sometimes embarrassing; we shall obviate this by modifying the statement of the principle.

To shorten our terms, let us denote by e the quantity

$$Q-\frac{W_e-W_1+W_0}{E}.$$

The principle of equivalence may be thus announced:

If a system describes a closed cycle, the quantity e, calculated for the whole of this closed cycle, is equal to zero.

Keeping this in mind, imagine a system to pass from a given initial state 0 to a final state 1, by a definite series of modifications; the quantity e, calculated for this first transformation, takes a certain value that we may call ε. Suppose now that the system returns from the state 1 to the state 0 by an equally definite series of modifications; the quantity e has, for this second transformation, a certain value η; for the total transformation, the quantity e has the value $(\varepsilon+\eta)$, but since this total transformation is a closed cycle, we have

$$\varepsilon+\eta=0.$$

Suppose now that the system passes from the same initial state 0 to the same final state 1, but by a different series of modifications than in the preceding case; for this transformation the quantity e will be equal to ε'; then bring it back from state 1 to state 0 by the same transformation as in the preceding case, for which the value of e is η; this time we shall have

$$\varepsilon'+\eta=0.$$

The two equalities require that

$$\varepsilon = \varepsilon',$$

so that the principle of the equivalence between heat and work leads to this consequence: in whatever manner a system passes from a given initial state to a definite final state, the quantity e keeps the same value.

If is clear, also, that this proposition includes also the principle of the equivalence between heat and work; thus, if the system describes a closed cycle, e will have, for this cycle, the same value as for any other transformation that takes the system in the same initial state and returns it to this state; or the same value as if the system were not modified at all, and this last value is evidently 0; so that, for any closed cycle, $e=0$; this is precisely the original statement of the principle of the equivalence between heat and work.

We are therefore authorized to state in the following way the PRINCIPLE OF THE EQUIVALENCE BETWEEN HEAT AND WORK:

The quantity

$$e = Q - \frac{W_e - W_1 + W_0}{I} \tag{3}$$

has a value which depends on the initial and final states of the system, but not on the nature of the modifications undergone by the system in order to pass from this initial state to this final state.

22. Internal energy.—It is now sufficient to follow a reasoning similar to that which gave us the idea of potential (Art. 10) to transform the preceding statement into the following:

For each state X of the system there may be made to correspond a quantity U_x such that, for every modification undergone by the system from the initial state 0 to the final state 1, this equality holds:

$$e = U_0 - U_1;$$

or again, replacing e by its value, defined by equation (3),

$$EQ - W_e = W_0 - W_1 + E(U_0 - U_1). \tag{4}$$

Such is the form given in 1850, by R. Clausius, to the principle of the equivalence between heat and work.

To the quantity U_x Clausius gave the name *Internal Energy* of the system in the state X.

The preceding equalities show us that U_x is a quantity of the same kind as e and Q, that is, it may be measured in heat-units.

Certain writers, instead of keeping a special name for the quantity U_x, prefer to consider the product EU_x, which they call the *potential energy* of the system in the state X; the potential energy is then a quantity of the same kind as W_e and W_1, or a quantity measured in units of work; we may say that *the potential energy is the equivalent of the internal energy.* The quantity

$$EU_x + W_x$$

may be called the *total energy* of the system in the state X.

23. Principle of the conservation of energy.—That an interest attaches itself to this nomenclature is seen from the following:

Imagine a system isolated in space and consequently deprived of the action of any external body.

No external body exerting force upon this system, every modification that it undergoes is accompanied by external work equal to 0:

$$W_e = 0.$$

No external body can give to or take heat from this system; whatever modification it undergoes, the quantity of heat set free can be neither positive nor negative; it is zero:

$$Q = 0.$$

Thus, for such a system, equation (4) becomes

$$EU_1 + W_1 = EU_0 + W_0,$$

and is stated:

Whatever modification an isolated system undergoes, this modification leaves the total energy of the system unchanged; what of the potential energy is lost is gained in kinetic energy, and conversely.

This proposition bears the name of the *principle of the conservation of energy.* It is quite unnecessary to assign to it a vague metaphysic sense or a mysterious origin; it is simply a special case of a *physical law,* the law of the equivalence of heat and work.

24. Gases which obey Mariotte's Law. Absolute temperature.—Before applying the principle of the equivalence of heat and work to the various problems of chemical calorimetry, we shall make an application which we shall employ in what follows.

All are acquainted with the statement of Mariotte's Law: at a given temperature, the product of the pressure supported by a mass of gas and the volume it occupies is a constant. Everybody knows also that this law does not apply rigorously to any gas, but that the gases distant from the conditions in which they liquify obey this law approximately.

It results from this law, as is taught in elementary classes, that the two coefficients of expansion, under constant pressure and constant volume, of a gas obeying this law have the same value, and this with whatever thermometer used; besides, according to the observations of Charles and of Gay-Lussac, this value is the same for all gases which follow sensibly Mariotte's Law; finally, if the thermometer chosen is one constructed with one of these gases, the value in question evidently does not depend upon the temperature.

If the temperature chosen is a centigrade temperature, the coefficient of expansion of gases which obey Mariotte's Law is sensibly, according to Regnault's observations, $\alpha = \frac{1}{273}$.

At two different temperatures t and t', and under two different pressures P and P', the same mass of gas will occupy the volumes V and V', joined by the relation

$$\frac{PV}{1+\alpha t} = \frac{P'V'}{1+\alpha t'}. \tag{6}$$

This relation may also be written

$$\frac{PV}{\frac{1}{\alpha}+t} = \frac{P'V'}{\frac{1}{\alpha}+t'}.$$

The expressions $T = \frac{1}{\alpha} + t$ and $T' = \frac{1}{\alpha} + t'$, which appear in this equation, are called, for reasons which we cannot go into here, the *absolute temperatures* corresponding to the centigrade temperatures

t and t'. The absolute temperature which corresponds to a given centigrade temperature is thus obtained by adding to the latter the number $\frac{1}{\alpha}$, equal to about 273. The difference between two absolute temperatures is equal, by definition, to the difference between the two corresponding centigrade temperatures:

$$T' - T = t' - t. \tag{7}$$

Making use of absolute temperatures, equation (6) may be written

$$\frac{PV}{T} = \frac{P'V'}{T'}. \tag{6'}$$

25. Expansion of a gas in vacuo. Gay-Lussac's experiment. —The gases which obey Mariotte's Law approximately, obey also, approximately, another law illustrated by an old experiment of Gay-Lussac, repeated by Regnault, by W. Thomson, and by Joule.

Two reservoirs R, R' (Fig. 17), one of volume v, the other of

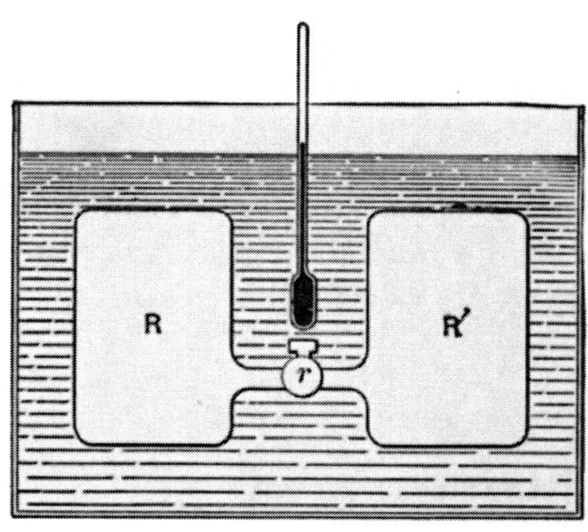

FIG. 17.

volume v', are immersed in a calorimeter; a tube fitted with a stop-cock r joins the two reservoirs.

At the beginning of the experiment, the cock r is closed, the reservoir R is full of gas and the reservoir R' is empty; the water in the calorimeter has a certain temperature.

When the cock r is opened, a part of the gas that filled the reservoir R passes into the reservoir R'.

At the end of the experiment the two reservoirs R and R' are filled with gas at the same density; it is observed that the water in the calorimeter has the same temperature as at the beginning.

Let us apply to this transformation the principle of the equivalence of heat and work, expressed by equation (4).

At the beginning, as at the end of the transformation, the gas is at rest; the reservoirs do not move; the kinetic energy is zero:

$$W_0=0, \qquad W_1=0.$$

The exterior forces applied to this system, the pressures exerted on the external surface of the reservoirs, have done no work, for this surface has not moved:

$$W_e=0.$$

Finally, since the water in the calorimeter returns to its initial temperature, the system considered has neither given out nor absorbed heat:

$$Q=0.$$

Then equation (4) gives us

$$U_1=U_0.$$

The final value of the internal energy of the system is equal to its initial value.

The system studied consists of two parts: the matter which forms the walls of the reservoirs, and the gas enclosed by these reservoirs; the matter which forms the walls of the reservoirs has undergone no modification, so that its internal energy has not changed in value; thus the internal energy of the gas should have, at the end of the transformation considered, the same value as at the start. Nevertheless, this gas has undergone a change: its temperature, it is true is the same at the beginning and at the end of the transformation; but its volume, which was v, has become $(v+v')$; we are thus led to announce the following law:

If he volume of a mass of gas is varied, while its temperature is kept constant, the internal energy of this body undergoes no change in value.

26. Perfect gases.—This law has the same limitations as Mariotte's Law; no gas obeys it rigorously, but certain gases taken in proper conditions obey it very closely, and these gases are precisely those that follow Mariotte's Law most exactly; such are, under the ordinary conditions of temperature and pressure, hydrogen, oxygen, nitrogen, carbonic oxide, nitrous oxide; we give to these gases the name of *gases near to the ideal state,* keeping the name *perfect gas* for an ideal gas which would obey exactly the two laws just cited.

Although a perfect gas does not exist any more than does an absolutely rigid body, the study of ideal gases in thermodynamics is as legitimate and as useful as the study of rigid bodies in mechanlcs; it furnishes an image simplified and approximate, it is true, but, in a great number of cases, practically sufficient, of the properties of real gases.

27. Specific heat at constant volume. Internal energy of a perfect gas.—Bring a gas from temperature t_0 to temperature t, keeping it in a reservoir of volume V. Let M be the mass of this gas expressed in grammes. It absorbs, during the transformation considered, a quantity of heat whose value q is expressed in gramme-calories or in small calories; other things being equal, q is proportional to M.

The ratio $\frac{q}{M(t-t_0)}=c$ is, by definition, the mean value, between the temperatures t_0 and t, *of the specific heat at constant volume* of the gas con idered.

Let us apply equation (4) to the transformation considered.

The volume of the gas remains constant; then, by formula (6) of the preceding chapter, the work done by the pressure the foreign bodies exert on the gas is equal to 0:

$$W_e=0.$$

The gas is at rest at the beginning and at the end of the transformation, so that $W_0=0$, $W=0$.

The quantity Q of heat set free is equal to $-q$, so that

$$Q=-Mc(t-t_0).$$

If then U_0 is the internal energy of the mass of gas at volume V and temperature t_0, and if U is the internal energy of this same mass at the same volume and at temperature t, equation (4) becomes

$$U-U_0=Mc(t-t_0). \tag{8}$$

The same gas might be heated between the same limits of temperature, but keeping it at a volume V' different from V. From what we have seen in the preceding article, if the gas is a perfect one, the values of U_0 and U will undergo no change; and similarly for c. Thus *the mean value between two given temperatures of the specific heat at constant volume of a perfect gas does not depend upon the value of this volume at which the gas is kept.*

If we suppose the temperatures t_0 and t read on a thermometer constructed of a sensibly perfect gas, we shall have, by equation (7)

$$t-t_0=T-T_0,$$

and equation (8) becomes

$$U=U_0-McT_0+McT. \tag{9}$$

28. Specific heat at constant pressure. Robert Mayer's Relation.—Let us take now the same mass M of the same gas, always expressed in grammes; in equilibrium at the temperature t_0, under a certain pressure P, it occupies a certain volume V_0; in equilibrium at the temperature t, and under the same pressure P, it occupies a volume V. By heating in this way at constant pressure, from the temperature t_0 to the temperature t, it absorbs a certain quantity of heat q'. By definition, the ratio $\frac{q'}{M(t-t_0)}=C$ is the mean value, between the temperatures t_0 and t, of *the specific heat at constant pressure* of the gas considered.

Let us apply equation (4) to the transformation considered.

The kinetic energy of the gas is zero at the beginning and at the end of the transformation:

$$W_0=0, \qquad W=0.$$

The pressure being kept constant, the external work is given by equation (5) of the preceding chapter:

$$W_e = P(V_0 - V).$$

The quantity of heat set free has the value $Q = -q'$, or

$$Q = -MC(t - t_0).$$

If, then, we denote by U_0 the initial value and by U the final value of the internal energy of the gas, equation (4) becomes

$$U - U_0 = MC(t - t_0) - \frac{P}{E}(V - V_0).$$

This equality is general.

Suppose now that we have to deal with a gas near to the perfect state; the value of the internal energy at a given temperature is constant; if then the temperatures t_0 and t are the same in the preceding equation as in equation (8), the values U_0 and U will be also the same in these two equalities; equating them gives

$$M(C - c)(t - t_0) = \frac{P}{E}(V - V_0).$$

A very simple transformation will allow us to put this equation into a celebrated form.

Suppose the temperature read upon a thermometer of a sensibly perfect gas; we have $t - t_0 = T - T_0$, T and T_0 being the absolute temperatures which correspond to the centigrade temperatures t and t_0; from this the preceding equation becomes

$$M(C - c)(T - T_0) = \frac{P}{E}(V - V_0). \tag{10}$$

Further, let

θ be the absolute temperature of melting ice, about 273°;

π the normal atmospheric pressure in the system of units chosen;

σ the volume occupied by one gramme of the gas studied, at the temperature of melting ice and under normal atmospheric pressure.

In the same conditions, M grammes of this gas occupy a volume $M\sigma$, such that the equation (6′) may be written

(11) $$\frac{PV}{T}=\frac{PV_0}{T_0}=\frac{M\pi\sigma}{\theta}.$$

These equations in their turn may be written somewhat differently.

Whatever be the nature and mass of the gas considered, the quotient

(12) $$\frac{\pi}{\theta}=R$$

has the same value, which depends solely on the mechanical units employed; thus in the system in which the unit of length is the metre and the unit of force the gramme-weight, we have

$$R=\frac{10333000}{273}=37848.$$

In the C.G.S. system, in which the unit of length is the centimetre and the unit of force the dyne, we have

$$R=\frac{1033.3\times 981}{273}=3713.$$

By means of equation (12) the equations (11) may be written

(13) $$PV=MR\sigma T. \qquad PV_0=MR\sigma T_0,$$

and equation (10) becomes simply

(14) $$C-c=\frac{R\sigma}{E}.$$

This is Robert Mayer's Relation, given by the illustrious Hielbronn physician at the dawn of thermodynamics, and of which we shall see many applications.

29. Influence of temperature on the specific heats of perfect gases. Clausius' Law.—According to the definition which has been given of the two quantities c and C, the value of each of them may very well depend upon the two extreme temperatures t and t_0; besides, we know that the value of c does not depend upon the constant volume at which the gas is heated; but we are ignorant if the value of C depends on the constant pressure supported by the gas while it is carried from the temperature t_0 to the temperature t.

Mayer's relation teaches us that the difference $(C-c)$ has, for a given gas, an absolutely definite value; it follows then, since for a given gas the value of c can depend only upon the temperatures t_0 and t, that it is the same with C; whence this first proposition: *The specific heat at constant pressure of a given perfect gas does not depend upon the value of the constant pressure under which the gas is heated.*

Besides, if we determine in what manner one of the two quantities C, c, depends on the two temperatures t_0, t, Mayer's relation will show us immediately how the other depends on these same temperatures.

Regnault measured, under constant atmospheric pressure, the mean specific heats of various gases at different temperatures.

For air the following numbers were found:

Between $t_0 = -30°$ C.	and $t = +10°$ C.,	$C = 0.23771$
0°	100°	0.23741
0°	200°	0.23751

More recently Witkovski has obtained for the same gas at the same pressure:

Between $t_0 = +20°$ C.	and $t = +98°$ C.,	$C = 0.2372$
$-77°$	$+16°$	0.2374
$-102°$	$+17°$	0.2372
$-170°$	$+18°$	0.2427

According to Regnault, the mean specific heat of hydrogen, under atmospheric pressure, has the same value between 0° and 200° as between $-30°$ and $+110°$.

These observations, joined to Mayer's relation, justify the following law, called Clausius' Law.

The specific heat at constant pressure and the specific heat at constant volume, for a given gas near to the perfect state, have fixed values.

The truth of this law for very high temperatures has been questioned, notably by Mallard and Le Chatelier; but their very complex experiments can be interpreted only by means of a certain number of hypotheses, some of which are in contradiction with known facts; thus these authors suppose that carbonic acid gas is undecomposable by heat up to 1800° and water vapor up to 2300°, which is contrary to the direct observations of H Sainte-Claire Deville. We think then that Clausius' law may be conserved until further notice, even for very high temperatures.

30. Evaluation of the mechanical equivalent of heat.—It is evident that Mayer's equation (14) may also be written

$$(14') \qquad E=\frac{R\sigma}{C}\cdot\frac{\frac{C}{c}}{\frac{C}{c}-1}.$$

All the quantities in the second member are accessible to experiment.

We have seen how R may be calculated.

We have mentioned Regnault's experiments which gave C.

If u is the volume occupied by 1 gramme of air under normal conditions of temperature and pressure, if Δ is the density of the gas considered with respect to air, we have $\sigma=\frac{u}{\Delta}$, so that σ may be determined.

Finally, under the pressure P, at the temperature t, sound is propagated in the gas considered with a velocity V, which, by Laplace's formula, has the value

$$V=\sqrt{R\sigma T\frac{C}{c}},$$

$T=273+t$ being the absolute temperature which corresponds to the centigrade temperature t. The experimental determination of the velocity V allows the computation of $\frac{C}{c}$.

We see then that equation (14′) gives a means of calculating the mechanical equivalent of heat; it is the method which led Robert Mayer to the first evaluation ever published of this quantity; before Mayer, Sadi Carnot had obtained a value of the mechanical equivalent, probably by the same method.

CHAPTER III.

CHEMICAL CALORIMETRY.

31. The quantity of heat set free by a system which undergoes a transformation does not depend solely upon the initial and final states.—The chemists who occupied themselves with calorimetry, from the time of Lavoisier and Laplace down to the time of the foundation of thermodynamics, have all admitted and made use of the following law:

The quantity of heat set free by a system which undergoes a transformation depends solely upon the initial and final states of the system and not at all upon the intermediate states.

This law may also be stated as follows:

When a system passes through a closed cycle, the liberation and absorption of heat are so compensated that the total quantity of heat set free is equal to 0.

A reasoning similar to that used in Art. 21 would prove the equivalence of these two statements.

It is easy to see that this law is not compatible with the principle of the equivalence between heat and work.

Let us cause a system to pass from an initial state 0 to a final state 1 and suppose it at rest in both of these states; we shall then have $W_0=0$, $W_1=0$, and, by equation (4) of the preceding chapter, the quantity of heat set free will be

$$Q=U_0-U_1+\frac{W_e}{E}. \tag{1}$$

The difference (U_0-U_1) has a value which depends exclusively upon the initial and final states of the system. But in general this is not so for W_e, nor for $\frac{W_e}{E}$, because (Arts. 9 and 10) the external forces which act on the system do not in general admit a potential. We must then, contrarily to the preceding law, announce the following proposition:

The quantity of heat set free by a system that undergoes a transformation does not depend solely upon the initial and final states, but also upon all the steps of the transformation.

32. **Example from the study of perfect gases.**—Let us consider immediately an example of this.

M grammes of a perfect gas are taken at the temperature t_0, under the pressure P; they occupy a volume V_0. At the constant pressure P this mass of gas is heated to the temperature t_1, higher than t_0; it then occupies a volume V_1 greater than V_0; at the same time there is set free the quantity of heat

$$Q = -MC(t_1 - t_0).$$

This gas may be brought from the same initial to the same final state by another process, as follows:

1°. It is heated, at the constant volume V_0, from the temperature t_0 to the temperature t_1, during which operation it absorbs a quantity of heat $Mc(t_1 - t_0)$.

2°. The reservoir of volume V_0, containing the gas, is put in communication with an empty reservoir of volume $(V_1 - V_0)$ and the temperature is allowed to return to t_1; according to Gay-Lussac's experiments, this operation occasions neither liberation nor absorption of heat.

The second transformation therefore produces a total liberation of heat,

$$Q' = -Mc(t_1 - t_0).$$

Although these two transformations bring the system from the same initial to the same final state, they do not give the same liberation of heat; we have, in fact,

$$Q' - Q = M(C - c)(t_1 - t_0),$$

or, according to Robert Mayer's relation [equation (14) of the preceding chapter],

$$Q' - Q = \frac{MR\sigma}{E}(t_1 - t_0).$$

33. Case in which the quantity of heat set free by a system depends solely upon the initial and final states.—The law stated at the beginning of Art. 31 is then in general false; nevertheless there are particular cases in which it is true.

Let us take equation (1), which gives us the quantity of heat set free by a system when it passes from a state 0 in which its kinetic energy is zero to a state 1 in which its kinetic energy is also zero. In order that the value of this quantity may depend solely upon the initial and final states and in no wise upon the intermediate states, it is necessary and sufficient that the same is true of W_e; in other terms (Arts. 9 and 10), *in order that the quantity of heat set free by a system which undergoes transformation may depend solely upon the initial and final states, it is necessary and sufficient that the external forces which act upon the system admit a potential.*

If Ω is this potential, we have [chap. I, eq. (12)]

$$W_e = \Omega_0 - \Omega_1,$$

and equation (1) becomes

$$Q = U_0 + \frac{\Omega_0}{E} - U_1 - \frac{\Omega_1}{E}. \tag{2}$$

The systems studied by the chemist may be regarded, in most cases, as undergoing a single external action, that of a normal, uniform pressure; this pressure (Art. 9) does not in general admit a potential; nevertheless, we may impose upon the transformations of the system studied restrictions such as were admissible in Art. 12; it is thus in the two following particular cases:

1°. *The external pressure remains a constant value P.* The potential of the external actions is then [Chap. I, eq. (14)] $\Omega = PV$, and equation (2) becomes

$$Q = U_0 + \frac{PV_0}{E} - U_1 - \frac{PV_1}{E}. \tag{3}$$

2°. *The volume occupied by the system rests constant.* The external pressure then admits the potential (Art. 12) $\Omega = 0$ and equation (2) becomes

$$Q = U_0 - U_1.$$

The chemist has then the right to use the law that the early thermo-chemists regarded as general, when they make use of one or the other of the two particular cases that we have just defined; they are, happily, the cases which are the most common in investigations.

1°. Very often all the transformations of the system studied take place in an open calorimeter, that is to say, under atmospheric pressure; this last being practically constant, the first of the above conditions is satisfied.

2°. Very often, also, all the modifications of the system studied are produced inside of the same combustion-chamber or within the same calorimetric bomb; during such transformations the volume occupied by the system does not change, so that the second condition is fulfilled.

34. Utility, in chemical calorimetry, of the preceding law.— In all cases in which the law stated in the preceding article is applicable it renders very great services to chemical calorimetry.

Suppose that a system sets free a quantity of heat q during a certain modification m which brings it from the state 0 to the state 1, a quantity of heat Q during a modification M which brings it from the state 1 to the state 2, finally a quantity of heat Q' during a modification M' that brings it from the state 0 to the state 2.

Suppose, besides, that the modifications m, M, M' are accomplished in such conditions that the external forces admit a potential; for example, if the external forces reduce to a normal, uniform pressure, suppose the three modifications m, M, M' accomplished either under the same pressure or at the same *volume*.

The series of modifications m and M on the one hand, the modification M' on the other hand, bring the system from the same initial state 0 to the same final state 2; each should set free the same quantity of heat

$$q+Q=Q', \tag{5}$$

or

$$q=Q'-Q, \tag{6}$$

Now it may happen that the modification M does not readily admit of calorimetric determinations, while the modifications M and

M', on the contrary, are easily produced within a calorimeter. The measurement of the two quantities of heat Q and Q' joined to equation (6) will give the data for the determination of the heat q.

Similarly, if the modifications m, M are adapted to calorimetric measurements while the modification M' is not so adapted, equation (5) allows the determination of the quantity of heat Q' from the measurement of the quantities q and Q.

This remark was first made by Berthollet, who gave the following application:

To determine the quantity of heat $(-Q')$ absorbed when a certain quantity of salt melts a certain quantity of ice at atmospheric pressure. This modification M' causes the system to pass from the state 0, in which the ice exists separately, at 0°, to the state 2, composed of a solution brought also to 0°.

Under atmospheric pressure let us melt the mass of ice considered, the salt remaining isolated; this modification m causes the system to pass from the state 0 to the state 1, formed of liquid water and of salt separated from one another and both at 0°; we can measure the quantity of heat that the ice absorbs.

Still at atmospheric pressure, dissolve the salt in water; this modification M causes the system to pass from the state 1 to the state 2; it may be easily produced in a calorimeter, so that the quantity of heat absorbed $(-Q)$ may be measured.

Whence, by equation (5), the quantity of heat sought $(-Q')$ may be found; the slowness of the modification M' would not have permitted its determination directly.

Let us take an application of equation (6); it applies to cases in which all the modifications of the system take place in a reservoir of constant volume.

Suppose that 12 grammes of diamond (C) at 0° are in the presence of 16 grammes of oxygen (O_2); this is the state 0 of the system.

By an incomplete combustion (modification m) the diamond combines with 16 grammes of oxygen, forming the mixture $CO+O$, brought to 0°, which is the state 1; it is desired to know the quantity of heat q set free by this reaction; this cannot be done directly, because it is impossible to regulate the combustion so that the product corresponds exactly to the preceding formula.

But, as was done by Berthelot and Matignon, the two following reactions may be realized in a calorimetric bomb:

1°. The complete combustion of the oxide of carbon (modification M) which causes the system to pass from the state 1, formed by the mixture of 28 grammes of carbon monoxide and 16 grammes of oxygen ($CO+O$), to the state 2, formed by 44 grammes of carbon dioxide (CO_2) brought to 0°. This modification sets free a quantity of heat

$$Q=68200 \text{ calories.}$$

2°. The complete combustion of the diamond (modification M') which causes the system to pass from the state 0 to the state 2. This modification sets free a quantity of heat

$$Q'=94300 \text{ calories.}$$

Equation (6) is here applicable and gives

$$q=26100 \text{ calories.}$$

35. Exothermic and endothermic reactions.—Here is another application of equation (6), made in 1852 by Favre and Silbermann.

The state 0 of the system is formed by 12 grammes of carbon (C) and 88 grammes of gaseous nitrous oxide ($2N_2O$), at the temperature 0°. Without change of volume, pass to state 1 (modification m) formed by 12 grammes of carbon in the presence of a mixture of 56 grammes of nitrogen and 32 grammes of oxygen ($4N+2O$), the whole brought to 0°. The modification m sets free a quantity of heat q which it is proposed to find.

For this we determine:

1°. The quantity of heat Q set free by the combustion (modification M) of 12 grammes of carbon in the mixture $4N+2O$; the state 2 of the system is formed by 56 grammes of nitrogen and 44 grammes of carbonic acid gas ($4N+CO_2$), brought to 0°.

2°. The quantity of heat Q' set free by the combustion (modification M') of 12 grammes of carbon in 88 grammes of nitrous oxide; the system passes from the state 0 to the state 2.

All the experiments of Favre and Silbermann were performed at atmospheric pressure.

q is then given by equation (6).

Now, the measurements of Favre and Silbermann have shown that the quantity Q' was greater than the quantity of heat Q; the quantity of heat q is therefore positive; thus, in the presence of a mass of carbon which takes no part in the modification m and of which, consequently, it is not necessary to take account, *the decomposition of nitrous oxide into nitrogen anh oxygen liberates heat.*

This result caused surprise at the epoch when Favre and Silbermann obtained it; up to that time, in fact, it was believed that every chemical *combination liberated* heat, and that every chemical *decomposition absorbed* heat.

Since this period, Favre and Silbermann's observation has been accurately repeated; by using the calorimetric bomb and operating, consequently, at constant volume, Berthelot has made the following measurements:

1°. Modification M':

$$CO+N_2O=CO_2+2N,$$
$$Q'=88800 \text{ calories.}$$

2°. Modification M:

$$CO+O+2N=CO_2+2N,$$
$$Q=68200 \text{ calories.}$$

It is concluded that the modification m,

$$CO+N_2O=CO+2N+O,$$

sets free a quantity of heat

$$q=20600 \text{ calories.}$$

Also, for reasons that would be too long to discuss here, it is admitted that, in this last reaction, the presence of the carbonous oxide may be neglected, as it takes no part in the reactions, so that the decomposition, at constant volume, of 44 grammes of nitrous oxide *liberates* 20600 calories.

There are many examples of the liberation of heat by chemical decomposition. Let us cite two taken from among the determinations made by Berthelot by means of the calorimetric bomb.

First example: Decomposition of nitric oxide.
Operation M:

$$2CN + 4O + 4N = 2CO_2 + 6N,$$
$$Q = 261800 \text{ calories.}$$

Operation M':

$$2CN + 4NO = 2CO_2 + 6N,$$
$$Q' = 349200 \text{ calories.}$$

Operation m:

$$2CN + 4NO = 2CN + 4N + 4O,$$
$$q = 87400 \text{ calories.}$$

Thus 30 grammes of nitric oxide, in decomposing at constant volume, set free $\frac{87400}{4} = 21850$ calories.

Second example: *Decomposition of acetylene.*
Operation M':

$$C_2H_4 + 5O = 2CO_2 + H_2O \text{ liquid,}$$
$$Q' = 314900 \text{ calories.}$$

Operation M:

$$2C + 2H + 5O = 2CO_2 + H_2O \text{ liquid,}$$
$$Q = 94300 \times 2 + 69000 = 257600 \text{ calories.}$$

Operation m:

$$C_2H_2 + 5O = 2C + 2H + 5O,$$
$$q = 57300 \text{ calories.}$$

Thus 26 grammes of acetylene, decomposing at constant volume, liberate 57300 calories.

When, in certain circumstances, the *formation* of a compound from its elements *liberates* heat, or when its *decomposition absorbs* heat, the compound is called an *exothermic compound* in the circumstances considered.

Water, hydrochloric acid, carbonous oxide, carbonic anhydride, are, in all circumstances, exothermic compounds.

When, in certain circumstances, a compound is *formed* with *absorption* of heat or is *decomposed* with *liberation* of heat, it is with an *endothermic compound* that one has to deal.

Nitrous oxide, nitric oxide, acetylene are endothermic in all circumstances.

36. Heats of formation at constant pressure and at constant volume.—Imagine that a mixture of two bodies A and B combine and furnish 1 gramme of the compound C. The quantity of heat set free by this combination depends upon the conditions in which it is produced.

Suppose, in the first place, that during the combination the temperature t remains constant and that the pressure also has a constant value P; let L be the heat set free during this combination; *L is, at the temperature t and constant pressure P, the heat of formation of the compound C.*

Suppose, in the second place, that during the combination the temperature t and the volume V occupied by the system are kept constant; let λ be the heat set free during the combination; *λ is called the heat of formation of the compound C, at the temperature t and at the constant volume V.*

It may happen that, in the conditions considered, the elements A and B cannot enter into combination, but on the contrary the compound C decomposes into its elements.

If, at the constant temperature t and at the constant pressure P, 1 gramme of the body C is decomposed into its elements A and B, and if this reaction *absorbs* a quantity of heat L or *liberates* a quantity $-L$ of heat, L is still called the heat of formation of the compound C, under the constant pressure P, at the temperature t; a similar remark may be made for the heat of formation at constant volume.

From these definitions, if a compound C is exothermic in the given conditions, its heat of formation in these conditions is positive; it is negative if the compound C is endothermic.

Often, instead of considering, in the preceding definitions, 1 gramme of compound C, w grammes are considered, w being the *molecular weight* of the compound C. The quantities L and λ are then replaced by other quantities L' and l, which are equal to wL and $w\lambda$ respectively. These quantities L' and l are, at the temperature t, the *molecular heats of formation* of the compound C, the one at constant pressure P, the other at constant volume V.

It is, in general, the values of L′ and l which are found in thermochemical tables.

37. Case in which the two heats of formation are equal to each other.—If the combination, taking place at constant temperature and under constant pressure, does not cause the volume of the system to change, case in which the combination takes place *without contraction or expansion,* it follows from the definition that *the two heats of formation L and λ are equal to each other.*

Thus, at a given temperature, the heat of formation of hydrochloric acid gas under constant pressure and the heat of formation of this gas under constant volume have an identical value.

38. General relation between the two heats of formation.—In general the above is not true if the combination, taking place at constant temperature and under constant pressure, causes the volume of the system to vary.

Suppose that 1 gramme of the mixture $A+B$, brought to the temperature t and supporting the pressure P, occupies the volume V_0; that, in the same conditions, 1 gramme of the compound C occupies the volume V_1; let, in these conditions, U_0 be the internal energy that 1 gramme of the mixture $A+B$ possesses, U_1 the internal energy of 1 gramme of the compound C.

Equation (3) gives

$$L=U_0-U_1+\frac{P}{E}(V_0-V_1). \tag{7}$$

Also, denote by u_1 the internal energy of 1 gramme of the compound C, at the temperature and the volume V_0. Equation (4) gives

$$\lambda=U_0-u_1. \tag{8}$$

Therefore we have

$$L-\lambda=u_1-U_1+\frac{P}{E}(V_0-V_1). \tag{9}$$

Such is, in general, the expression for the difference which exists between the heat of formation at constant pressure and the heat of formation at constant volume of the same compound, at the same temperature. This difference depends upon the diminution which the internal energy of 1 gramme undergoes when, without varying the temperature t, the volume changes from V_0 to V_1.

39. Case in which the compound is a perfect gas.—This variation of energy is in general not known; but there is a particular case in which we know how to evaluate this variation; it is the case in which the compound C is a gas near enough to the perfect state that we may apply the laws which characterize this state; in this case, the two internal energies u_1, U_1 refer to the same mass of gas, taken at the same temperature t, consequently (Art. 25) they are equal to each other and equation (9) reduces to

$$L-\lambda=\frac{P}{E}(V_0-V_1). \tag{10}$$

When the compound is a perfect gas, the excess of the heat of formation at constant pressure over the heat of formation at constant volume is equivalent to the external work done by the formation of a gramme of the compound under the constant pressure considered.

This proposition assumes nothing as to the nature of the components, which may be solid, liquid, or gas, the latter being or not near the ideal state.

40. The distinction between the two heats of formation has small importance in practice.—Let us apply the preceding formulas to the calculation of the difference between the two heats of formation of water vapor, both referred to 0°; the pressure P is taken equal to atmospheric pressure.

Let us take for unit of force the gramme-weight, for unit of length the centimetre, for unit of heat the small calorie; we shall then have

$$P=1033.3,$$
$$E=42500.$$

The volume of 1 gramme of water vapor, in the normal conditions of temperature and pressure, reckoned in cubic centimetres, has the value

$$V_1=\frac{1}{0.622\times 0.001293}.$$

Again, water vapor is formed with a contraction equal to $\frac{1}{3}$, so that we have

$$\frac{V_0-V_1}{V_0}=\frac{1}{3},$$

or

$$V_0-V_1=\frac{V_1}{2}.$$

We then find

$$L-\lambda=15.1 \text{ calories.}$$

If we observe that λ is equal to about 3220, we see that the difference between these two heats of formation of water vapor is negligible compared with each of these two heats of formation.

It is thus for the greater number of cases. The distinction between the heat of formation at constant pressure L and the heat of formation at constant volume λ, essential from a theoretical point of view, has, in general, a minimum of practical importance.

41. Influence of temperature on the heats of formation.— Of greater practical importance are the following remarks:

In defining the heats of formation of a given compound at constant pressure and at constant volume, we have stated the temperature at which the reaction takes place. This indication is essential, for the two quantities L and λ vary, in general, with the temperature, and we shall now discuss the laws of these variations.

Let us consider, for example, the heat of formation at constant pressure.

Take 1 gramme of the mixture $A+B$ under the constant pressure P at the temperature t and, without varying either pressure or temperature, cause the compound C to be formed; the system liberates a quantity of heat L; next bring the compound C from the temperature t to the temperature t'; it *absorbs* a quantity of heat $C(t'-t)$, C being the mean specific heat of the compound C, between the temperatures t and t', under the constant pressure P. The total quantity of heat liberated in the transformation considered has the value

$$L-C(t'-t).$$

Now take 1 gramme of the mixture $A+B$, under the constant pressure P; without letting it undergo any combination, bring it from the temperature t to the temperature t'; it absorbs a quantity $\Gamma(t'-t)$ of heat, Γ being the specific heat of the mixture at the constant pressure P; then at the temperature t' and the pressure P, cause the mixture to combine; it liberates a quantity of heat L′. The total quantity of heat liberated by this second transformation is

$$L'-\Gamma(t'-t).$$

The two transformations cause the system to pass from the same initial to the same final states; they take place at the same pressure P; they therefore liberate the same quantities of heat and we have

$$L'-L=(\Gamma-C)(t'-t). \tag{11}$$

A similar line of reasoning may be applied to the heat of formation at constant volume; if λ and λ' are the values of this heat at the temperatures t and t', if γ and c are the mean specific heats between the temperatures t and t', at constant volume, of the mixture $A+B$ and of the compound C, we have

$$\lambda'-\lambda=(\gamma-c)(t'-t). \tag{12}$$

42. **Heat of formation referred to a temperature at which the reaction considered is impossible.**—It often happens that in books on thermochemistry the heat of formation of water at 0° is spoken of, while at 0° oxygen and hydrogen will not combine and water is undecomposable; the definitions of the quantities L and λ are therefore illusory at this temperature and the words used seem void of sense.

We may interpret them in the following way:

Suppose one of the two reactions possible at the temperature t; at this temperature the two quantities L and λ have the experimental meaning that we have given them.

If, at the temperature t', the two reactions are impossible, we shall regard the two heats L′ and λ' at this temperature as purely algebraical quantities defined by equations (11) and (12).

43. Importance of the variations that changes of temperature cause in the heats of formation.—It follows from the formulæ (11) and (12) that changes of temperature may cause very considerable variations in the heats of formation of a compound. According to Berthelot the following variations exist in the heat of formation of water vapor under atmospheric pressure:

at +15° C.,	$L=3228$	calories
2000°	2811	"
4000°	2001	"

We see that changes of temperature, provided they are sufficiently extended, may cause the heat of formation of a body to vary by an amount comparable with the value of this heat.

We shall even meet cases in which a body, formed with absorption of heat at one temperature, is formed with liberation of heat at a higher temperature; the variations of temperature in such cases change the sign of the heat of formation of the compound.

44. Case of perfect gases which combine without condensation. Delaroche and Bérard's law. The heats of formation are independent of the temperature.—The general formulæ (11) and (12) take on a form easier to apply in the particular cases in which the two mixed bodies A and B are perfect gases. In this case the specific heat at constant pressure of the mixture is obtained by applying the classic *rule of mixtures* to the specific heats at constant pressure of the two mixed gases; an analogous rule applies, also, to the specific heats at constant volume of the mixture and of the mixed gases.

Let us suppose that a molecule of the compound C is formed by the union of n_a molecules of the body A and n_b molecules of the body B; let w_a and w_b be the molecular weights of these two bodies A and B, and w the molecular weight of the compound C. We have

$$w=n_aw_a+n_bw_b.$$

In order to form 1 gramme of the compound C, it would be necessary to take $\frac{n_aw_a}{w}$ grammes of the body A and $\frac{n_bw_b}{w}$ grammes of the body B. If we denote by C_a, c_a the two specific heats of the gas A, by C_b, c_b the two specific heats of the gas B, the rule that we have just recalled gives

$$\Gamma=\frac{n_a w_a}{w}C_a+\frac{n_b w_b}{w}C_b;$$

$$\gamma=\frac{n_a w_a}{w}c_a+\frac{n_b w_b}{w}c_b.$$

Equations (11) and (12) may be written

$$(13) \qquad w(L'-L)=(n_a w_a C_a+n_b w_b C_b-wC)(t'-t);$$

$$(14) \qquad w(\lambda'-\lambda)=(n_a w_a c_a+n_b w_b c_b-wc)(t'-t).$$

It is evident that these equations allow of determining very simply the variation of the heat of formation with the temperature, either at constant pressure or at constant volume, of a compound and of its gaseous components. Here is a remarkable example:

Suppose that *the body C is a sensibly perfect gas formed by the union, in equal volumes and without condensation, of two simple, diatomic, sensibly perfect gases, A and B.*

For this, hydrochloric acid gives us a fair example.

In this case a molecule of the compound includes a half-molecule of each of the component gases; n_a, n_b are both equal to $\frac{1}{2}$, and equations (13) and (14) may be written

$$(15) \qquad w(L'-L)=\frac{w_a C_a+w_b C_b-2wC}{2}(t'-t);$$

$$(16) \qquad w(\lambda'-\lambda)=\frac{w_a c_a+w_b c_b-2wc}{2}(t'-t).$$

Again, a very old law, discovered by Delaroche and Bérard, and since verified by many observers, shows that *for all the simple diatomic gases, near to the ideal state and for all gaseous compounds, formed without condensation and near to the ideal state, the product of the molecular weight by the specific heat at constant pressure has a single value:*

(17) $$w_aC_a = w_bC_b = wC.$$

Here are some examples of the truth of this law, taken from Regnault's observations:

1° Simple diatomic gases.	Value of $w \times C$
Oxygen.	3.4800
Nitrogen.	3.4112
Hydrogen.	3.4128

2° Gaseous compounds formed without condensation.	Value of $w \times C$
Nitric oxide.	3.4800
Carbon monoxide.	3.4128
Hydrochloric acid.	3.3744

If we denote by σ_a, σ_b, σ the volumes occupied, in the normal conditions of temperature and pressure, by one gramme of each of the gases A, B, C, respectively, Robert Mayer's relation [Chap. II, eq. (14)] gives

$$w_a(C_a - c_a) = \frac{R}{E} w_a \sigma_a;$$

$$w_b(C_b - c_b) = \frac{R}{E} w_b \sigma_b;$$

$$w(C - c) = \frac{R}{E} w \sigma.$$

Again, from Avogadro's Law, we have

$$w_a \sigma_a = w_b \sigma_b = w \sigma,$$

so that we may write

$$w_a(C_a - c_a) = w_b(C_b - c_b) = w(C - c).$$

From equation (17) it follows that

(18) $$w_a c_a = w_b c_b = wc.$$

The law of Delaroche *and Bérard applies also to the sp.cific heats at constant volume.*

By means of equations (17) and (18), equations (15) and (16) become

$$L-L'=0; \tag{19}$$

$$\lambda-\lambda'=0. \tag{20}$$

When a perfect gas is formed by the union in equal volumes and without condensation of two simple, diatomic gases, the heat of formation at constant pressure and the heat of formation at constant volume (equal to each other) are independent of the temperature.

CHAPTER IV.

CHEMICAL EQUILIBRIUM AND THE REVERSIBLE TRANSFORMATION.

45. Idea of chemical equilibrium. It differs from the idea of mechanical equilibrium.—In mechanics a system of bodies is said to be in equilibrium when each of these bodies and each of the parts that compose it keep an invariable form and position in space.

In chemistry a system is said to be *in chemical equilibrium* when no chemical reaction is taking place in it.

Chemical equilibrium is not a special case of mechanical equilibrium. It is possible to observe a chemical reaction in a system notwithstanding that each of the parts of the system keeps an invariable form and position. Such a system is therefore in mechanical equilibrium, but not in chemical equilibrium.

Let us take, for example, a homogeneous mixture of hydrogen and chlorine and let diffused light act upon it in a hermetically closed reservoir; this mixture remains perfectly at rest; each part, however small, as far as can be distinguished keeps an invariable position and form; yet the system is the seat of a chemical reaction; the hydrogen and chlorine combine to form hydrochloric acid gas.

46. Chemical equilibrium may be the common limit of two oppositely directed reactions. Phenomena of saponification.—In the example that we have just cited, the reaction does not cease until one of the gases, chlorine or hydrogen, which takes part in the reaction has entirely disappeared. The reaction does not stop until the instant that it would be absurd to suppose that it persisted; such a reaction is called an *unlimited reaction.*

It is not so with all reactions in chemistry. For example, Berthelot and Péan de Saint-Giles[1] mixed together masses of water and benzoic ether proportional to the molecular weights of these two substances; they heated them at 200° in a sealed tube; *saponification* was produced; that is to say, there was formation of benzoic acid and alcohol according to the equation

$$\underset{\text{benzoic ether}}{C_6H_5CO_2C_2H_5} + \underset{\text{water}}{H_2O} = \underset{\text{benzoic acid}}{C_6H_5CO_2H} + \underset{\text{alcohol}}{C_2H_5OH}.$$

If the reaction were unlimited, it would not stop so long as the mixture enclosed a quantity, however small, of benzoic ether and water.

This is not what is observed.

At the end of 24 hours' heating, the mixture still encloses a certain mass of non-saponified ether. This mass is a considerable fraction of the quantity of ether introduced originally into the sealed tube, the fraction being 0.664. The time of heating may be indefinitely prolonged without producing the slightest change in the composition of the system; the saponification has therefore ceased, and it has stopped when its continuation would be nowise in contradiction to the chemical formulæ, and while there still exists in the system bodies susceptible of taking part.

This fact is expressed by saying that the saponification of benzoic ether at 200° is a *limited reaction.*

Now let us mix, as did Berthelot and Péan de Saint-Giles, masses of benzoic acid and alcohol, proportional to the molecular weights of these two bodies; heat them at 200° in a sealed tube; we shall observe the inverse to saponification, a formation of benzoic ether and water represented by the equation

$$\underset{\text{benzoic acid}}{C_6H_5CO_2H} + \underset{\text{alcohol}}{C_2H_5OH} = \underset{\text{benzoic ether}}{C_6H_5CO_2C_2H_5} + \underset{\text{water}}{H_2O}.$$

This reaction, the inverse of the preceding, is, like it, a limited reaction; it stops before the benzoic acid and alcohol have been transformed entirely into benzoic ether and water; no matter how long the experiment is prolonged, the mass of benzoic ether obtained remains equal to a fraction of the mass that

---

[1] Berthelot and Péan de Saint-Giles, *Annales de chimie et de Physique*, v. 45, p. 485, 1862; v. 46, p. 5, 1862; v. 48, p. 225, 1863.

would be obtained if the reaction were unlimited; this fraction is 0.664.

Let us compare these two experiments.

In both, the starting-point was two mixtures whose elementary composition is the same; we may regard these two mixtures as the two extreme states of the same system; one, the mixture of ether and water, represents the extreme state of etherification; the other, the mixture of alcohol and acid, represents the state of extreme saponification.

Starting from these two different states, there is produced at 200° two reactions the inverse of each other; in the midst of the completely etherified system a saponification is produced; in the midst of the completely saponified system an etherification is produced. Each of these two reactions is limited. Each of them ceases when the mixture attains a certain composition intermediate between total etherification and total saponification. This composition for which chemical equilibrium is established is the same in the two cases. It is obtained when the mass of ether existing in the system is a fraction of the mass of possible ether, equal to 0.664.

Thus, at 200°, chemical equilibrium is established in the system considered when the mass of ether that it encloses is the fraction 0.664 of the possible mass of ether. *This state of chemical equilibrium is the common limit of two inverse reactions, etherification and saponification.*

47. Reciprocal action of two soluble salts within a solution.—The phenomena of etherification, studied by Berthelot and Péan de Saint-Giles, are not the only ones in which may be observed a state of chemical equilibrium, the common limit of two reactions which are the inverse of each other. Berthollet had predicted that such a state of equilibrium should be produced in a solution in which two soluble salts may, by double decomposition, produce two other soluble salts. Malaguti [1] has verified the truth of Berthollet's prediction in the following way:

In a known mass of water, dissolve a molecule of strontium acetate and two molecules of potassium nitrate; there will be

---

[1] MALAGUTI, *Annales de Chimie et de Physique*, 3me série, v. 37, 1853.

formed, in the solution, potassium acetate according to the equation

$$Sr(CH_3CO_2)_2+2KNO_3=2KC_2H_3CO_2+Sr(NO_3)_2.$$

To determine the composition of the mixture at a given instant, it is sufficient to treat it with an excess of alcohol mixed with ether; this alcohol and ether dissolve the acetates, but not the nitrates.

If this analysis is made after a long wait at ordinary temperature, it is seen that the double decomposition ceases before it is completed; the condition of equilibrium which limits this double decomposition corresponds nearly to the following composition of the mixture:

$\frac{2}{3}$ molecule of potassium acetate,
$\frac{4}{3}$ molecule of potassium nitrate,
$\frac{1}{3}$ molecule of strontium nitrate,
$\frac{2}{3}$ molecule of strontium acetate.

Suppose now that there is dissolved in the same quantity of water one molecule of strontium nitrate and two molecules of potassium acetate; by a reaction the inverse of the preceding, there is formed strontium acetate and potassium acetate, accord- to the equation

$$Sr(NO_3)_2+2KC_2H_3CO_2=Sr(CH_3CO_2)_2+2KNO_3.$$

At ordinary temperatures this reaction is limited and leads to the same state of equilibrium as the preceding reaction; this state of equilibrium is therefore again the common limit of two reactions that are the inverse of each other.

48. Many chemical systems seem incapable of possessing a state of equilibrium which is the common limit of two reciprocally inverse reactions.—In a great number of chemical systems, states of equilibrium similar to those we have just studied are met; each of these states of equilibrium is the common limit of two reactions the inverse of each other.

But a not less number of chemical systems show themselves, at first sight, incapable of possessing such states of equilibrium. Let us take, for instance, a system formed of oxygen and hydrogen;

if we are content to observe superficially the properties of this system, we shall be led to give the following description, for a long time considered the true one:

At low temperatures oxygen and hydrogen do not combine; water is not decomposed. At high temperatures oxygen and hydrogen combine; this combination is not limited, but total; water is not decomposable.

49. Grove's experiment. Water is decomposable by heat.— Notwithstanding, an old experiment contradicts this description of the properties of water. By letting fall into water a sphere of platinum brought to white heat, an explosion is produced; the water is decomposed by the contact of the platinum sphere; afterwards the oxygen and hydrogen recombine.

This experiment, due to Grove, was known long ago, but chemists were content, with Berzelius, to attribute it to the *catalytic force* of the platinum. This experiment was repeated on a larger scale by H. Sainte-Claire Deville and H. Debray; rejecting the explanation by catalytic force and accepting purely and simply the teaching of the experiment, they admitted the following proposition: *At a temperature inferior to the fusing-point of platinum, water vapor is decomposed into its elements, oxygen and hydrogen.*

Further, water is even decomposable at the temperature of the fusing-point of silver, that is, at a temperature less than 1000° C. When silver is melted in the presence of water vapor it absorbs oxygen and gives it out again only at the moment of solidification, this constituting the phenomenon of rochage; the rochage proves therefore that the elements of water are at liberty at the temperature of 1000°, unless one would attribute to melted silver a chemical action on the oxygen; this objection, however, may be obviated; it is sufficient to replace the silver by litharge, a substance chemically saturated with oxygen, incapable of further oxidation; the phenomenon of rochage, certain proof of the decomposition of water, is produced just as sharply.

50. Direct demonstration of the dissociation of water.— Other experiments, still more direct and more conclusive,[1] place

[1] H. Sainte-Claire Deville, *Comptes Rendus*, v. 56, pp. 195 and 322, 1863; *Lessons on Dissociation*, given before the Chemical Society, Mar. 18 and April 1, 1864; H. Debray, *Wurtz's Dictionary*, art. *Dissociation*.

beyond doubt the dissociation of water at temperatures easily attained in laboratories.

Use is made of an apparatus composed of a glazed porcelain tube VV' (Fig. 18), in the interior of which is another tube, PP',

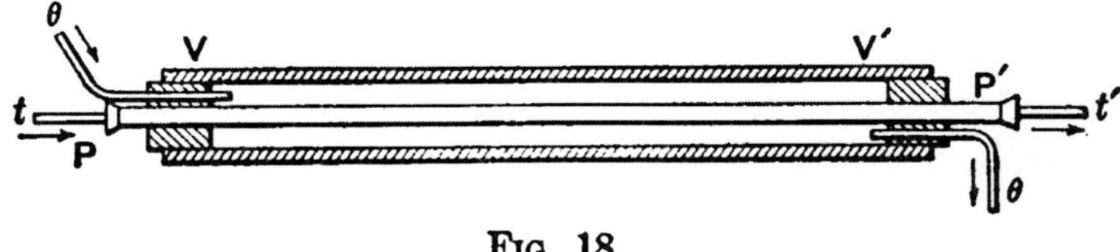

FIG. 18.

of porous substance; the two tubes being strongly heated in a furnace fed with coke or hard coal, water vapor is let into the interior tube of porous earthenware at t, and a current of carbonic acid is introduced by the tube θ into the annular space included between the porous and the porcelain tubes; the gas from the apparatus is received in éprouvettes over a bowl of potash solution to catch the carbonic acid. While the furnace is in activity, a strongly explosive gaseous mixture is obtained composed of the elements of water, oxygen, and hydrogen. Thus a part of the water vapor is spontaneously decomposed or dissociated in the porous earthenware tube; the hydrogen, following the ordinary laws of osmosis, has passed through the semi-permeable partition and is separated as if by the action of a simple filter from the oxygen which stays in the interior tube; furthermore, there is found with this oxygen a considerable quantity of carbonic acid coming from outside.

In general, when water vapor traverses, without special precaution, a strongly heated tube, only water vapor is collected at the outlet and not oxygen nor hydrogen; in fact, the water vapor, decomposed in the hotter parts of the apparatus, is totally re-formed in the cooler parts passed through by the gases resulting from this decomposition. If, however, the passage of the water vapor through a strongly heated tube is extremely rapid, and if the water vapor is mixed with a large excess of carbonic acid whose presence hinders the recombination of the oxygen and hydrogen, a small quantity of explosive gas may be collected at the outlet, and the dissociation undergone by water vapor at a high temperature may be shown in a very simple manner.

51. Dissociation of carbonic acid gas.—This simple apparatus serves also to show another decomposition, one not less remarkable than that of water vapor: the decomposition of carbonic acid gas at high temperatures.

It is sufficient to pass a current of quite pure carbonic acid through a small-bore porcelain tube filled with pieces of porcelain and heated in a combustion-furnace to the highest temperature possible (1200° to 1300°).

The gases coming from the tube pass through long tubes filled with potash solution, where they are separated from the excess of carbonic acid.

The carbonic acid is decomposed by heat into carbonous oxide and oxygen, and if these gases do not totally recombine on reaching the colder parts of the apparatus, this is probably due to the difficulty with which their mixture ignites when disseminated through a large mass of inert gas, as carbonic acid.

52. These decompositions are not complete but limited; at the temperatures at which they are produced, the inverse reaction also takes place.—Must we conclude from these observations that at the temperatures attained in the investigations of H. Sainte-Claire Deville water is *totally* decomposed into oxygen and hydrogen, carbonic acid *totally* broken up into oxygen and carbonous oxide? If this were so, we should encounter this incomprehensible paradox: Water does not exist at the temperature of melting silver, and nevertheless oxygen and hydrogen in combining produce a temperature such that their flame fuses iridium; how is it that this flame melts platinum and that melted platinum decomposes water?

It is clear that the decomposition of water at a given temperature should not be total, but partial: this decomposition should stop when the gaseous mixture formed by the water vapor and by the oxygen and hydrogen which comes from its decomposition has a certain composition; this composition, for which the system is in equilibrium, would naturally depend upon the temperature.

Conversely, when a mixture of hydrogen and oxygen is brought to a temperature sufficient to ignite it, the combination of the two gases should not be complete; it should cease for a certain proportion of water vapor, variable with temperature; this is the conclusion reached by H. Sainte-Claire Deville from the analysis of the properties of the oxyhydrogen flame.

If, as was the case before the researches of H. Sainte-Claire Deville, we accept the hypothesis that above 500° oxygen and hydrogen combine entirely to form water vapor, it is easy to calculate the temperature reached in the oxyhydrogen flame. The computation requires a knowledge only of the specific heat of water vapor and the heat of formation of water. In this way is found the excessively high temperature of 6800°.

Now this temperature seems quite improbable. It is true that the oxyhydrogen flame melts platinum-iridium, but its temperature should not much exceed the point of fusion of this alloy, for the alloy placed in the flame is hardly brighter than at its fusing-point.

This temperature may even be determined approximately: by pouring into cold water considerable masses of melted platinum or iridium brought to the highest possible temperature given by oxygen and hydrogen which combine in equal equivalents, and observing the maximum rise of temperature produced in the water, it is found that the fixed point of the combination of these two gases cannot exceed 2500°, and is perhaps less.

How explain these results? Evidently they are due to the fact that in the oxyhydrogen flame the gases which burn do not totally combine; a part of these gases escapes combustion.

When oxygen and hydrogen, mixed in equal equivalents, burn there is formed a certain quantity of water vapor, but a certain quantity of hydrogen and of oxygen remain in a state of liberty. With a special arrangement, the gases which feed the oxyhydrogen flame may be drawn partly off, and it is found that the hottest parts always contain uncombined oxygen and hydrogen.

If this experiment is repeated with a burner of oxygen and carbonous oxide, it is readily seen that the flame is far from uniquely made of carbonic acid; in the hottest part of the flame two thirds at the most of the gases oxygen and carbonous oxide are combined; it is only in the coolest part of the flame that the combustion is total.

These various experiments put beyond doubt the following propositions:

At a high temperature water and carbonic acid are decomposed but the decomposition is not complete; an equilibrium is set up when the mixture formed by the compound and its component gases has reached a certain composition; the mixture in equilibrium contains a less proportion of the compound considered in the proportion that the temperature is higher.

Conversely, at a high temperature oxygen and hydrogen combine, carbonous oxide and oxygen form carbonic acid, but the combination is not total; it tends towards a state of equilibrium at which it stops; in this equilibrium state the proportion of the gases which have escaped combination is the greater as the temperature is higher.

53. **Example of a state of equilibrium which is the common limit of two reactions the inverse of each other. Action of water vapor on iron and the inverse action.**—The preceding experiments show that in the same system which includes a molecule of oxygen and one of hydrogen, at the same temperature, two inverse reactions may be observed: decomposition of water vapor and formation of water vapor; they show us that each of the two inverse reactions stops when the system has reached a certain state of equilibrium; but they do not show us that these two states of equilibrium are identical with each other. We shall see farther on, when we study the states of false equilibrium (Chap. XVIII), that it is not useless to demonstrate experimentally this equality.

Here is a case in which experiment shows states of equilibrium of which each is the common limit of two inverse reactions, and for which the laws governing these states of equilibrium may be completely analyzed.

At a high temperature iron reduces water vapor, giving the magnetic oxide of iron; inversely, by passing a current of hydrogen over the magnetic ferrous oxide, iron and water vapor are obtained; H. Sainte-Claire Deville[1] and Debray[2] have tried to determine the conditions in which these two inverse reactions are produced.

A porcelain tube containing iron and the magnetic oxide were immersed in a bath brought to a fixed temperature; hydrogen

---

[1] H. Sainte-Claire Deville, *Comptes Rendus*, v. 70, pp. 1189 and 1201, 1870; v. 71, p. 30, 1871.

[2] H. Debray, *Comptes Rendus*, v. 88, p. 1341, 1879.

could be admitted to this tube, which also received water vapor coming from a flask filled with cold water; in accordance with Watt's principle, the tension of the water vapor in the whole apparatus was equal to the tension of the saturated water vapor at the temperature of the flask, and consequently had a known value; a manometer gave the pressure of the mixture of hydrogen and water vapor and thus, by difference, the pressure of the hydrogen.

Suppose, for example, the flask brought to the temperature at which the pressure of the saturated water vapor is 4.6 mm., and heat the porcelain tube to 200°; so long as the pressure of the hydrogen is less than 95.7 mm., the water vapor attacks the iron, reaction having for effect the increase of the hydrogen pressure; when, on the contrary, this pressure becomes greater than 95.7 mm., it diminishes, because a part of the hydrogen is used to reduce the ferrous oxide; when the temperature is 200° and the vapor pressure of the water 4.6 mm., the system is in a state of equilibrium corresponding to the value 95.7 mm. for the pressure of hydrogen; the system, removed from this state in the one or the other direction, undergoes a chemical reaction which brings it back; this state is therefore one of stable equilibrium.

This state varies with the temperature; the vapor pressure of water being always 4.6 mm., the hydrogen pressure at the instant of equilibrium has, at various temperatures, the values given in the following table:

Temperature.	Hydrogen Pressure.
200°.	95.7 mm.
Mercury boiling-point.	40.5 "
Sulphur. "	25.8 "
Cadmium "	12.9 "
Zinc "	9.2 "
About 1000°.	5.1 "

At a given temperature this state of equilibrium changes with the vapor pressure of water; the hydrogen pressure, at each temperature, is approximately proportional to the vapor pressure of the water vapor; thus at 200° when the water vapor has a pressure of 4.6 mm., the hydrogen pressure at the instant of equilibrium has the value 95.7 mm., whose ratio to the vapor pressure of the water is 20.8; at the same temperature, when the water-vapor has a pressure of 9.7 mm., the pressure of the hydrogen, at the instant of equilibrium, is 195 mm., giving the ratio of 20.1.

These observations show us, for high temperatures, what the phenomena of etherification and the double decomposition of salts showed us for ordinary temperatures: namely, the existence, in a chemical system, of a state of equilibrium, the common limit of two reactions the inverse of each other.

54. Changes of physical state give rise to equilibrium conditions each of which is the common limit of two modifications the inverse of each other. Saturation of solutions.—Similar conclusions hold for changes of physical state as were furnished by chemical reactions.

Take, at 0°, an aqueous solution of sodium chloride in the presence of a crystal of this salt. If the solution contains less than 36 grammes of salt to 100 grammes of water, it dissolves the new pieces of salt until the concentration corresponds to 36 grammes of salt dissolved in 100 of water; then the transformation considered ceases and the solution is *saturated.* If the solution contains more than 36 grammes of salt to 100 grammes of water, salt is precipitated and the solution attains, without exceeding it, the concentration 36/100.

A system formed of water and salt is then in equilibrium, at the temperature 0°, when the solution contains 36 grammes of salt dissolved in 100 grammes of water. This state of equilibrium is the common limit of two inverse modifications, solution and precipitation.

55. Another example. Tension of saturated vapor.—Another example is furnished by the vaporization of water. A vessel containing water and water vapor is brought to 100° C. Whatever may be the form and size of the vessel and the respective masses of water and vapor, the following facts are observed:

If the pressure in the vessel is inferior to the pressure of 1 atmosphere, the water is changed into vapor; the vaporization ceases when the pressure becomes 1 atmosphere.

If the pressure is greater than one atmosphere, the vapor condenses; the condensation ceases at atmospheric pressure.

At 100°, liquid water and water vapor are in equilibrium if the pressure in the vessel is equal to the pressure of 1 atmosphere; this state of equilibrium is the common limit of two changes of state the inverse of each other, vaporization and condensation.

These facts are included in the well-known law:

At a given temperature a liquid of definite composition is in equilibrium with its own vapor when the pressure supported by these fluids has a certain value; this value does not depend upon the size or the form of the containing vessel, of the masses of the liquid and vapor; it depends solely upon the nature of the liquid and upon the temperature; it is called the *tension of saturated vapor of the given fluid at the temperature considered.*

The tension of saturated vapor of a given fluid increases with the temperature.

At a given temperature the liquid vaporizes if the pressure is less than the tension of saturated vapor for the temperature considered; the vapor condenses, on the contrary, if the pressure is greater than the tension of saturated vapor.

This law is susceptible of a common geometrical representation.

Take two coordinate axes OT, OP (Fig. 19), of which OT is the axis of temperatures and OP of pressures.

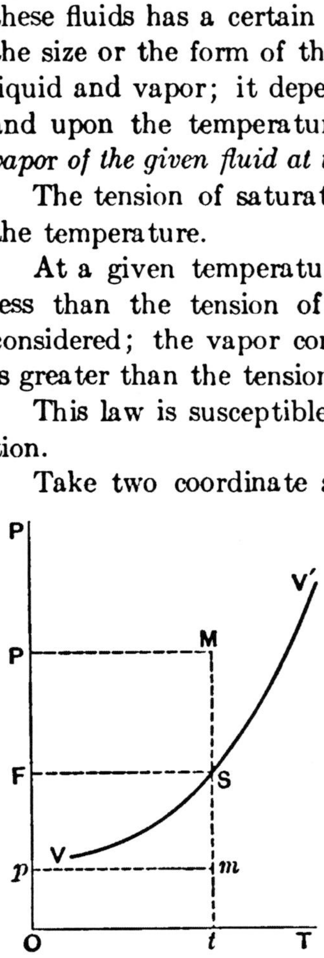

FIG. 19.

Let F be the tension of saturated vapor of a certain fluid at the temperature t; the point S whose abscissa is t and whose ordinate is F shows the conditions in which the liquid will be in proper equilibrium with its own vapor.

When the temperature t varies, the tension F varies also and the point S describes a certain curved line VV' which is the *vapor-pressure curve for saturated vapor* of the fluid considered.

Since the tension of saturated vapor F increases at the same time as the temperature t we see that the vapor-pressure curve rises from left to right.

At the temperature t take a pressure P, greater than the tension F of the saturated vapor; the point M, of abscissa t and ordinate P, will be above the point S belonging to the curve VV'. Take likewise a pressure p, less than the tension F of saturated vapor; the point m, of abscissa t and ordinate p, will be below the point S.

The law stated above may be expressed in the following way:

The curve of tensions for a saturated vapor, each point of which represents a state of equilibrium of the system formed by the liquid and its vapor, is the boundary of two regions.

Every point in the region situated above the vapor-pressure curve represents a state of the system in which the liquid vaporizes.

Every point in the region below this curve represents a state in which the vapor condenses.

56. Dissociation of calcium carbonate. Tension of dissociation.—Guided by the intuitions of H. Sainte-Claire Deville, H. Debray[1] has proved that the laws of the vaporization of a liquid, laws which we have just recalled, may be applied almost literally to the chemical decomposition of certain bodies, notably to the dissociation of carbonate of calcium into lime and carbonic acid gas.

Heat, at a known temperature t, carbonate of lime in a vessel which communicates with a mercury pump; this pump serves either to remove the carbonic acid produced or to force in carbonic acid, and at the same time gives a measure of the gaseous pressure at each instant.

At a given temperature t the decomposition of the carbonate of lime stops when the pressure of the carbonic acid reaches a certain value F; if, keeping the temperature constant, the carbonic acid produced is removed by the pump, a new decomposition takes place, which again stops when the pressure of the carbonic acid retakes the value F; if carbonic acid gas is forced into the apparatus, this gas combines with the free lime until the pressure returns to the value F. These results, similar to those that would be had if the vessel contained a solid or liquid in the presence of its vapor, may be expressed by saying that the space containing the carbonate of lime is *saturated* with carbonic acid gas when this gas is at the pressure F.

---

[1] H. DEBRAY, *Comptes Rendus*, v. 64, p. 603, 1867.

At a given temperature t, this pressure has a value which does not depend upon the various peculiarities which may characterize the experiment; in particular, it does not change if at the beginning of the experiment there is put into the apparatus not only Iceland spar, but also an excess of quicklime. Depending solely upon the temperature t, this pressure may be called the *dissociation tension of calcium carbonate at the temperature t.*

The dissociation tension of carbonate of calcium at a temperature t varies with this temperature and increases with it. H. Debray found the following values for this tension:

Temperature.	Dissociation Tension.
Mercury boiling-point.......	Zero
Sulphur "	Hardly sensible
Cadmium "	85 mm.
Zinc "	520 "

The decomposition of calcium carbonate is not the only reaction showing a dissociation tension, fixed at each temperature and similar in all points to the tension of saturated vapors; H. Debray found the same law in studying [1] the decomposition of a certain number of hydrated salts into water vapor and anhydrides.

We may evidently construct, for each of the cases of which we have spoken, a *curve of dissociation tensions*, which possesses all the properties of the vapor-pressure curve for saturated vapors. In each of the reactions studied by Debray the determinations of this investigator give us a certain number of points on the curve of dissociation tensions; but these points are too few and too widely separated from each other to allow us to draw the curve.

Isambert has tried to fill this gap. He made use of [2] the compounds that certain metallic chlorides, bromides, or iodides form with ammonia gas.

At a given temperature there is emission or absorption of ammonia gas according as the pressure of this gas is less or greater than a certain dissociation tension. The dissociation tension at a given temperature depends exclusively upon this

[1] H. DEBRAY, *Comptes Rendus*, v. 66, p. 194, 1868.

[2] ISAMBERT, *Comptes Rendus*, v. 66, p. 1259, 1868; *Annales de l'Ecole Normale Supérieure*, v. 5, p. 129, 1868.

temperature and upon the ammonia compound which is destroyed or formed under pressures near to this tension. Isambert was able to determine the curves of dissociation tension for a certain number of ammonia compounds.

Since the time when Isambert published this work several chemists have made known a great number of dissociation tension curves. These curves have exactly the appearance and properties of those for the tension of saturated vapors of solids or liquids.

57. The study of chemical reactions and the study of physical changes of state depend on the same theory, that of chemical mechanics.—These observations clearly show that chemical reactions and physical changes of state sometimes obey exactly similar laws; consequently every theory applicable to chemical reactions in general should include also physical changes of state.

From the beginning of its development, thermodynamics was applied successfully to vaporization, fusion, and solution. It is therefore natural to seek to extend it to chemical reactions.

This attempt, crowned with success, has given birth to chemical mechanics founded upon thermodynamics, the subject of these lessons.

58. Idea of reversible transformation.—It is to thermodynamics that we must look for the constitution of a truly rational chemical mechanics. We shall see this chemical mechanics developed from the union of the principle of the equivalence between work and heat with a new principle, discovered by Sadi Carnot, transformed and enlarged by Clausius.

In order to state this principle, it is necessary for us to become acquainted with one of the most delicate principles in all thermodynamics, that of *reversible transformation.*

Take a system, acted upon by certain forces, in a given state 1 and suppose that, under the action of these forces. the system passes to the state 2. It has undergone a modification in the course of which it has passed through various states which succeed each other in a continuous manner. As soon as the system, during this modification, is in any one of these states, it immediatiely leaves it to pass into the state following, which proves this state passed through by the system not to have been one of equilibrium.

When the forces which act on the system and the conditions in which it is placed are given, the nature and the direction of the modification which it undergoes are necessarily fixed. If, for example, these forces and these conditions cause the system to go from the state 1 to the state 2 passing through the intermediate states *A*, *B*, *C*, D . . ., it cannot be admitted that, placed in the same conditions and acted upon by the same forces, the system can repass from the state 2 to the state 1 by traversing in inverse order precisely the same states . . . D, *C*, *B*, *A*; this is expressed by saying that a *real transformation is never reversible.*

It is not *reversible*, but it may be *renversable;* it will be renversable if, by modifying the forces which act upon the system and the conditions in which it is placed, the system may be made to repass from the state 2 to the state 1; but in general, in this transformation, the inverse of the preceding, the system will pass through the states . . . *D'*, *C'*, *B'*, *A'* different from the states . . . D, *C*, *B*, *A*, either in the properties possessed by the body in each of these states or in the forces which act upon it.

Take an example borrowed from mechanics.

In an Atwood's machine (Fig. 20) let a weight π pass from the level *A* to the level *Z* under the action of the counterweight π', less than π; at the moment that the weight π passes the level *N*, between *A* and *Z*, its velocity of fall is not zero; it is not therefore in equilibrium.

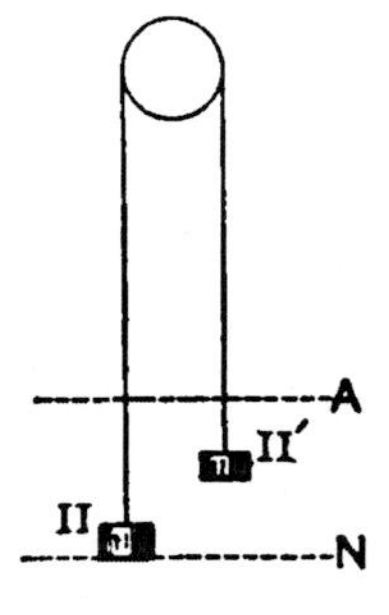

FIG. 20.

Take the weight π at the level *Z*; give to it as initial velocity the velocity directed from above downwards that it had at the end of the preceding fall; under the action of the same counterweight, the weight π will not remount from the level *Z* to the level *A*; it will continue, on the contrary, to descend; the modification considered is not reversible.

It is, however, renversable; by taking a counterweight π' greater than π, the weight π may be brought back from the level *Z* to the same level *A*; but at the instant during this ascension when it repasses the level N, it will be no longer acted upon by the force which acted upon it when, in descending, it passed through the same level and its velocity, instead of being directed downwards, will be directed upwards.

Take two states 1 and 2 of a system; suppose that certain forces can cause the system to pass from state 1 to the state 2, passing through a series of intermediate states A, B, C, D . . . ; that other forces bring it from the state 2 to the state 1, passing through other states . . . D′, C', B', A'. Between the states 1 and 2 arrange a series of *equilibrium positions* α, β, γ, δ, . . . following each other in a continuous manner. The system, placed in any one of these positions, would remain there indefinitely. This series of positions of equilibrium cannot then be passed through by the system either in one direction or in the other; *it does not correspond to a realizable transformation of the system.*

Take the transformation ABCD . . . which brings the system from state 1 to state 2. Change slowly the forces which act upon the system during this transformation so that they approach gradually the forces which would assure the system's equilibrium in each of the positions through which it passes; suppose that, by the effect of this operation, the state that we denote by A tends towards the equilibrium position α, that the state we call B tends towards the equilibrium position β. . . . We shall have thus constructed a continuous series of realizable transformations suitable to make the system pass from state 1 to state 2; and this series of transformations will have for limiting form the chain of equilibrium positions α, β, γ, δ . . . , which is not a realizable modification.

Suppose, for example, that in our Atwood's machine we cause to approach π the counterweight π' ($\pi' > \pi$), under whose action the weight π falls from A to Z; the velocity of this weight π, at the instant it passes the level N, will approach 0; the force acting upon it will approach 0; the state of the weight π, at the instant it passes through N, will approach a limiting state in which it would be kept in equilibrium at the level N, by an equal counterweight; the actual descents which we studied, but taken more and more slowly, will have for l miting form a series of equilibrium positions of the weight π under the action of an equal counterweight, positions succeeding each other with continuity from the level A to the level Z; this series of equilibrium positions is the limit of a series of falls; but it does not constitute a fall; the weight π cannot really pass by this series of states, which does not prevent us from turning our attention, if we wish, to each one of these states.

Similarly, if we modify gradually the forces which act upon the system to make it pass from the state 2 to the state 1 passing through the states . . . D′, C', B', A', we could cause the real transformation . . . D′, C', B', A' to vary in a continuous manner and admit for limiting form the series of limiting states of equilibrium . . . δ, γ, β, α, which is not a realizable transformation.

In our Atwood's machine, for example, we might lift the weight π from Z to A under the action of the counterweight π', greater than π; by varying in a continuous manner the counterweight π', we may modify this lifting in a continuous manner also; if the counterweight π' approaches π as a limit, the ascent of the weight π, becoming slower and slower, will have for limiting form a series of equilibrium states of the weight π, extended in a continuous manner from the level Z to the level A.

This series of equilibrium states α, β, γ, δ . . . which is passed over by no modification of the system is, in some sort, the common boundary of the real transformations that bring the system from the state 1 to the state 2 and of the real transformations that bring the system from state 2 to state 1; change, however slightly, in a certain direction, the conditions that maintain the system in equilibrium in each of these states and the system will undergo a real transformation which will conduct it from state 1 to state 2; change, however slightly, these conditions in the opposite direction, and you obtain a real transformation from state 2 to state 1; this series of equilibrium states is called a *reversible transformation.*

Thus the reversible transformation is a continuous series of equilibrium states; *it is essentially unrealizable;* but we may give our attention to these equilibrium states successively either in the order from state 1 to state 2, or in the reverse order; this purely intellectual operation is denoted by these words: *to cause a system to undergo the reversible transformation considered, either in the direction 1–2, or in the reverse direction.*

59. Example of a reversible transformation furnished by the vaporization of a liquid.—This idea of reversible transformation is of great importance in all branches of thermodynamics; one cannot insist too much upon it; let us illustrate it by several examples taken from physical changes of state and changes of chemical constitution.

At the temperature of 100° the vapor tension of water is equal to 1 atmosphere; at 200° it is 15 atmospheres.

Take a mass of water, entirely liquid, at the temperature of 100° and under atmospheric pressure; this will be the state 1 of our system; next take the same mass of water, entirely vapor, at 200° and under 15 atmospheres pressure; this will be the state 2 of our system.

We may, by a real transformation, cause the system to pass from the state 1 to the state 2; heat it so as to raise the temperature gradually from 100° to 200°; at the same time increase the pressure, but so slowly that at each instant its value is less than the tension of saturated water vapor for the temperature of the system at the given instant; under these conditions the liquid water will be changed into vapor until it has all passed over to this latter state.

Similarly, by a real transformation, we may cause the system to pass from state 2 to state 1; lower gradually the temperature from 200° to 100°; at the same time decrease the pressure, but so slowly that its value, at each instant, is greater than the tension of saturated water vapor for the temperature at which the system is at the same instant; in these conditions the water vapor will continue to condense until it has all passed over into the liquid state.

Change gradually the conditions in which the first transformation is produced, so that the pressure at each instant is nearer and nearer the tension of saturated vapor for the temperature of the system at this same instant. Change also the conditions under which the second transformation is produced, so that the pressure exceeds less and less the tension of saturated vapor for the temperature of the system at this same instant. Our two real transformations, the inverse of each other, will approach the same limiting form; this limiting form will be composed of a series of systems in which the water will be partly liquid and partly vapor; from one system to the following the temperature increases, the liquid mass decreases, the mass of vapor increases;

each of these systems will support a pressure which is precisely the tension of saturated water vapor for the temperature to which the system is brought, so that *each of these systems is in equilibrium;* it will be the seat neither of a vaporization nor of a condensation.

According as one passes mentally over this series of states of equilibrium in a certain order or in the reverse order, one will pass by a *reversible transformation* from the state 1 to the state 2 or from state 2 to state 1.

60. Example of a reversible transformation furnished by the dissociation of cupric oxide.—The vaporization of a liquid has just furnished us with an example of a reversible transformation; the dissociation of a solid compound, such as the carbonate of calcium or of cupric oxide,[1] will give us a second example quite analogous to the preceding.

Take a certain mass of cupric oxide (CuO) at the temperature t_1 and under the pressure P_1, equal to the pressure of dissociation for the temperature t_1; next, take at the temperature t_2 and pressure P_2, equal to the dissociation tension relative to the temperature t_2, cuprous oxide (Cu_2O) and the oxygen furnished by the dissociation of this mass of cupric oxide. From the state 1 to the state 2 we may pass through a real transformation, by varying the temperature from t_1 to t_2 and constantly keeping the system under a pressure less than the dissociation tension of the cupric oxide at the actual temperature of this system. We might also pass from the state 2 to the state 1 through a real transformation by varying the temperature from t_2 to t_1, and in maintaining the system constantly at a pressure greater than the dissociation tension of the cupric oxide at the actual temperature of this system. These two groups of inverse transformations will have a common boundary; this boundary will be composed of a series of systems containing cupric oxide, cuprous oxide, and oxygen, at a temperature t, comprised between t_1 and t_2, and under a pressure P equal to the dissociation pressure at the temperature t; all these systems will be in equilibrium, and this series of states of equilibrium will form a *reversible transformation* joining the state 1 to the state 2.

[1] Debray and Joannis, *Comptes Rendus*, v. 99, pp. 583 and 688, 1884.

61. Example of a reversible transformation furnished by the dissociation of water vapor.—Here is another example of the reversible transformation, taken from the phenomena of dissociation.

A vessel of constant volume contains water vapor and the mixture of hydrogen and oxygen produced by the decomposition of this vapor; at a given temperature t such a system is in equilibrium when the ratio of the mass of non-decomposed water vapor to the total mass of the system has a value x; this value x depends on the temperature t and diminishes when the temperature rises. If, in the vessel brought to the temperature t, the proportion of water vapor is greater than x, the water dissociates, while the oxygen and hydrogen cannot combine; if, on the contrary, the proportion of water vapor is less than x, the oxygen and hydrogen combine while the water vapor cannot dissociate.

Take the two temperatures $t_1=1200°$ and $t_2=1500°$; let x_1, x_2 be the values of x which correspond to these temperatures; x_1 will be greater than x_2.

From a state 1, where the temperature is $t_1=1200°$ and where the proportion of water vapor is x_1, to a state 2, where the temperature is $t_2=1500°$ and where the proportion of water vapor is x_2, we may pass through a real transformation, in the course of which the temperature will constantly rise while the water vapor will dissociate unceasingly; it will be necessary for this that at each instant the proportion of water vapor in the system is greater than the value of x which corresponds to the temperature of the system at the same instant.

We may also pass through a real transformation from the state 2 to the state 1, in the course of which the temperature will fall constantly, while the oxygen and hydrogen will combine; it will be necessary for this that at each instant the proportion of water vapor in the system is less than the value of x which corresponds to the temperature of the system at the same moment.

These two groups of inverse transformations will admit a limiting form which is common to both; this limiting form will consist of a series of systems in equilibrium; in each of them the temperature will have a value t, included between t_1 and t_2, and the proportion of water vapor will have the value x which corresponds to the temperature t; this series of systems in equilibrium will form a *reversible transformation* permitting passage from state 1 to state 2 and inversely.

From now on, we shall consider a state of chemical equilibrium in the light of its ability to take part in reversible transformations; its essential character will no longer consist in the absence of all change; it will consist rather in the separation of states which are the seat of a transformation of definite direction from those states which are the seat of a transformation in the opposite direction; we may then characterize such a state of chemical equilibrium as one where two reactions, the inverse of each other, limit each other.

CHAPTER V

THE PRINCIPLES OF CHEMICAL STATICS.

62. Sadi Carnot's principle. Generalization of this principle by Clausius.—In 1824 Sadi Carnot [1] published a short work on the mechanical effects of heat; depending on the one hand upon the impossibility of perpetual motion, on the other hand upon the principle, then accepted without question, that around a closed cycle a system undergoes losses and absorptions of heat which exactly compensate each other, he demonstrated a theorem of the greatest importance both for the theory of heat and for the applications of this science to heat-engines.

The discovery of the principle of the equivalence between heat and work, by annihilating one of the two postulates on which Carnot's demonstration was based, seemed as if it would crush his theorem by a single blow. Happily this was not so; Carnot's theorem was not incompatible with the new principle; by combining the principle of the equivalence between heat and work with a postulate analogous to the impossibility of perpetual motion, this theorem may be proved in a manner more precise than that given it by its discoverer.

It was Clausius who in an imperishable memoir [2] reconciled Carnot's theorem with the principle of equivalence. But Clausius did not limit himself to the realization of this work, which alone would have assured him the admiration of physicists. He

---

[1] Sadi Carnot, *Réflexions sur la puissance motrice du feu et sur les machines propres à augmenter cette puissance;* Paris, 1824. Reprinted 1872 *et seq.;* also in English in several forms.

[2] R. Clausius, *Poggendorf's Annalen,* Bd. 79, pp. 368 and 500, 1850.

generalized and transformed Carnot's[1] theorem so as to make of it one of the most important and fruitful principles of natural philosophy. It is with justice that the name of *Principle of Carnot and Clausius* is given to the great law that they established.

This general law, which we shall regard as one of the PRIMARY HYPOTHESES on which thermodynamics is based, and which therefore we place in the same rank as the *Principle of the equivalence between heat and work*, may be stated in a very simple manner.

Let us first define what Clausius calls the *value of transformation* of a modification.

This modification may be *isothermal;* this means that the system undergoing a change has the same temperature at all points and that this temperature remains constant during the whole modification; in this case, if Q is the quantity of heat set free by the system which undergoes this change and if T is the constant value of the absolute temperature, the *value of transformation* for the change is

$$\varepsilon=\frac{Q}{T}. \tag{1}$$

In general, even if the temperature is supposed to keep the same value at all points of the system, this value varies from one instant to the next, and that in a continuous manner; the definition of the value of transformation is then somewhat more complicated.

Let T_0 be the initial value and T the final value of the absolute temperature.

Separate the total change into n partial modifications, M_0, $M_1, \ldots, M_{n-1}$. Let $T_0, T_1, \ldots, T_{n-1}$ be the values of the temperature at the beginning of the various modifications. Let $Q_0, Q_1, \ldots, Q_{n-1}$ be the quantities of heat set free by the system during these modifications. Form the sum

$$\frac{Q_0}{T_0}+\frac{Q_1}{T_1}+\ldots+\frac{Q_{n-1}}{T_{n-1}}.$$

Next increase without limit the number n of the partial modifications into which the total change is divided, so that each of

[1] R. CLAUSIUS, *Poggendorf's Annalen*, Bd. 93, p. 481, 1854; Bd. 116, p. 73, 1862; Bd. 125, p. 353, 1865.

these partial modifications tends to disappear. Seek the limit approached by the preceding sum. By definition this limit is the value of transformation of the total modifications considered:

$$(1') \qquad \varepsilon = \lim. \left(\frac{Q_0}{T_0} + \frac{Q_1}{T_1} + \ldots \frac{Q_{n-1}}{T_{n-1}}\right).$$

The statement of the CARNOT-CLAUSIUS PRINCIPLE is then the following:

Suppose that a system undergoes a series of modifications forming a closed cycle. If all the modifications which compose the cycle are reversible, the value of transformation for the cycle is zero:

$$(2) \qquad \varepsilon = 0.$$

If all the modifications which compose the cycle are real, or if certain of them are real and others reversible, the value of transformation for the cycle is positive:

$$(3) \qquad \varepsilon > 0.$$

63. The absolute temperature is always positive.—These statements cause the *absolute temperature* to play an important rôle; let us consider this.

We have seen (Chap. II, Art. 24) that when a centigrade-scale thermometer constructed with a perfect gas indicated the temperature t, the absolute temperature had the value $T = \frac{1}{\alpha} + t$, α being the coefficient of expansion of a perfect gas. We have seen also that if the same mass of gas, at the centigrade temperatures t and t', and under the pressures P and P', occupied the volumes V and V', the following equation holds [Chap. III, eq. (6')]:

$$\frac{PV}{T} = \frac{P'V'}{T'}.$$

Suppose, for instance, that the absolute temperature T' is that which corresponds to 0° C., that is, to melting ice; we shall then have $t' = 0$, $T' = \frac{1}{\alpha}$; let P_0 and V_0 be the pressure and volume of this mass of gas at this temperature; at an absolute temperature

T the same mass of gas, at the same value V_0, would exert a force P and we should have

$$P = P_0 \alpha T.$$

Imagine that this mass of gas is cooled until the corresponding absolute temperature becomes zero, then negative; the pressure of this gas whose volume is supposed invariable would also become zero, then negative, which is incompatible with the properties of a gas.

This contradiction may be resolved in two ways: it may be considered as a proof that it is impossible to sufficiently cool a body so that the absolute temperature of this body becomes zero or negative; it may also be taken as a proof that no body can remain in the state of a perfect gas when the temperature becomes sufficiently low.

As, in fact, the gases approaching the ideal state differ more and more from this state when the temperature is lowered, the second statement may be easily accepted and the reasoning that we have just developed cannot suffice to establish the first affirmation.

It is to other considerations, which bear upon the definitions of absolute temperature and which we cannot explain here, that we must look for the justification of the follow'ng proposition:

It is impossible to sufficiently cool any system whatever so that the absolute temperature of this system becomes zero or negative.

The *absolute zero of temperature* would thus appear to be a lower limit of cold; the most energetic refrigerating methods that can be imagined will allow us to approach nearer and nearer this limit; they will never let us reach it.

64. Property of a real isothermal cycle.—These remarks will be useful to us on several occasions. Let us consider a first application.

Imagine that a system actually describes a closed cycle and that during the cycle the absolute temperature keeps the constant value T; we then have a real *isothermal* cycle, for which we may write the equation

$$\varepsilon = \frac{Q}{T}.$$

The inequality (3) may then be written

$$\frac{Q}{T}>0.$$

But, from what has just been said, the factor $\frac{1}{T}$ is essentially positive; the preceding inequality then becomes

$$Q>0.$$

When a system is put through a real isothermal cycle it always sets free more heat than it absorbs.

Similarly, if we denote by W_0 the kinetic energy that the system possesses at the instant it began to describe the cycle, by W_1 the kinetic energy it possessed at the moment it completed this cycle, by W_e the work done during the description of the cycle, we shall have, by the principle of the equivalence between heat and work [Chap. II, eq. (2′)],

$$W_e - W_1 + W_0 = EQ.$$

The preceding inequality gives, therefore,

$$W_e > W_1 - W_0. \tag{4}$$

65. Impossibility of perpetual motion.—Several conclusions may be drawn from this inequality.

Suppose, in the first place, that it is question of a system relieved from all external action; W_e in this case becomes zero; the inequality (4) becomes

$$W_1 - W_0 < 0.$$

When a system at constant temperature, relieved of the action of external forces, undergoes a transformation which returns it to its initial state, it returns having a less kinetic energy than at the start.

Thus a simple machine, whose temperature cannot vary, and which must pass periodically through the same state, will so pass with a slower motion at each return; this is the *impossibility of perpetual motion,* a subject of engrossing study of mechanics toward the close of the eighteenth century and at the beginning of the nineteenth.

66. Continuously acting machines.—Let us now imagine *a continuously acting machine*, that is to say, a system which periodically repasses through the same state and whose various parts retake periodically the same velocity, which restores to the system the same kinetic energy. If the machine is kept at a constant temperature, each of these periods will constitute an isothermal cycle, to which will apply also the equality $W_1 = W_0$. The inequality (4) will then become

$$W_e > 0.$$

In order to keep a constant temperature machine in continuous operation, it is necessary that the external forces applied to this machine do positive work for each period.

Such a machine would not constitute a *motor;* a motor, in fact, is a machine which compels the external forces to act in a sense contrary to their natural tendency—which obliges, for instance, a weight to rise; each period of a motor in continuous action must correspond to negative work of the external forces; it is thus evident that as long as isothermal machines only are at our disposal we have not a true motor; to realize a motor capable of permanent action, it is necessary to consider a system which, during each period, undergoes variations of temperature; it is necessary to construct a *heat-engine.*

67. Statement of the Carnot-Clausius principle for an open reversible transformation. Entropy.—These propositions suffice to give a clue to the importance of the principle of Carnot and Clausius in the theory of heat-engines; but the study of the motive power of heat is not our object; so let us return to the principle itself and try to put it into such a form that it may be easily applied to the problems of chemical mechanics.

In the form that we have stated it, this principle applies to a closed cycle only; this is a troublesome condition for its application. The principle of the equivalence of work and heat, in the first form stated, possessed the same inconvenience (Chap. II, Art 21); we transformed it in such a way as to remove this inconvenience. and it is this transformation which introduced into our reasonings the quantity called *internal energy*. We shall transform the theorem of Carnot and Clausius in an analogous manner, and this transformation will introduce into our reasonings another quantity, *entropy*, which will not be of less importance than internal energy.

Let us consider in the first place, quite exclusively, reversible transformations.

The principle of Carnot and Clausius then takes the following form: If a system describes a reversible closed cycle, the transformation value ϵ, calculated for the whole cycle, is equal to 0.

This stated, suppose that a system passes from any initial state 0 to a final state 1 by a definite series M of reversible transformations; the transformation value, calculated for this first modification, has the value e; next suppose that the system returns, by a series μ of reversible transformations, from the state 1 to the state 0; the transformation value calculated for this second modification has the value η; the totality of these two series of transformations M and μ forms a reversible cycle for which the transformation value is $(e+\eta)$; the principle of Carnot and Clausius requires that

$$e+\eta=0. \tag{5}$$

Now imagine that the system passes from the same initial state 0 to the same final state 1 by a series of reversible transformations M', differing from the series M; for this series of transformations the transformation value will be e'; next suppose that the system returns to the state 0 from the state 1 by the series μ of reversible transformations, for which the transformation value is η; we shall have this time

$$e'+\eta=0.$$

But the two equalities obtained require that

$$e=e',$$

so that the principle of Carnot and Clausius involves the following proposition:

By whatever reversible path a system passes from a given initial state to a given final state, the transformation value remains constant.

It is clear, besides, that this proposition involves in its turn the exactness of the principle of Carnot and Clausius for a reversible closed cycle; in fact, if the system describes a reversible cycle, the transformation value will have, for this cycle, the same value as for every other reversible modification taking the system in the same initial state and returning it to this; consequently it has the same value as if the system was not modified at all, and this value is visibly 0; thus, for any reversible cycle whatever, $\varepsilon=0$ which is the statement of Carnot and Clausius's principle for such a cycle.

Thus, *so long as reversible cycles alone are considered*, the principle of Carnot and Clausius may be regarded as the exact equivalent of the following proposition:

The transformation value of a reversible modification depends on the initial and final states of the system, but not on the reversible modification chosen by which the system passes from this initial to this final state.

It suffices now to resume a reasoning similar to that which gave us the notion of potential (Chap. I, Art. 10) in order to transform the preceding statement into the following:

For every state X of the system there corresponds a quantity S_x, such that for every reversible modification which causes the system to pass from the state 0 to the state 1 the transformation value is given by the equality

$$\varepsilon=S_0+S_1.$$

Clausius, to whom is due these considerations, gave to the quantity S_x the name *entropy* of the system taken in the state X.

By means of this nomenclature equation (5) may be stated in the following way:

The transformation value of any reversible modification whatever is equal to the decrease in the entropy of the system brought about by this modification.

This proposition includes as a particular case the proposition relative to a reversible closed cycle. A closed cycle is in fact a transformation in which the final state 1 is identical with the initial state 0; S_1 is equal to S_2 and equation (5) gives equation (2).

68. Entropy of a perfect gas.—As an example, let us calculate the variation in entropy of a perfect gas of mass M when this gas passes from a state in which its volume is V_0 and its absolute temperature T_0 to a state in which its volume is V and its absolute temperature T.

To do this we may bring the gas from the initial state to the final state by whatever reversible process we please; we shall choose the following:

1°. The gas passes from the volume V_0 to the volume V, the temperature keeping the constant value T_0.

2°. At the constant volume V the gas passes from the temperature T_0 to the temperature T.

The first modification is isothermal; according to equation (1) its transformation value is $\frac{Q}{T_0}$, Q being the quantity of heat set free by the gas during this modification.

It is easy to calculate this quantity of heat Q.

The temperature being constant, the internal energy of the gas does not change (Art. 25). The modification being reversible, the gas is constantly in equilibrium and the kinetic energy is constantly zero. Equation (4) of Chap. II gives

$$Q=\frac{W_e}{E}.$$

Finally, the work of exterior pressure is given by equation (11) of Chap. I, which may be written

$$W_e=2.325P_0V_0(\log V_0-\log V).$$

The first modification furnishes thus a transformation value:

$$2.325\frac{P_0V_0}{ET_0}(\log V_0-\log V).$$

From equation (13) of Chap. II we have

$$P_0V_0=MR\sigma T_0,$$

and the preceding quantity becomes

$$(6) \qquad 2.325\frac{MR\sigma}{E}(\log V_0-\log V).$$

Let us now consider the second modification.

Separate it, as was done in Art. 62, into n partial heatings, all accomplished at the constant volume V, these heatings causing the system to pass first from the temperature T_0 to the temperature T_1, then from the temperature T_1 to the temperature T_2, finally from the temperature T_{n-1} to the temperature T.

If we denote by c the specific heat at constant volume of the gas considered, these various heatings set free respectively the following quantities of heat:

$$Q_0 = -Mc(T_1 - T_0),$$
$$Q_1 = -Mc(T_2 - T_1),$$
$$\cdot\ \cdot\ \cdot\ \cdot\ \cdot\ \cdot\ \cdot\ \cdot$$
$$Q_{n-1} = -Mc(T - T_{n-1}).$$

Form the sum

$$\frac{Q_0}{T_0} + \frac{Q_1}{T_1} + \ldots + \frac{Q_{n-1}}{T_{n-1}} = -Mc\left(\frac{T_1 - T_0}{T_0} + \frac{T_2 - T_1}{T_1} + \ldots + \frac{T - T_{n-1}}{T_{n-1}}\right)$$

and find the limit approached by this sum when the number n of partial heatings is indefinitely increased, by making each of them approach zero.

Evidently it is sufficient for this to find the limit of the sum:

$$\frac{T_1 - T_0}{T_0} + \frac{T_2 - T_1}{T_1} + \ldots + \frac{T - T_{n-1}}{T_{n-1}} = \Sigma.$$

Let us take (Fig. 21) two coordinates axes OT, Oy. Upon the axis of abscissæ OT lay off the values of the absolute temperature T. For each abscissa let there correspond a point M whose coordinate y is equal to $\frac{1}{T}$. The locus of the point M is a certain curve CC'; y being smaller as T is greater, this curve descends from left to right; it is one of those curves we studied in Art. 7 and that the geometers call *equilateral hyperbolas.*

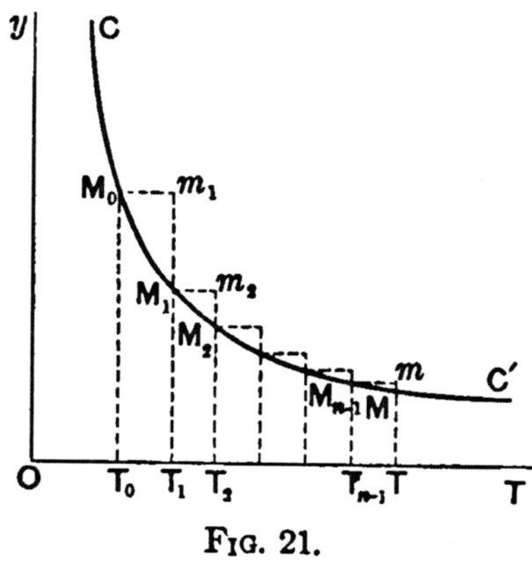

Fig. 21.

The quantity $\frac{T_1 - T_0}{T_0}$ is measured by the rectangle $T_0T_1M_0m_1$.

The sum

$$\frac{T_1 - T_0}{T_0} + \frac{T_2 - T_1}{T_1} + \ldots + \frac{T - T_{n-1}}{T_{n-1}} = \Sigma$$

is measured by the area included between the straight line T_0T, the two ordinates T_0M_0, T_0m, and the broken line $M_0m_1M_1m_2M_2 \ldots M_{n-1}m$.

When the points of division $T_1, T_2, \ldots, T_{n-1}$, marked between T_0 and T_1 become more and more numerous and nearer and nearer together, all the points on this broken line approach the points of the arc M_0M of the equilateral hyperbola CC'. The limit of the sum Σ therefore measures the area included between the straight line T_0T, the two ordinates T_0M_0 and TM, and the arc M_0M of the equilateral hyperbola.

Now we said in Art. 7 that the geometers can evaluate such an area, which has the value (denoting by *log* the common logarithm)

$$2.325 \log \frac{T}{T_0} = 2.325(\log T - \log T_0).$$

Thus our second modification has the transformation value

$$(7) \qquad 2.325Mc(\log T_0 - \log T).$$

Uniting the results of (6) and (7), keeping in mind equation (3) we may state the following:

When a perfect gas of mass M passes from a state in which it occupies the volume V_0 and in which its absolute temperature is T_0 to a state in which its volume is V and its temperature T, its entropy passes from the value S_0 to the value S, and we have

$$(8) \quad S = S_0 + 2.325M\left(\frac{R\sigma}{E}\log V + c\log T - \frac{R\sigma}{E}\log V_0 - c\log T_0\right).$$

69. Statement of the principle of Carnot and Clausius for a real, open transformation. Compensated and non-compensated transformations.—Let us now consider a real transformation M which causes the system to pass from the initial state 0 to the final state 1; what can we say concerning the transformation value ε for this change?

Consider a reversible transformation M' which returns the system from the state 1 to the state 0; according to equation (5), the modification M' has for transformation value $(S_1 - S_0)$. The ensemble formed by the modification M followed by the modification M' has, therefore, for transformation value

$$\varepsilon+S_1-S_0.$$

But this ensemble of changes brings back the system to its initial state; it is a closed cycle composed in part of real transformations, in part of reversible transformations; from this, according to the inequality (3), its transformation value must be positive. If, therefore, we denote by P a quantity essentially positive, we shall have

$$\varepsilon+S_1-S_0=P,$$

or

$$\varepsilon=S_0-S_1+P. \tag{9}$$

This equality characterizes any realizable modification whatever bringing the system from the initial state 0 to the final state 1.

This quantity P, of which we know nothing except that it is positive, has received from Clausius the name of *non-compensated transformation* relative to the realizable change which causes the system to pass from the state 0 to the state 1.

We see that *the non-compensated transformation relative to any realizable modification whatever is positive.* If we compare equation (5) with equation (9), we see that *a reversible modification is accompanied by a non-compensated transformation equal to zero.*

We shall give to the quantity (S_0-S_1), which reduces to zero for every closed cycle, reversible or not, the name of *compensated transformation* relative to the modification which causes the system to pass from the state 0 to the state 1.

A fundamental property distinguishes the compensated transformation from the non-compensated transformation. In order to know the compensated transformation which accompanies a modification undergone by a system, it is sufficient to know the initial and final states of the system; on the contrary, to know a non-compensated transformation, it is necessary, at least in general, to know the whole series of intermediary states that the system has passed through and the actions to which it was submitted while it passed through them.

70. Principle of the increase in entropy of an isolated system. —The laws that we have just stated lead to very important consequences.

Imagine that a material system be absolutely isolated in space; around it there is nothing, neither ponderable matter nor ether; nothing that may furnish it with heat nor take away heat.

This system undergoes a certain modification which brings it from the state 0 to the state 1; what, for this modification, is the transformation value? As was done in Art. 63, we may decompose this modification into a certain number of partial modifications $m, m', m'' \ldots$; we shall denote by $q, q', q'' \ldots$ the quantities of heat set free by the system in these various modifications and by $T, T', T'' \ldots$ the values of the absolute temperature at the beginning of each of them; we shall form the sum

$$\frac{q}{T}+\frac{q'}{T'}+\frac{q''}{T''}+\ldots,$$

and we shall find the limit approached by this sum when the changes $m, m', m'' \ldots$ become more and more numerous.

Now, the system being able neither to gain nor to lose heat, each of the quantities $q, q', q'' \ldots$ is equal to zero; it is the same for the preceding sum constantly and therefore for its limit.

Hence, when an isolated system undergoes a real modification, the transformation value for this modification is zero:

$$\varepsilon=0.$$

This equality taken with equation (9) gives

$$S_1-S_0=P,$$

or, since P is essentially positive,

$$S_1>S_0.$$

The modification considered thus causes the entropy of the isolated system to increase.

On the other hand, we have seen (Chap. II, Art. 23) that such a modification left the *total energy* of the system unchanged. We are therefore led to the following statement, which was given by Clausius: [1]

---

[1] R. Clausius, *Poggendorf's Annalen*, Bd. 125, p. 400, 1865.

If we consider a system absolutely isolated in space, every real modification of this system has the following properties:

1°. *It leaves its total energy constant.*

2°. *It causes the entropy to increase.*

71. Use of this principle in chemical statics.—The proposition that we have just established is of use in chemical statics.

The chemical system A is in a certain state and is submitted to the action of other bodies B; it is proposed to determine if, in these conditions, the system A remains in equilibrium.

The system A and all the bodies B which act upon it may compose a system isolated in space; let C be the total system; we shall find the entropy S of the system C.

We then consider all the modifications by which the system A may be imagined to change from the given state, and we calculate the change in value that they would give to the entropy S of the total system C. If none of these modifications causes the entropy S to increase, none of them can be realized and the system A remains forcibly in equilibrium in the state considered.

Such is the principle of chemical statics proposed by Horstmann [1] and by J. Willard Gibbs.[2] In its first form it has to be used somewhat carefully on account of the obligation to reckon in the calculation of the entropy, not only with the system A that is studied, but also with all the foreign bodies B which act on it.

We shall substitute another principle, theoretically equivalent, but more convenient to use.

72. Compensated and non-compensated work **in an isothermal modification.**—Let us turn to equation (9) and see what it becomes for an isothermal change.

If Q is the quantity of heat set free in this change taking place at the absolute temperature T, the transformation value is, accord- to equation (1),

$$\varepsilon=\frac{Q}{T},$$

[1] Horstmann, *Theorie der Dissociation* (*Liebig's Annalen der Chimie und Pharmacie*, Bd. 170, p. 192, 1873).

[2] J. Willard Gibbs, *On the equilibrium of heterogeneous substances* (Transactions of the Acad. of Sci. and Arts of Conn., v. 3, 1875 to 1878).

so that equation (9) becomes, for an isothermal modification

$$\frac{Q}{T}=S_0-S_1+P,$$

or

$$Q=T(S_0-S_1)+TP. \tag{10}$$

The first member of this equation (10) being a quantity of heat, the two terms which compose the second member must be quantities of the same kind as a quantity of heat; it is natural to call the first, $T(S_0-S_1)$, the *quantity of compensated heat* set free by the isothermal change, and the second, TP, the *quantity of non-compensated heat* set free by the same change.

The first, the quantity of compensated heat, has a definite value when we know the initial state 0 and the final state 1 between which the modification is taken. This is not so for the second; like the quantity P, it cannot be determined if the initial and final states alone are known, but only when all the intermediate states through which the system has passed and the external forces which have obliged the system to pass from each to the following are known.

These two quantities of heat are distinguished by a still more essential property.

The sign of the quantity of compensated heat has nothing arbitrary about it; if an isothermal change which brings the system from the state 0 to the state 1 liberates a positive quantity of compensated heat, a change leading from the state 1 to the state 0 will liberate a negative quantity of heat equal in absolute value to the preceding; in other words, it will absorb as much compensated heat as the first liberated.

On the contrary, the absolute temperature T is always positive; the non-compensated transformation P, equal to 0 for a reversible transformation, is positive for every realizable transformation; therefore *the quantity of non-compensated heat liberated by any realizable isothermal change is a quantity essentially positive; it is equal to* 0 *for a reversible change.*

Multiply the two members of equation (10) by the mechanical equivalent of heat E; this equality becomes

$$EQ = ET(S_0 - S_1) + ETP. \tag{11}$$

The first member of this equation, product of a quantity of heat by the mechanical equivalent of heat, is a quantity of the same kind as work; it is therefore the same for each of the two terms composing the second member. We shall give to the first,

$$\vartheta = ET(S_0 - S_1), \tag{12}$$

the name of *compensated work* accomplished in the isothermal change considered, and to the second,

$$\tau = ETP, \tag{13}$$

the name of *uncompensated work.*

Each of these two quantities of work possesses naturally the same characteristics as the quantity of heat to which it is equivalent. In particular we may state this fundamental proposition:

Every real isothermal change engenders positive non-compensated work; this work is zero for a reversible isothermal modification.

73. First form of the equilibrium condition of a system kept at a given temperature.—This proposition will furnish us with a criterion by which we may recognize if a system is in equilibrium at a given temperature in a given state.

Suppose that we examine all the infinitely small isothermal modifications which would be susceptible, if they were realized, of making the system leave its actual state; calculate the non-compensated work that each of these *virtual* modifications would cause; if all these modifications correspond to a zero or negative non-compensated work, none of them is realizable; the system cannot quit the state in which it is; it remains there forcibly in equilibrium; hence the following theorem:

A system taken in a given state, at a given temperature, is in equilibrium if all the isothermal modifications by which it may be conceived to quit this state correspond to a zero or negative non-compensated work.

74. Expression for non-compensated work accomplished in an isothermal modification.—Let us compare the equations (11) and (13); we find

$$\tau = EQ - ET(S_0 - S_1).$$

But if we denote by W the kinetic energy of the system in a given state, by U the internal energy in this same state, by W_e the work done by the external forces during the modification considered, the principle of the equivalence of heat and work gives [Chap. II, eq. (4)]

$$EQ = E(U_0 - U_1) + W_e + W_0 - W_1.$$

These two equations give us the following expression for the non-compensated work produced by a system which undergoes an isothermal modification:

$$\tau = E(U_0 - TS_0) - E(U_1 - TS_1) + W_e + W_0 - W_1. \tag{14}$$

This equation (14) gives rise to a great number of important remarks.

75. Characteristic function of a system.—In the first place equation (14) shows that, in order to form the expression for non-compensated work engendered in an isothermal change, it is not necessary to seek separately the variation of internal energy and the variation of entropy; it suffices to calculate the variation undergone by a certain quantity, entirely determined when the state of the system is known; it is the quantity

$$\mathcal{F} = E(U - TS). \tag{15}$$

In thermodynamics this quantity plays a fundamental rôle. Every equation furnished by thermodynamics is, in the last analysis, the statement of the quantity $\mathcal{F}$. When the value that this quantity takes in each state of the system is known, one may, by ordinary mathematical operations, determine the value, in each state, of the internal energy and entropy of the system, the external forces that must be applied to keep it in equilibrium in this state, the quantity of heat that must be furnished to raise its temperature or to produce any modification; it is possible, in a word, to determine all the mechanical and thermal properties of the system. The knowledge of the value that the quantity $\mathfrak{F}$ assumes for each state of the system therefore *characterizes* this system. Hence this quantity has been called *the characteristic function of the system* by Massieu,[1] who first treated it and established its principal properties.

76. Characteristic function of a perfect gas.—Let us calculate, as an example, the characteristic function of a perfect gas of mass M occupying the volume V at the temperature T. Let us choose arbitrarily an initial state in which the gas occupies the volume V_0 at the temperature T_0; let U_0, S_0 be the internal energy and entropy of the gas in this state; let U, S be the values of the same quantities when the gas occupies the volume V at the temperature T. Equation (9) of Chap. II and equation (8) of the present chapter give

$$U = U_0 - McT_0 + McT,$$

$$S = S_0 - 2.325M\left(\frac{R\sigma}{E}\log V_0 + c\log T_0\right) + 2.325M\left(\frac{R\sigma}{E}\log V + c\log T\right).$$

Denote by

$$A = E(U_0 - McT_0),$$

$$B = 2.325EM\left(\frac{R\sigma'}{E}\log V_0 + c\log T_0\right) - ES_0,$$

two quantities which do not depend upon V and T, and which therefore remain constant as V and T vary; hence equation (15) will give the following expression for the characteristic function of a perfect gas:

$$(16)\quad \mathfrak{F} = A + BT + EMcT(1 - 2.325\log T) - 2.325MR\sigma T\log V.$$

[1] F. Massieu, *Sur les fonctions caracteristiques*, Comptes Rendus, v. 69, pp. 858 and 1057, 1869; *Mémoire sur les fonctions caracteristiques*, Savants étrangers, v. 22, 1876.

$$(17) \qquad W_1 - W_0 - W_e = \mathfrak{F}_0 - \mathfrak{F}_1 - \tau.$$

Let us see what this equation teaches.

The change considered may be used to *produce work,* that is, to constrain the external forces to do negative work; W_e being this negative work of the external forces, the *work produced,* W_e', is the absolute value of this negative work: $W_e' = -W_e$. This change may also be used *to increase the kinetic energy* of a part of the system, as in the bow when the spring of the cord sends forth the arrow. In a general way, if the change considered is used as source of motive power, it may be said that its *useful effect* is measured by the quantity

$$(18) \qquad \mathcal{E} = W_1 - W_0 - W_e.$$

In virtue of this equation, we may write equation (17) as

$$(19) \qquad \mathcal{E} = \mathfrak{F}_0 - \mathfrak{F}_1 - \tau.$$

If we remember that the non-compensated work τ, equal to zero for every reversible isothermal change, is positive for every realizable isothermal change, we obtain without difficulty the following proposition:

If we consider all the isothermal modifications susceptible of causing a system to pass from a given initial state 0 *to a final state* 1, *also known, it may be shown that the useful effect of these changes is always less than the decrease* ($\mathfrak{F}_0 - \mathfrak{F}_1$) *in the quantity* $\mathfrak{F}$ *when the system passes from the state* 0 *to the state* 1. *The useful effect approaches this higher limit, without reaching it, when the realizable isothermal change approaches a reversible change.*

This fundamental proposition seems to have been appreciated for the first time by Maxwell;[1] in the first editions of his book Maxwell gave to the quantity $\mathfrak{F}$ the name *entropy of the system,* already employed by Clausius in a different sense; in the fourth

[1] J. Clerk Maxwell, *Theory of Heat,* p. 186 (London, 1871).

edition he calls it *available energy,* which had been suggested by Gibbs;[1] Helmholtz[2] gave it the name of *free energy;* to-day the name ordinarily used is the *internal thermodynamic potential of the system;* this name results from the rôle that this quantity plays in the study of equilibrium conditions in a system.

78. Definite form of the equilibrium condition of a system kept at a given constant temperature.—Referring to equation (17), we may write it in the form

$$\mathfrak{F}_0-\mathfrak{F}_1+W_e=\tau+W_1-W_0. \tag{17'}$$

Suppose the system placed in the state 0 without any initial velocities being imparted to its various parts; will it remain in equilibrium in this state or will it undergo some isothermal change, any reaction that will disturb this equilibrium?

Imagine that an isothermal reaction brings it from the state 0 to the state 1, near together; the initial kinetic energy was 0; the kinetic energy at the instant when the system is in the state 1 can be only positive or zero, and similarly for (W_1-W_0); as to the non-compensated work τ, it can only be positive; therefore every isothermal modification capable of removing the system from the state 0, where it was without initial kinetic energy, corresponds to a positive value of the quantity $(\tau+W_1-W_0)$ and consequently, by equation (17'), of the quantity $(\mathfrak{F}_0-\mathfrak{F}_1+W_e)$.

Whence the following proposition:

A system placed with no initial velocity in the state 0 *remains of necessity there in equilibrium if all the virtual isothermal modifications by which it might be brought from the state* 0 *to a neighboring state* 1 *verify the conditions*

$$\mathfrak{F}_0-\mathfrak{F}_1+W_e \leqq 0. \tag{20}$$

79. Internal thermodynamic potential.—Referring to what was said in Art. 14, we see that the equilibrium condition just

---

[1] J. Willard Gibbs, *A method of geometrical representation of the thermodynamic properties of substances by means of surfaces,* Trans. Conn. Acad., v. 2, p. 400, 1873.

[2] H. von Helmholtz, *Zur Thermodynamik chemischer Vorgänge,* Sitzungsberichte der Berliner Akademie, p. 2, 1882.

stated has exactly the form that, in mechanics, the equilibrium condition of the system constituted as follows would have:

This system is acted upon by the same external forces as the preceding; it is, besides, the seat of *internal forces* whose potential is $\mathfrak{F}$ (Art. 10).

Effectively, in a modification of such a system, while the external forces would perform work equal to W_e, the internal forces would do the work $(\mathfrak{F}_0 - \mathfrak{F}_1)$; equation (20) would therefore express the fact that the work of all the forces acting upon the system is negative or zero for every virtual modification capable of removing the system from the state 0. We come therefore to the conclusion:

There is an absolute analogy between the laws of equilibrium in thermodynamics and the statics of a mechanical system in which the internal forces admit a potential; the same rôle is played in the latter theory by the potential of the internal forces, and in the former by the internal thermodynamic potential.

80. Total thermodynamic potential under constant pressure or at constant volume.—In general the external forces acting upon a system do not admit a potential; but there are always special cases which admit a potential; in Art. 12 we cited two that are of especial interest for the chemist:

1°. When the external forces reduce to a *constant, uniform pressure* π they admit a potential of value

$$\Omega = \pi V, \tag{21}$$

denoting by V the total volume of the system.

2°. When the external forces reduce to a uniform pressure, constant or variable, if the *total volume of the system remains fixed*, the external forces have the potential

$$\Omega = 0. \tag{22}$$

Let us consider one of those particular cases in which *the external forces applied to the system have a potential*, and let Ω be this potential; when the system passes from the state 0 to the state 1 the external forces will do work given by the equation

$$W_e = \Omega_0 - \Omega_1 \tag{23}$$

and the first member of the condition (20) becomes

$$\mathfrak{F}_0+\Omega_0-(\mathfrak{F}_1+\Omega_1).$$

It is then equal to the increase undergone, from the state 0 to the state 1, by a certain quantity

$$\Phi=\mathfrak{F}+\Omega \tag{24}$$

which has, for each state of the system, a definite value. *This quantity, the sum of the internal thermodynamic potential and of the potential of the external forces, is called total thermodynamic potential.*

If we have to deal with a system kept at a constant uniform pressure π, this quantity becomes, by equations (21) and (24),

$$\Phi=\mathfrak{F}+\pi V. \tag{25}$$

This is the *total thermodynamic potential at constant pressure.*

If the system is kept at constant volume and at uniform pressure, by equations (23) and (24) this quantity becomes

$$\Phi=\mathfrak{F}. \tag{26}$$

This is the *total thermodynamic potential at constant volume.*

For the case in which the system studied admits a total thermodynamic potential the condition (20) becomes, by equations (23) and (24),

$$\Phi_0-\Phi_1\leqq 0. \tag{27}$$

Placed without initial kinetic energy in a state 0, *a system which admits a total thermodynamic potential remains in equilibrium in this state if none of the virtual isothermal modifications which may displace it do not cause the total thermodynamic potential to decrease.*

Hence a system is surely in equilibrium in a state in which the total thermodynamic potential has a minimum value among those that it may assume at the same temperature.

81. Stability of equilibrium.—If to this proposition we add the complement that we cannot demonstrate here: *Such a state is one of stable equilibrium,* we shall have finished noting the complete analogy between the thermodynamic statics of a system which admits a total thermodynamic potential and the statics of a mechanical system acted upon by forces admitting a potential (see Art. 16).

82. Interpretation of non-compensated work.—Consider again the equation

$$Q=T(S_0-S_1)+TP, \tag{10}$$

from which we have already drawn so many conclusions. We shall find an interesting interpretation of the non-compensated heat TP set free in this modification.

Let us suppose the kinetic energy of the system constantly negligible, which is ordinarily true for the changes of state that the chemist studies; equation (17), in which now $W_0=0$, $W_1=0$, gives

$$W_e=\mathfrak{F}_1-\mathfrak{F}_0+\tau,$$

or from equation (13), which defines τ,

$$W_e=\mathfrak{F}_1-\mathfrak{F}_0+ETP \tag{28}$$

The real isothermal change considered is composed of a series of states of the system; imagine that by means of external actions suitably chosen, whose nature it is not necessary to specify, we may keep the system in equilibrium in each of these states. The series of these states of equilibrium would form a reversible isothermal modification bringing the system from the same initial state 0 to the same final state 1 as the real modification considered; for this reversible modification the non-compensated transformation P would have the value 0; if, therefore, we denote by W_e' the work done by the external forces which act upon the system during this reversible modification, equation (28) would give, for this change,

$$W_e'=\mathfrak{F}_1-\mathfrak{F}_0. \tag{28'}$$

Equations (28) and (28′) give the equation

$$ETP=W_e-W_e'. \tag{29}$$

The non-compensated work which accompanies a real isothermal modification is the excess of the work done during this modification by the external forces really acting upon the system over the work that would be done during the same change by the external forces capable of maintaining the system constantly in equilibrium.

83. Intensity of reaction; slow reactions.—We often speak in chemistry of vigorous, intense, or energetic reactions without giving these words an exact meaning; what precedes allows us to fill this gap; we may take the difference $(W_e - W_e')$, which is always positive, as a measure of the *intensity* of the isothermal reaction considered; when the reaction is produced under conditions very near to those which would assure equilibrium at every instant, this intensity has a very small value; to have a large value, it is necessary that the reaction be produced under conditions extremely different from those assuring equilibrium, that we have a system very much out of equilibrium.

By means of equation (29), equation (10) becomes

$$Q = T(S_0 - S_1) + \frac{W_e - W_e'}{E}. \tag{30}$$

Let us suppose, in the first place, that the reaction be not at all intense; the quantity $(W_e - W_e')$ will have a very small value; the sign of the quantity Q will be that of $T(S_0 - S_1)$; but this last sign is not arbitrary; the quantity $T(S_0 - S_1)$ may be either positive or negative.

Suppose, for instance, that for this change the quantity $T(S_0 - S_1)$ and consequently the quantity Q are positive; the isothermal change considered liberates heat. This very slightly intense modification is very nearly reversible; that is to say, in conditions very slightly different from those that determine the modifications considered, there is produced a modification in the opposite direction, absorbing about as much heat as the first sets free.

Hence a *physical or chemical change of very slight intensity may either absorb or liberate heat.*

For example, accomplished at a constant temperature, the vaporization of water absorbs heat and the condensation of water vapor liberates heat.

The dissociation of the carbonate of calcium into lime and carbonic acid gas absorbs heat, and the combination of carbonic acid gas with lime sets free heat.

The oxidation of iron by water vapor liberates heat; the reduction of the magnetic oxide by hydrogen absorbs heat.

84. Very intense reactions; principle of maximum work.— Let us now consider the case, the opposite of the preceding, in which the modification studied is of great intensity; the difference $(W_e - W_e')$, which is always positive, has a very great value; it therefore gives its sign to the quantity Q. Whence the following proposition:

Every isothermal change of state, of great intensity, is accompanied by a liberation of heat.

In 1854 J. Thomsen[1] stated the following proposition, which Berthelot has called the PRINCIPLE OF MAXIMUM WORK:

In order that a chemical reaction may be produced at a temperature kept constant, it is necessary that this reaction be accompanied by a liberation of heat.

Thomsen took care to limit the application of this law to purely chemical reactions; it is but too clear in fact that it may not be applied to physical changes of state; at a temperature kept constant, one may observe the vaporization of a liquid, the fusion of a solid, the solution of sea-salt in water, yet all these changes of state absorb heat.

The principle of maximum work may not even be regarded as a principle applicable to all chemical reactions; at a fixed temperature, carbonate of calcium dissociates, hydrogen reduces magnetic iron oxide, in spite of the fact that these reactions absorb heat; we might cite an immense number of exceptions to the principle of maximum work, all chosen from among reactions of feeble intensity.

The principle of maximum work should therefore be limited to reactions of great intensity, for which it is applicable from the preceding deductions.

Thus restricted, this principle may, in a great number of cases, indicate to the chemist the direction of possible isothermal reactions.

Let us consider some examples:

The combination of hydrogen with chlorine, forming a mole-

---

[1] J. THOMSEN, *Poggendorf's Annalen der Physik und Chemie*, Bd. 92, p. 34, 1854.

cule of hydrochloric acid, liberates 22000 calories; the formation of a molecule of sulphurous acid gas from sulphur and oxygen sets free 69260 calories; also, there is observed the direct combination of hydrogen and chlorine, of sulphur and oxygen.

Chlorine and oxygen, in forming a molecule of hypochlorous acid gas, would absorb 15100 calories; three atoms of oxygen, in uniting to form one molecule of ozone, would absorb 30700 calories; oxygen, in combining with water to form a molecule of hydrogen peroxide with a great excess of water, would give rise to an absorption of 21700 calories; also, hypochlorous acid gas, ozone, hydrogen peroxide are substances whi h decompose spontaneously.

The reaction

$$Ag+Cl=AgCl$$

sets free 29000 calories, while the reaction

$$Ag+Br=AgBr$$

sets free only 27100 calories. It follows from this that chlorine will displace bromine from its combination with silver, for the reaction

$$AgBr+Cl=AgCl+Br$$

liberates $29000-27100=1900$ calories.

The reaction

$$2AgNO_3\,(\text{dilute})+Cu=Cu(NO_3)_2(\text{dilute})+2Ag$$

sets free 25300 calories; the reaction

$$Cu(NO_3)_2\,(\text{dilute})+Zn=Zn(NO_3)_2(\text{dilute})+Cu$$

liberates 61700 calories. It follows that copper should precipitate silver from a dilute solution of silver nitrate, and zinc should precipitate copper from a dilute solution of copper nitrate. Experiment confirms these conclusions.

A compound formed with absorption of heat cannot, according to the principle of maximum work, be formed in a direct and isolated manner; but an endothermic compound may be formed if its formation is the necessary consequence of reactions whose ensemble corresponds to a liberation of heat.

Thus hydrogen peroxide, which is, as we have just seen, an endothermic compound, may be formed when dilute hydrochloric acid acts on barium dioxide, for under these conditions the reaction

$$BaO_2+2HCl=BaCl_2+H_2O_2$$

sets free 22000 calories.

These examples, which might be indefinitely multiplied, will show the variety of cases to which the principle stated by Thomsen may be applied. We must not, however, entertain any illusions as to the generality of this principle.

85. Available mechanical effect of an adiabatic change.—All the propositions stated from Art. 72 to the preceding (with the exception of Arts. 75 and 76) suppose that the changes imposed upon the system studied are *isothermal*, that is to say, accomplished without any change of temperature. Without this condition these propositions may lead to false results. We shall give an example.

Instead of treating an isothermal change, let us consider an *adiabatic* change; such a change is so called in thermodynamics when the system neither loses nor gains heat.

The equation which expresses the principle of equivalence between heat and work [Chap. II, equation (4)],

$$EQ=E(U_0-U_1)+W_e+W_0-W_1,$$

becomes in the present case, for which $Q=0$,

$$W_1-W_0-W_e=E(U_0-U_1). \tag{31}$$

Now the first member of this equation is what we have called the available mechanical effect of the change considered; hence we have this proposition:

The available mechanical effect of an adiabatic change is obtained by multiplying the mechanical equivalent of heat by the decrease sustained in the internal energy of the system during this change.

This rule, evidently, is very different from the one obtained for an isothermal change (Art. 77).

Suppose that the adiabatic change considered causes the system to pass from the state 0 to the state 1 in which it has the same temperature as in the state 0; we may then pass by an isothermal change from the same state 0 to the same state 1; let $\mathcal{E}_Q$ be the available mechanical effect of the adiabatic modification and $\mathcal{E}_T$ the available mechanical effect for the isothermal change; let us compare the two effects.

Equation (31) gives

$$\mathcal{E}_Q = E(U_0 - U_1), \tag{31'}$$

while equation (19) gives

$$\mathcal{E}_T = \mathfrak{F}_0 - \mathfrak{F} - \tau.$$

Taking account of the equation

$$\mathfrak{F} = E(U - TS), \tag{15}$$

it is clear that we have

$$\mathcal{E}_Q - \mathcal{E}_T = \tau + ET(S_0 - S_1) = ETP + ET(S_0 - S_1).$$

But by equation (11), denoting by Q the quantity of heat set free in the isothermal change. this equality becomes

$$\mathcal{E}_Q - \mathcal{E}_T = EQ. \tag{32}$$

If a system may be brought from the same initial to the same final state, on the one hand by an isothermal change, on the other hand by an adiabatic change, the available mechanical effect produced in the latter case exceeds the available mechanical effect produced in the former by a quantity exactly equal to the quantity of heat liberated by the isothermal change.

According as the isothermal reaction is *exothermic* $(Q>0)$ or *endothermic* $(Q<0)$, its available mechanical effect is smaller $(\mathcal{E}_T<\mathcal{E}_Q)$ or greater $(\mathcal{E}_T>\mathcal{E}_Q)$ than the available mechanical effect of the corresponding adiabatic change.

Furthermore, the available mechanical effect $\mathcal{E}_Q$ of the adiabatic change may exceed the limit $(\mathfrak{F}_0-\mathfrak{F}_1)$ that the mechanical effect $\mathcal{E}_T$ of the isothermal change could not reach. We have in fact

$$\mathcal{E}_Q - (\mathfrak{F}_0 - \mathfrak{F}_1) = (U_0 - U_1) - (\mathfrak{F}_0 - \mathfrak{F}_1),$$

or, by equation (15),

$$\mathcal{E}_Q - (\mathfrak{F}_0 - \mathfrak{F}_1) = ET(S_0 - S_1).$$

The mechanical effect of an adiabatic change for which the final temperature is identical with the initial temperature exceeds the limit $(\mathfrak{F}_0-\mathfrak{F}_1)$, *which could not be reached by the mechanical effect of an isothermal change, by a quantity equal to the compensated work which accompanies this latter change.*

This compensated work may also be positive.

86. Application to the theory of explosives.—These considerations are of great importance in the theory of explosives.

The charge which detonates in pushing the ball in the bore of a cannon undergoes such rapid reactions that we may regard as quite feeble the exchanges of heat between this change and the metal of the cannon and projectile. For a first approximation we may neglect these exchanges and treat as adiabatic the changes which are produced in the bore of the piece.

The charge is taken at the ordinary temperature, which we shall call T_0; at the moment it explodes, the temperature of the products formed increases enormously; then these products expand, expelling the projectile, and in this expansion they are cooled; the temperature and pressure of these products at the moment when the base of the projectile leaves the mouth of the piece depend evidently upon the peculiarities of the combustion of the powder and of the path of the passage of the bullet through the bore of the piece. Denote by 1 this final state. The mechanical effect produced is

$$\mathcal{E}_Q=E(U_0-U_1), \tag{31'}$$

U_0 being the internal energy of the charge in the initial state and U_1 the internal energy of the products of the combustion taken in the final state.

Suppose that at the moment the projectile quits the bore of the piece the combustion of the powder is complete; the products of the combustion at this moment will be entirely in the gaseous state, and we may without serious error consider them as perfect gases. Then to know the internal energy U it will be sufficient to know the mass M of the gaseous mixture, equal to the mass of the charge, the chemical nature of this mixture, and the temperature T_1 to which it is brought; it will not be necessary to know the pressure (see Art 25).

The temperature T_1 differs from the external temperature T_0; in all firearms it is notably higher:

(33) $$T_1 - T_0 > 0.$$

Let u_0 be the internal energy, at the temperature T_0, of the gaseous mixture furnished by the combustion of the explosive in the bore of the piece; let c be the specific heat at constant volume of this mixture; we shall have [Chap. II, eq. (9)],

$$U_1 = u_0 + Mc(T_1 - T_0).$$

Equation (31′) will then become

(34) $$\mathcal{E}_Q = E(U_0 - u_0) + EMc(T_1 - T_0).$$

Taken together with equation (33), this equality shows that the mechanical effect of the explosion, supposed adiabatic, is less than the quantity

(35) $$P = E(U_0 - u_0).$$

It is the nearer to this higher limit as the temperature of the gases in the bore of the piece, at the moment the ball quits it, is nearer to the ordinary temperature.

This quantity P represents the mechanical effect that the charge of explosive considered would produce in detonating under the most favorable conditions; in ballistics it is called *the potential of this explosive charge.*

This quantity may be determined by means of data furnished by chemical calorimetry.

Suppose, for instance, that the given explosive charge, taken at the ordinary temperature T_0 (or +15° C.), undergoes at constant volume a series of modifications *having for last stage a gaseous mixture of the same composition as th· mixture contained in the piece at the departure of the ball;* suppose, besides, that at the end of this modification the temperature becomes T_0. This change liberates a quantity of heat λ. The kinetic energy is zero at the beginning and at the end of the change. The volume of the system studied has remained constant, so that the external forces have done no work. The principle of the equivalence of work and heat gives, therefore,

$$\lambda = U_0 - u_0,$$

or, by equation (35),

$$P = E\lambda. \tag{36}$$

The computation of the explosive potential of the charge is thus reduced to the determination of λ, that is to say, to a problem of chemical calorimetry. The whole difficulty of the problem, from the experimental point of view, consists in assuring oneself that the reaction produced in the calorimeter and the reaction produced in the bore of the piece really give rise to the same gaseous mixture.

Below are the potentials of some common explosives expressed in gramme-metres.[1]

POTENTIALS OF SOME COMMON EXPLOSIVES.

No.	Name.	Potential.
1	War-powder (black powder)	270×10^3
2	Fulminate of mercury	173 "
3	Ammonium nitrate	267 "
4	Picric acid	323 "
5	Ammonium picrate	254 "
6	Enduanitric coton (gun-cotton)	457 "
7	Octonitric coton (collodion)	313 "
8	Nitroglycerine	669 "

It is evident that it is useful to be informed in an exact manner on the intensity of the mechanical effect that may be attained with a given explosive under the most favorable conditions.

[1] SARRAU, *Poudres de guerre, ballistique intérieur,* Cours de l'École d'application de l'artillerie et du génie, 1893.

CHAPTER VI.

THE PHASE RULE.

87. The number of independent components of a chemical system of given kind.—After having discussed the principles of thermodynamics which are the basis of modern chemical mechanics, we come to the second part of our task: the exposition of the consequences that may be deduced from these principles. It is naturally by the aid of algebra that this passage from principles to consequences is made; but, setting aside in this work the mathematical developments that the reader, if he desires, may get elsewhere, we shall give merely the results and compare them with the teachings of experiment.

When we have reduced the representing of a problem by an equation to be no more than an algebraic expression, the first point we have to examine is the following: How many distinct quantities are there in the equations of the problem? And the examination of this point is immediately followed by the study of this other point: Among these distinct quantities, how many independent relations does algebra furnish? In making this double enumeration for the problems of chemical mechanics, J. Willard Gibbs was led to the propositions whose ensemble constitutes the *phase rule.*

Two numbers characterize a system: the number of *independent components* which form it and the number of *phases* into which it is divided.

A chemical system is always composed of a certain number of simple substances; but ordinarily, by reason of conditions imposed upon the system studied, conditions that give in a sense the *definition* of this system, determining the *kind,* the masses of these various elements cannot be taken arbitrarily; there exist certain relations among them. Thus, a system containing calcium carbonate, lime, and carbonic acid gas consists of calcium, carbon, and oxygen; but the masses of these three elements cannot be taken arbitrarily; in saying that these simple substances were grouped in the system so as to form exclusively calcium carbonate, lime, and carbonic acid gas, we have imposed upon these three masses a certain condition; we may choose arbitrarily two of these masses, the mass of calcium and that of carbon, for example; but the third mass, that of oxygen, is then determined without ambiguity.

We may always group the simple bodies which form a system of a given kind into a certain number of components or of residues of components, so that the mass of each of these groupings may be chosen in an arbitrary way, without contradicting the definition itself of the kind of systems that we study; thus, in saying that the systems are to be composed of calcium carbonate, lime, and carbonic anhydride, we have defined their *kind;* by taking arbitrary masses of calcium, carbon, and oxygen, we could not in general compose from them a system of the kind studied; but in taking arbitrary masses of lime and of carbonic anhydride, we may always compose such a system; the lime and carbonic anhydride are, for the kind of systems studied, *independent components.*

In many cases it is possible to choose in many different ways the independent components of the systems of a given kind; for instance, take systems formed of hydrated crystals of sodium acetate and of an aqueous solution of sodium acetate; we may take as independent components of the systems of this kind water and hydrated sodium acetate; we may take also, for independent components, water and anhydrous sodium acetate. But if the *nature* of the independent components of a definite kind of chemical systems may, in certain cases, show a certain indefiniteness, the *number* of these independent components can show none; it is easy to demonstrate the following theorem: *The number of independent components of systems of a given kind is always the same, whatever the manner of grouping the independent components of the elements which form the systems of the kind considered.*

Thus in the systems that we have taken for illustration the

independent components may be chosen in two different ways; but whichever way is adopted, their number remains equal to 2.

It is in general a very easy matter to determine the number of independent components for a kind of chemical system when the formulæ of the bodies taking part are known; hereafter we shall denote by the letter c the *number of independent components* which form the systems of the kind studied.

88. Number of phases into which a given system is divided.—The heterogeneous systems studied by the chemist are divided into a certain number of homogeneous masses; a system formed of water and water vapor is divided into a homogeneous mass of liquid water and a homogeneous mass of water vapor.

Several of the homogeneous masses which compose a heterogeneous system may have the same nature, the same chemical and physical properties; in a bath where a saturated solution of sodium chloride deposits the salt that it contains, the crystals of salt adhere to various parts of the walls of the bath; although separated from each other, these crystals have all the same composition and the same properties; if they were joined together they would form a homogeneous solid. These different masses, which, although separated from each other, have the same composition and properties, belong to the same phase; in the crystal bath of which we have just spoken, the salt crystals constitute one phase; the saline solution constitutes another phase; the system has two phases.

It is in general very easy to enumerate the distinct phases which are encountered in a given system.

A homogeneous mixture contains a single phase.

A system formed of liquid water and water vapor has two phases, the liquid and the vapor; a system containing salt crystals and a solution of this salt has likewise two phases, the solid salt and the solution.

A system formed of calcium carbonate, of lime, and of carbonic acid gas has three phases: solid calcium carbonate, solid lime, carbonic acid gas.

We shall denote by the letter ϕ the *number of phases* possessed by any system.

89. Fundamental hypothesis.—Let there be a system having the ϕ phases $1, 2, 3, \ldots, \phi$; let $M_1, M_2, \ldots, M_\phi$ be the masses of these various phases; the analysis of J. Willard Gibbs is based on the following principle:

The internal thermodynamic potential, $\mathfrak{F}$, of the system is given by the following formula:

$$\mathfrak{F} = M_1\mathfrak{F}_1 + M_2\mathfrak{F}_2 + \ldots + M_\phi\mathfrak{F}_\phi. \tag{1}$$

In this formula $\mathfrak{F}_1$ is a quantity whose value depends upon the temperature, upon the nature of the phase 1, of its composition, of its density; we may say that $\mathfrak{F}_1$ is the internal thermodynamic potential which would characterize a unit mass of the phase 1, if we considered separately this mass at the same temperature, in the same state, with the same composition and density that it has in the midst of the system. Analogous considerations apply to $\mathfrak{F}_2 \ldots \mathfrak{F}_\phi$.

Is the truth of this principle evident and absolute? Not at all. It may be said that this principle regards as negligible the forces that the different parts of the system exercise on each other. This hypothesis may, in a great number of cases, be near enough to the truth so that the consequences deduced from it accord in a sufficiently satisfactory manner with the facts; but it is to be expected that this may not always be the case; we shall see, in fact, that certain phenomena do not accord with the laws that we are going to develop; to take account of these phenomena it will be necessary to renounce the too simple principle that we have just stated (see Chap. XVII).

Nevertheless, in an immense number of cases, this principle will lead us to laws which will be in most satisfactory accord with experiment; these are the cases of which we shall now treat.

90. Variance of a system.—From the point of view of simplified chemical statics to which we shall be led by making use of the hypothesis implied in formula (1), a system is characterized by the number c of its independent components which form it and by the number ϕ of phases into which it is divided; in a still more precise manner, *the form of the law of equilibrium for a chemical system depends exclusively upon the number*

$$V = c + 2 - \phi, \tag{2}$$

which we shall call the VARIANCE *of the system.*

91. Systems of negative varian.e.—Let us enumerate the forms that the law of equilibrium takes according to the value of this variance.

In the first place, *if the variance of the system is negative, the system can be in equilibrium at no temperature and under no pressure;* thus it would be impossible to have, at any temperature or pressure, in a state of equilibrium a system which would simultaneously include sulphur vapor, liquid sulphur, the orthorhombic form and the monoclinic form; such a system would enclose a single component divided into four phases; from equation (2) the variance would be -1.

92. Invariant systems.—When the variance of a system is equal to zero the system is called INVARIANT; *there exists but one temperature and one pressure for which an invariant system may be in equilibrium;* the composition and the density of each of the phases which compose the system in equilibrium are, besides, determined; but this is not so for the mass of each phase; even if the total mass of each of the independent components which form the system is given, it would be possible to divide in an infinite number of different ways these components into phases having the composition proper for equilibrium.

Consider, for example, a system which includes water simultaneously in each of its three states of ice, liquid, and vapor; this system is formed of a single independent component ($c=1$), separated into three phases ($\phi=3$); it is invariant; it can therefore be in equilibrium only at a definite temperature and at a definite pressure; very careful investigations have shown that this temperature, very near to 0° C., has the value +0°.0076 C., and that this pressure is the tension of saturated water vapor for this temperature, or about 4.60 mm. In these conditions the state of ice, liquid water, and vapor is determined without ambiguity, but this is not so for the mass of each of these three phases; one may without disturbing the equilibrium give to these three masses any values, provided that the sum of these three values remains constantly equal to the total mass of the system.

93. Monovariant systems. Transformation tension at a given temperature. Transformation point under a given pressure.—A system whose variance is equal to unity is called a MONOVARIANT system. In order to have a monovariant system in equilibrium either the pressure or the temperature may be arbitrarily chosen. *At a given temperature the pressure for which the system is in equilibrium has a definitely determined value which is called the* TRANSFORMATION TENSION *at the temperature considered.* The composition and the density of each of the phases which form the system in equilibrium are likewise determined; as with the transformation tension, they do not depend upon the masses of the independent components which constitute the system; on the contrary, the masses of the various phases are not entirely determined, not even when the masses of these independent components are given.

Under a given pressure the temperature for which the system is in equilibrium has a definite value, which is called the transformation point under the pressure considered; this equilibrium temperature does not depend upon the masses of the independent components making up the system, and it is the same for the composition and density of each of the phases of the system into which the system is divided, at the instant of equilibrium; besides this, the masses of these phases are not entirely determined, not even when the masses of the independent components are given.

94. Examples of monovariant systems.—The most simple type of monovariant system is furnished by a liquid in the presence of its vapor; a single component ($c=1$) is divided into two phases ($\phi=2$).

For a given temperature, equilibrium corresponds to a completely determined value of the pressure, which is the *tension of saturated vapor* at the given temperature; the densities of the liquid and of the vapor have definite values called the *density of saturated liquid* and the *density of saturated vapor* at this temperature; on the other hand, the mass of liquid and the mass of vapor that the system includes are not determined; we may impose upon these two masses all the variations that have their sum constant, equal to the total mass of the system.

Under a given pressure there is an equilibrium temperature which is entirely determined by the knowledge of this pressure; it is the *boiling-point* under the pressure considered.

A system which includes a single substance at once in the solid and liquid states is also a monovariant system; under a given pressure equilibrium corresponds to a definite temperature which is the *fusing-point* under the pressure considered; and the fusing-point depends upon this pressure alone.

A system which contains a gas such as cyanogen and a polymerous solid from this gas, such as paracyanogen, is also a monovariant system; similarly, at a given temperature, it is necessary for equilibrium that the gas reach a definite tension; this tension, which depends upon the temperature alone, is the *transformation tension* for this temperature; this is, in fact, the law found to be true by Troost and Hautefeuille in their classic investigations.

A system containing calcium carbonate, lime, and carbonic acid gas consists of two independent components ($c=2$) divided into three phases ($\phi=3$); it is a monovariant system; at a given temperature the system is in equilibrium for a definite value of the pressure, called the *dissociation tension* of calcium carbonate at the given temperature; this tension depends exclusively upon the temperature; it depends in no wise upon the masses of the independent components, lime, and carbonic anhydride, which make up the system; this is the celebrated law predicted by Henri Salnte-Claire Deville, demonstrated by Debray for the case that we have just taken as example, and confirmed by Debray and by G. Wiedemann when studying the dissociation of hydrated salts, and by Isambert from a study of the dissociation of compounds that ammonia gas forms with certain metallic chlorides.

The number of monovariant systems is very great. An anbydride or a hydrated salt is taken in the presence of a water solution of this salt in the presence of water vapor; two independent components ($c=2$), the salt and the water, exist in three phases ($\phi=3$), the solid salt, the solution, the vapor; thc system is monovariant; for every temperature there is an equilibrium state for the system; the temperature once given, the tension of the water vapor and the concentration of the solution for the system in equilibrium have definite values.

Chlorine is dissolved in water; crystals of chlorine hydrate are deposited from the solution, which has above it a gaseous atmosphere which is a mixture of chlorine and water vapor; two independent components ($c=2$), water and chlorine, exist in three phases ($\phi=3$), the solution, the chlorine hydrate crystals, and the gas mixture; the system is therefore monovariant; for to each temperature corresponds a state of equilibrium of the system; for this state of equilibrium the tension of the gas mixture is determined, as is also the composition of the liquid mixture; this law has been verified by Isambert and by H. Le Chatelier for mixtures of chlorine and water; it has been verified by Wroblewski, Bakhuis Roozboom, and P. Villard for various other mixtures with gases which form hydrates.

A mixture of ether and water separates into two layers; the one, more rich in ether and therefore the lighter, floats, while the other, richer in water, occupies the bottom of the vessel; a mixed vapor is above the liquids; two independent components ($c=2$), ether and water, have three phases ($\phi=3$), the two liquid layers and the mixed vapor; the variance of the system has the value 1; for each temperature the system may be observed in equilibrium; and it is sufficient to state the temperature in order to know the tension of the mixed vapor at the instant of equilibrium, the composition of the vapor, and the two layers of liquid.

These examples, that might be multiplied, suggest the infinite variety of types of monovariant systems; and nevertheless, in spite of the diversity of these types, the value of the variance common to them all imposes upon them all the same form of the law of equilibrium; in all we find a *transformation tension* depending solely upon the temperature.

95. **Rôle of monovariant systems in the history of chemical mechanics.**—The rôle that the monovariant systems have played in the history of chemical mechanics is well known; it is because they appealed to monovariant systems that Debray, Isambert, Troost, and Hautefeuille found, in the study of chemical decompositions, in the study of allotropic modifications, a *dissociation tension*, a *transformation tension*, analogous to the tension of saturated vapors; it is in showing the analogy between the dissociation tension, the transformation tension, and the tension of saturated vapor that they have made even the most skeptic accept the far-reaching thought of Henri Sainte-Claire Deville: There is no chemical mechanics distinct from physical mechanics; all

physical changes of state and changes of chemical composition depend upon the same general laws.

96. Bivariant systems.—The importance of monovariant systems should not make us forget the not less important BIVARIANT systems.

Such are called, evidently, those systems whose variance is 2; they are therefore the systems existing in ϕ phases equal in number to the independent components c of which they are composed.

A bivariant system may be in equilibrium at any pressure and at any temperature; when the temperature and pressure are given, the density and composition of each phase are known; they depend in no wise upon the masses of the independent components of the system; furthermore, if these masses are known, the mass of each of the phases into which the system is divided is in general determined.

A very simple case of a bivariant system is furnished by a solid salt in presence of an aqueous solution of this salt; two independent components, the salt and the water, have two phases, the solid salt and the solution. For every temperature and pressure such a system is in equilibrium; the solution is then *saturated* with the salt; the concentration of the saturated solution depends upon the temperature to which it is brought and upon the pressure it supports; but it is independent of the masses of salt and water that the system contains. Also, if to a knowledge of the temperature and pressure we join the knowledge of the total mass of the salt and water in the system, the masses of the solution and of the undissolved salt are determined.

97. Remark on the law of equilibrium of bivariant systems.—Here we must guard against a possible confusion. We have said that, when the temperature and pressure were stated, the concentration of the saturated solution was *determined;* we understand by this that it is impossible, at a given temperature and pressure, to find a series of saturated solutions such that the concentration varies in a continuous manner from one solution to the following; but it is not to be understood that the constitution of the saturated solution is determined *without ambiguity;* it may happen, in fact, that to a given temperature and pressure correspond *two distinct concentrations* of the saturated solution; if, for example, we studied a system where such a solid hydrate is in presence of a liquid mixture of water and anhydride, we could, for a given temperature and pressure, obtain two saturated solutions of distinct compositions, the one containing more water than the solid hydrated salt, the other containing less water than this salt. If, besides the pressure and temperature, the total mass of the anhyhride and that of the water which compose the system are given, the ambiguity will be removed and the equilibrium state of the system will be completely determined.

A similar remark may be made relative to the composition of each of the phases of a bivariant system in equilibrium at a given pressure and temperature; it is a remark whose importance we shall see while studying in Chap. XI the *indifferent states* of a bivariant system.

The water solution of a salt, in the presence of this solid salt gave us a first example of a bivariant system. For another, consider a definite mass of ether into which it poured increasing quantities of water; the first quantities of water poured in mix completely with the ether; but beyond a certain point the mixture divides into two layers, an upper one richer in ether and a lower richer in water; we therefore have to do with a system formed of two independent components, ether and water, and divided into two phases, the two superposed liquid layers; such a system is bivariant; and if the temperature and pressure rest constant, the composition of the two liquid layers will remain invariable; as water is little by little added to the mixture, we see the upper mass decrease and the lower mass increase, but neither the concentration of the upper nor of the lower layer undergoes any change up to the moment when enough water has been added to cause the upper layer to disappear; the system will then cease to be bivariant.

A great number of important problems in chemical statics are dependent upon the study of bivariant systems. The theory of the solubility of gases is the theory of a bivariant system; for the two independent components, the gas and the solute, exist in two phases, a liquid solution and a gaseous atmosphere, mixed or simple according as the solute is volatile or not. The theory of the vaporization of the mixture of two volatile liquids, the theory of the liquefaction of the mixture of two gases also depends upon the study of a bivariant system, for two fluids which play the rôle of two independent components are divided into two phases, the liquid and vapor mixtures.

98. There are contradictions to the phase rule.—What we have just said, and also what we shall say in the following chapter, show the extreme importance of the phase rule in chemical mechanics.

Does it follow that the phase rule is absolutely true and that it never encounters an experimental contradiction? It is not so. Observation shows us a considerable number of facts which are irreconcilable with the phase rule or with the various laws that depend upon the same principles as this rule.

Let us consider the following example, which later we shall consider more at length:

A glass tube encloses two independent components, selenium and hydrogen; there are two phases; a liquid phase, occupying the lower part of the tube, consists of selenium which has dissolved selenium hydride; a gaseous phase, filling the upper portion of the tube, contains hydrogen, the vapors of selenium and of selenium hydride gas.

This system, consisting of two components, and divided into two phases, is a bivariant system; according to the phase rule it may be in equilibrium at every pressure and temperature, but once the temperature and pressure fixed, the composition of each of the two phases of the system is determined. This is, in fact, the law obeyed by the system when the temperature is sufficiently high—when, for instance, this temperature exceeds 350° *C.*

But it is not the same when the temperature is lower—when, for instance, it is equal to 200° or 250° *C.* When the pressure and temperature are given, the composition of the two phases in equilibrium is not at all determined. At a given temperature and pressure the gaseous mixture may have all possible compositions between two limiting values, the one rich in selenium hydride, the other poor in this substance. Thus there may be an infinity of systems in equilibrium which should not be the case according to the phase rule.

99. J. Moutier's rule concerning these contradictions.—A great number of analogous examples might be cited; the examination of all these cases would lead to the following conclusion, first announced by J. Moutier:[1]

In all cases that thermodynamics, with the aid of the principles and hypotheses mentioned previously, announces that a certain state will be, for the system studied, one of equilibrium, experiment shows that the system, put in this state, will remain there actually in equilibrium. But when thermodynamics announces that the system studied, when placed in a certain state, will undergo a definite modification, it may happen that the system, put in this state, will remain there in equilibrium.

In other words, experiment always recognizes the existence of all the equilibrium states predicted by thermodynamics; but it recognizes, besides, the existence of a great number of equilibrium states which contradict the predictions of thermodynamics. For these equilibrium states that observation recognizes, while thermodynamics, such as we have developed it to the present time, may not predict the existence, we shall give the name of STATES OF FALSE EQUILIBRIUM.

The states of false equilibrium will be treated at length towards the end of this work; but it was necessary to point out their existence now; indeed, we shall have constantly to note the existence of these states when we wish to compare the indications of theory with experimental results.

---

[1] J. MOUTIER, *Bulletin de la Société philomathique,* 7th Series, v. 4, p. 86, 1880; *Sur quelques relations de la Physique et de la Chimie* (Frémy's Chemical Encyclopedia, v. 2, 1881.)

CHAPTER VII.

MULTIVARIANT SYSTEMS.

I. Trivariant Systems.

100. Multivariant systems. Trivariant systems.—We have just described briefly the monovariant and bivariant systems. These systems are at once the simplest and the most important in chemical mechanics; hence several chapters will be devoted to them.

For the moment we are going to demonstrate the utility of the phase rule by studying the multivariant systems, that is to say, the systems whose variance is at least equal to 3. The complexity of these systems is in general so great that they could hardly be comprehended without the aid of the rule for which we are indebted to Gibbs.

When the number of independent components exceeds by unity the number of phases into which the system is divided ($c=\phi+1$), the variance is equal to 3; the system is Trivariant.

In order to know completely the composition of the phases into which a trivariant system in equilibrium is divided, it does not suffice to know the temperature and the pressure; it is necessary to add a third quantity.

101. Theory of double salts.—Let us consider an example of a trivariant system.

Two anhydrides, that we may denote by 1 and 2, have been dissolved in water, denoted by 0; the liquid mixture is present with a solid body formed by the combination, in definite proportions, of the three substances 0, 1, 2 or of some of them; this solid body may be merely ice or one of the anhydrides 1 and 2; it may be a simple hydrated salt formed by the salt 1 or by the salt 2; it may be, finally, a double salt, anhydride or hydrate; in all cases the system is formed of three independent components, water 0 and the salts 1 and 2, divided into two phases, the solid and the liquid.

We have supposed that the solid state was a definite compound; the composition of the liquid phase may therefore alone vary; we may represent this composition by means of the two concentrations

$$s_1=\frac{M_1}{M_0}, \quad s_2=\frac{M_2}{M_0},$$

M_0, M_1, M_2 being the three masses of water and of the salts 1 and 2 contained in the solution.

It will not suffice, in order to know the values of the two concentrations s_1, s_2 of a solution in equilibrium with the solid, to know the temperature and the pressure. At a given temperature and pressure an infinity of values of the concentrations s_1, s_2 may be obtained, for which there is equilibrium between the solid body and the solution; if it is desired to have the concentrations s_1, s_2 determined, it is necessary to join to the temperature and pressure a third quantity conveniently chosen; one may, for instance, take for the supplementary known quantity one of the two concentrations s_1, s_2; the other is then entirely determined.

102. Surface of solubility of a double salt at a given pressure.—Imagine that it is desired to study a similar system under the constant pressure π, equal, for example, to the atmospheric pressure; take three rectangular axes OT, Os_1, Os_2 (Fig. 22), on which we lay off lengths proportional respectively to T, s_1, s_2; besides the pressure π, let us assume a value of the temperature T and of the concentration s_1; these values determine a point m in the plane TOs_1. It follows, when π, T, and s_1 are given, that the value assumed by the concentration s_2 of the solution in equilibrium with the solid body is determined; through the point m draw a line mM parallel to Os_2 and whose length is proportional to s_2; the point M, which has the coordinates T, s_1, s_2, represents the composition of

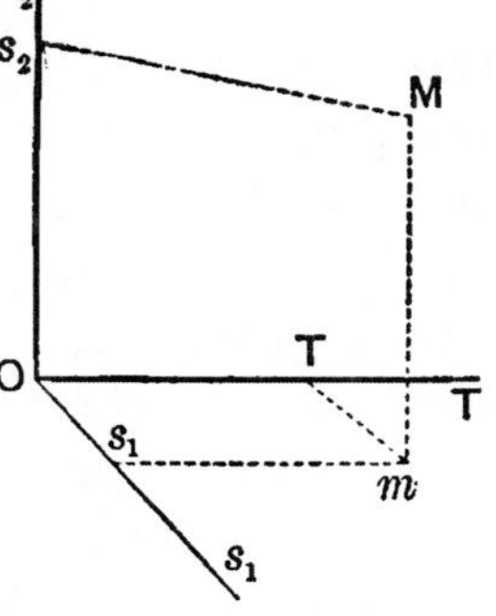

FIG. 22.

a solution susceptible of being in equilibrium at the pressure π, at the temperature T, with the solid phase considered. For every system of values of T and s_1, or, in other words, for every point m in the plane TOs_1 corresponds similarly a point M; these points taken together form a certain surface S; each point in this surface represents the temperature and the concentrations of a solution that may be observed, under the pressure π, in equilibrium in contact with the solid body. This surface S separates the space into two regions; every point situated in one of these regions represents, by its coordinates, a temperature and concentrations such that, under the pressure π at this temperature, a solution having these two concentrations deposits a certain quantity of solid; every point in the other region represents, by its coordinates, a temperature and concentrations such that, under the pressure π at this temperature, a solution having these two concentrations is capable of dissolving a new quantity of the solid substance. The surface S is the *surface of solubility* of the solid substance considered.

It is also easy to distinguish the region of space whose various points represent non-saturated solutions of the solid C from the region whose various points represent supersaturated solutions. Thus the different points of the axis OT correspond to $s_1=0$, $s_2=0$; they represent therefore pure water, which evidently could not be saturated with respect to the substance C except in the particular case where the substance C is ice; the region in which the axis OT is situated is therefore the region which represents non-saturated solutions.

103. Case in which the solution may furnish two distinct salts.—It generally happens that a solution of the substances 1 and 2 in water 0 may, according to circumstances, deposit different solid bodies: simple salts of different bases or acids, salts of the same base and of the same acid, but differently hydrated, distinct double salts, etc. Let C and C' be two different solids. For each of them there will be a surface of solubility; S will be this surface for C, and S' for C'.

Suppose that these two surfaces S and S' intersect in a certain line L (Fig. 23). The line L will separate the surface S into two parts S_1 and S_2, and the surface S' into two parts S_1' and S_2'.

Take a point on the surface S; T, s_1, s_2 are its coordinates; at the temperature T, a solution for which s_1 and s_2 are the concentrations may neither dissolve a new quantity of the solid C nor let this solid be deposited. Does it follow that a system consisting of the solid C and the solution is in equilibrium? This is

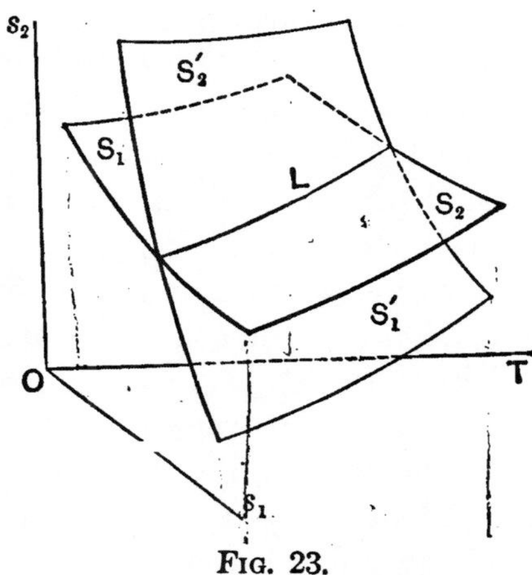

FIG. 23.

not assured, for it may happen that the solution deposits a certain quantity of the solid C'; before affirming that the system is in equilibrium, we must be sure that this last phenomenon does not take place.

The surface S' separates the space into two regions; the points in the first region represent by their coordinates T, s_1, s_2, the conditions in which the solution may dissolve the solid C' but not deposit it; the points in the second region represent conditions in which the solid C' cannot stay in solution but may be precipitated from it.

Of the two parts S_1 and S_2 into which the line L divides the surface S, one of them, S_1, is in the first of these regions and the other, S_2, is in the second.

Take a point of coordinates T, s_1, s_2, on the surface S_1; we know that at a temperature T a solution of concentrations s_1, s_2 cannot—

1°. Dissolve the solid C;

2°. Precipitate the solid C;

3°. Precipitate the solid C'.

The point considered represents therefore conditions where there is necessarily equilibrium in a system which contains only the solution and the solid C.

Take, on the contrary, a point on the surface S_2; let T, s_1, s_2 be its coordinates. At the temperature T a solution of concentrations s_1, s_2 can neither dissolve nor precipitate the substance C; but it precipitates the substance C'; a system which includes only the substance C and the solution is not in equilibrium in the conditions represented by the point considered; there is formed a precipitate of substance C'.

We should find similarly that of the two parts S_1', S_2' into which the line L divides the surface S', there is one, as S_1', for which each point represents a state of equilibrium for a system enclosing only the solution and the solid C', while the second, S_2', does not represent equilibrium states for such a system.

To sum up, if it is desired to represent the conditions (temperature and concentrations) in which may be observed in equilibrium the solution and *one only* of the two solid deposits C, C', all of the points of the two surfaces of solubility S, S' should not be kept, but only (Fig. 24) the points of one part, S_1, of the surface

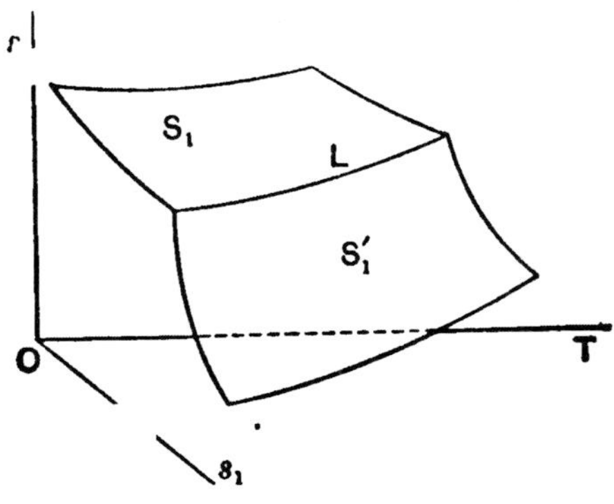

FIG. 24.

S and of a part, S_1', of the surface S', these two parts having for common boundary the line L; if the point chosen is on the surface S_1, the solid deposit will be formed exclusively of the substance C; if the point chosen is on the surface S_1', the solid deposit will consist exclusively of the substance C'.

104. Conditions in which the two precipitates are simultaneously in equilibrium with the solution.—May we observe a system in equilibrium enclosing at the same time the two solid precipitates, C and C' in the presence of the solution? For this it is necessary and sufficient that the solution cannot—

1°. Dissolve the substance C;

2°. Precipitate the substance C;

3°. Dissolve the substance C';

4°. Precipitate the substance C'.

The first two conditions require that the representative point be on the surface S, and the last two that this point be on the surface S'; all of these conditions teach us that the representative point is on the line L, the intersection of the surfaces S and S'. The line L is therefore the locus of the points which represent the conditions in which the solution may be observed in contact with the *two* solid deposits C and C'

When the solution exists in the presence of two solid deposits C, C', the system, formed of *three* components, is divided into *three* phases; it is therefore *bivariant;* let us show that the results we have just obtained are in accord with the properties of bivariant systems.

Let us take arbitrarily a pressure π and a temperature T. Construct (Fig. 25) the two surfaces S, S', which correspond to the pressure π, and let L be their line of intersection. On the axis of temperatures take a point T whose abscissa OT corresponds to the given temperature, and through this point draw a plane T_1TT_2 parallel to the plane s_1Os_2; this plane cuts the line L in a certain point P whose coordinates are T, s_1, s_2; under the pressure π, at the temperature T, the solutions of concentrations s_1, s_2 will remain in equilibrium in contact with the two deposits C, C'. We see therefore that at each pressure and temperature our bivariant system may have an equilibrium state; when the pressure π and the temperature T are given, the composition of each phase at the instant of equilibrium is determined.

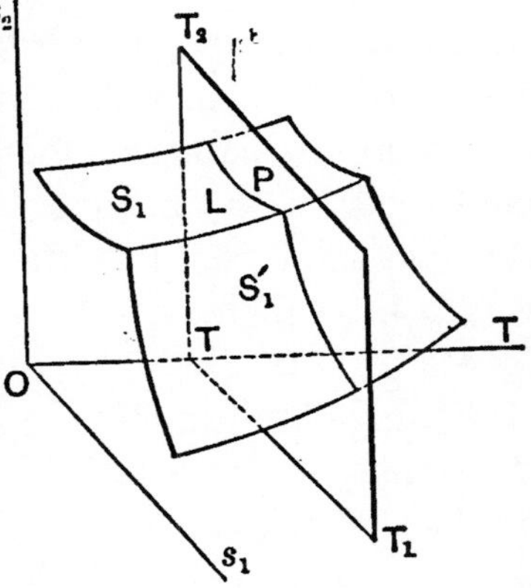

FIG. 25.

105. Case in which the solution may give three distinct salts.—It may be that the solution considered may precipitate not only two but *three* solids, C, C', C''. We shall therefore have to distinguish three kinds of equilibrium states of the system:

1°. Equilibrium of the system formed by the solution and a *single* solid precipitate.

At the pressure considered, the conditions in which such a state of equilibrium may be observed are represented by the three co-ordinates of a point situated on one of the three surfaces S, S', S'' (Fig. 26), which are parts of the solubility surfaces of the substances C, C', C'', respectively; according as the representative point is situated on the surface S, S', or S'', the solid deposit consists exclusively of the substances C, C', or C''.

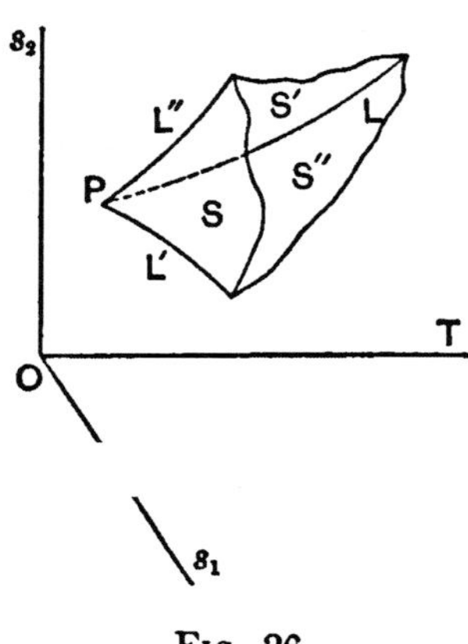

Fig. 26.

2°. Equilibrium of the system formed by the solution of *two* solid deposits.

The representative point will be on one of the three lines L, L′, L″, boundaries of the surfaces S, S', S''; if the two solid deposits are the substances C', C'', the representative point will be on the line L, boundary of the two surfaces S', S''; if the two solid deposits are the substances C'', C, the representative point will be on the line L′, boundary of the surfaces S'', S; if the two solid deposits are the substances C, C', the representative point will be on the line L″, boundary of the two surfaces S, S'.

3°. Equilibrium of the system formed by the solution and the *three* solid deposits.

The representative point will be the point P common to the three surfaces S, S', S'', and hence of the three lines L, L′, L″. Under the pressure considered, there is thus a single temperature and a single composition of the solution for which such an equilibrium is possible, a condition that should not astonish us; the system, formed of three independent components, is then divided into four phases; it is *monovariant.*

106. The Alloy: Lead, tin, bismuth. Charpy's Researches. —A chemical system which gives a very clear illustration of the preceding considerations has been studied recently by G. Charpy.[1] The liquid phase consists of a mixture of the three metals, lead, tin, and bismuth, in the state of fusion, a mixture that may be compared to the solution of which we have just spoken when we attribute the three indices 0, 1, 2 to the three metals studied, taken in any order.

The representative points of the states of equilibrium between the mixture in fusion and the solid lead form a surface S; the representative points of equilibrium states between the fused mixture and the solid tin form a surface S'; finally, the representative points of the equilibrium states between the fused mixture and the solid bismuth form a surface S''; these three surfaces S, S', S'' have been constructed by G. Charpy.

The points on the line L represent the conditions in which the liquid mixture may be in equilibrium with solid bismuth and solid tin; the points on the line L′ represent the conditions in which the liquid mixture may be in equilibrium with solid lead and solid bismuth; the points on the line L″ represent the conditions in which the liquid mixture may be in equilibrium with solid lead and solid tin.

Finally, the coordinates of the point P represent the values that must be given to the temperature and composition of the liquid mixture in order that the latter may remain in equilibrium in contact with the three metals in the solid state. According to Charpy's researches the value of this temperature is $+96°$ C. and the liquid mixture corresponding to the point P has the following composition:

Lead	0.32
Tin	0.16
Bismuth	0.52

107. Mixture of three melted salts.—Hector R. Caweth[2] has studied an analogous system; but here the liquid mixture, instead

[1] G. Charpy, *Comptes Rendus*, v. 126, p. 1569, 1898.

[2] H. R. Caweth, *Journal of Physical Chemistry*, v. 2, p. 209, 1889.

of being formed of three melted metals, consisted of three nitrates in fusion, the nitrates of potassium, sodium, and lithium.

108. Domain of a precipitate.—In a great number of cases the distinct solid precipitates that are encountered are much greater than three in number. Nevertheless the properties of the system may be studied and represented in accordance with the principles just developed.

Let C, C', C'' . . . be the precipitates that may be observed. Under the given pressure π the states of equilibrium between the liquid mixture of the three independent components and the single solid precipitate C, states in which the system is trivariant, are represented by the various points on a limited surface S that is called the *domain of the precipitate C*.

This surface S touches other surfaces, S', S'' . . ., which are the domains of the precipitates C', C'' . . . Thus the possible equilibrium states, under the pressure considered, between the liquid mixture and *a single* solid precipitate are represented by the various points of a polyhedral surface with curved faces which has as many *faces* as there are distinct solid precipitates able to exist within the liquid mixture.

The points on the intersections of these faces represent a state of equilibrium, under the given pressure, between the mixture and two distinct solid precipitates; for these states the system has become bivariant; the bounding line between the domain of the precipitate C and that of C' represents all the possible states of equilibrium between the liquid mixture, the precipitate C, and the precipitate C'.

Each of the cusps of this surface represents a state of equilibrium for which the system, become monovariant, is composed of a liquid mixture and of three solid precipitates C, C', C'', which are those whose domains S, S', S'' join at the cusp considered.

Finally, it is not necessary to consider states of equilibrium for which the liquid mixture exists in the presence of three solid precipitates; the system would then be invariant; it could be in equilibrium only at a single temperature and pressure; the case in which this pressure would be the one given is evidently exceptional.

It is easy to see what services such a system of representation may render; it makes easy the prediction which precipitates may, in the given conditions, exist in presence of a liquid mixture of three independent components.

109. System composed of water, ferric chloride, hydrochloric acid. Researches of Bakhuis Roozboom and Schreinemakers.—H. W. Bakhuis Roozboom and Schreinemakers[1] have used this method to represent the various states of saturation of a liquid mixture formed of the three following components:

Water:	H_2O
Hydrochloric acid:	HCl
Ferric chloride:	Fe_2Cl_6.

The solid precipitates which were observed in these investigations are twelve in number, as follows:

Ice:	H_2O
3 hydrates of hydrochloric acid:	$HCl \cdot 3H_2O$
	$HCl \cdot 2H_2O$
	$HCl \cdot H_2O$
Anhydric ferric chloride:	Fe_2Cl_6
4 hydrates of ferric chloride:	$Fe_2Cl_6 \cdot 12H_2O$
	$Fe_2Cl_6 \cdot 7H_2O$
	$Fe_2Cl_6 \cdot 5H_2O$
	$Fe_2Cl_6 \cdot 4H_2O$
3 tertiary compounds:	$Fe_2Cl_6 \cdot 2HCl \cdot 12H_2O$
	$Fe_2Cl_6 \cdot 2HCl \cdot 8H_2O$
	$Fe_2Cl_6 \cdot 2HCl \cdot 4H_2O$

It is evident how difficult it would have been to unravel the possible equilibrium states from such a system without the aid of the theoretical principles previously developed.

110. System: water, potassium sulphate, and magnesium sulphate. Investigations of Van der Heide.—These principles have been of service in the study of not less complicated systems.

Van der Heide[2] has applied them to the study of systems whose three independent components are:

---

[1] H. W. Bakhuis Roozboom and Schreinemakers, *Zeit. für physikalische Chemie*, v. 15, p. 588, 1894.

[2] Van der Heide, *Zeit. für physikalische Chemie*, v. 12, p. 416, 1893.

Water:	H_2O
Potassium sulphate:	K_2SO_4
Magnesium sulphate:	$MgSO_4$

At a temperature less than 100° *C*. the six following precipitates may be obtained:

Ice:	H_2O	
Anhydrous potassium sulphate:	K_2SO_4	
2 hydrated sulphates of magnesium:	$MgSO_4 \cdot 7H_2O$	
	$MgSO_4 \cdot 6H_2O$	
2 double salts:	$MgK_2(SO_4).6H_2O$	(Schœnite)
	$MgK_2(SO_4) \cdot 4H_2O$	(Léonite)

The various equilibrium states that may be had at temperatures less than 100° are represented by the surfaces whose general ap-

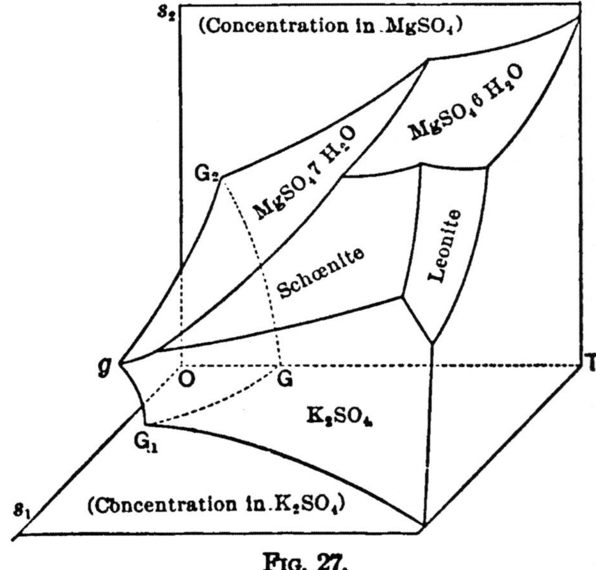

FIG. 27.

pearance is given in Fig. 27. The hidden face GG_1G_2g is the domain of ice.

111. System: water, potassium chloride, and magnesium chloride. Researches of Van't Hoff and Meyerhoffer.— Van't Hoff and Meyerhoffer[1] have studied in the same way

[1] VAN'T HOFF and MEYERHOFFER, *Sitzungsberichte der Berliner Akademie*, 1897, p. 487; *Zeit. für phys. Chemie*, v. 30, p. 64, 1899.

the system formed of the following three independent components:

Water:	H_2O
Potassium chloride:	KCl
Magnesium chloride:	$MgCl_2$

The studies of Van't Hoff and Meyerhoffer extended to about 185°. In these conditions one may obtain the following eight distinct precipitates:

Ice:	H_2O	
Anhydrous potassium chloride:	KCl	
5 hydrates of magnesium chloride:	$MgCl_2.12H_2O$	
	$MgCl_2\cdot 8H_2O$	
	$MgCl_2\cdot 6H_2O$	
	$MgCl_2\cdot 4H_2O$	
	$MgCl_2\cdot 2H_2O$	
A double salt:	$MgKCl_3\cdot 6H_2O$	(Carnallite)

Fig. 28 gives a general idea of the aspect of the surface which

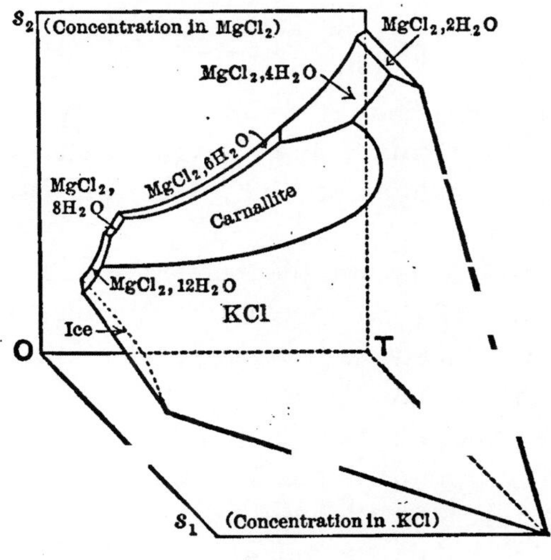

FIG. 28.

represents the states of equilibrium of such a system.

II. Quadrivariant Systems.

112. Quadrivariant systems formed of four components divided into two phases.—If four components exist in two phases, we have $c=4$, $\phi=2$,

$$V=c+2-\phi=4,$$

and the system is Quadrivariant. When the values of the pressure π and of the temperature T are given, the quadrivariant system may be observed in equilibrium; but the composition of the two phases in equilibrium is far from being determined by the knowledge of the values of π and T alone; to these quantities it is necessary to add *two* others in order to fix the composition of the system in equilibrium.

Let us take an example. A system is composed of four independent components: A solute 0, water, for instance, and three salts, 1, 2, and 3. This system is separated into two phases, one liquid and one solid phase; the liquid phase is composed of a mass M_0 of water and of masses M_1, M_2, M_3 of salts 1, 2, 3; the three concentrations of this solution are

$$s_1=\frac{M_1}{M_0},\quad s_2=\frac{M_2}{M_0},\quad s_3=\frac{M_3}{M_0}.$$

The solid phase is a body of definite composition formed at the expense of the components 0, 1, 2, 3: ice, simple salt—anhydrous or hydrous—double anhydrous or hydrous salt; we shall denote this solid by the letter C.

When the pressure π and the temperature T are given, the constitution of the solution capable of remaining in equilibrium in contact with the substance C is not yet determined; an infinity of different solutions may, under this pressure and at this temperature, remain in equilibrium in contact with the substance C, without dissolving it and without letting a new mass of this body be precipitated. To the knowledge of the pressure π and the temperature T let us join the values of *two* of the concentrations s_1, s_2, s_3; then, and then only, shall we know the value which the third concentration should have in order that there may be equilibrinm between the solution and the substance C.

113. Three salts dissolved in water. Solubility surface of a precipitate at a given pressure and temperature.—Let us try to represent all the compositions for which the solution may be in equilibrium with the substance C under the pressure π whose value is chosen once for all, atmospheric pressure for example, and at a temperature whose value is also fixed, as $+15°$ C.

On three axes of rectangular coordinates Os_1, Os_2, Os_3 (Fig. 29) lay off lengths proportional to the three concentrations s_1, s_2, s_3 respectively.

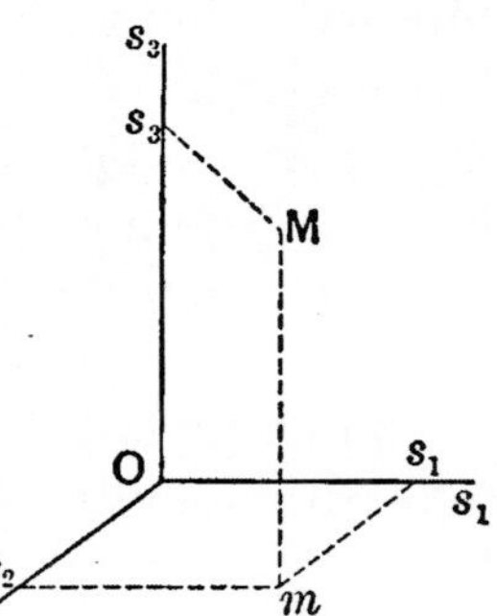

FIG. 29.

Suppose given two of these concentrations, as s_1 and s_2; and let m be the point in the plane s_1Os_2, whose coordinates Os_1, Os_2 represent these two concentrations. At these two concentrations it is necessary and sufficient, in order to assure equilibrium between the solution and the substance C, under atmospheric pressure and at the temperature 15° C., to add a well-determined value of the concentration s_3; through the point m draw a parallel mM to Os_3 whose length $mM = Os_3$ is measured by this value Os_3; the three coordinates of the points M represent the three concentrations of a solution capable of remaining in equilibrium in the presence of the substance C at atmospheric pressure and a temperature of 15°.

s_1 and s_2 may be taken arbitrarily, or, in other words, the point m may be taken anywhere in the plane s_1Os_2; for every position of the point m corresponds a point M on the surface S.

Each point M of the surface S represents, by its three coordinates, the three concentrations of a solution susceptible of remaining in equilibrium, under atmospheric pressure and at 15°, in contact with the substance C.

This surface S divides the space into two regions. Every point of one of these two regions represents, by its three coordinates, the three concentrations of a solution capable of dissolving a certain quantity of the substance C under atmospheric pressure and at 15° temperature.

Each point in the other region represents by its coordinates the three concentrations of a solution capable of precipitating a certain quantity of the substance C under the same conditions.

The origin of the coordinates for which $s_1=0$, $s_2=0$, $s_3=0$, represents pure water. If the solid C is not ice, pure water could not be saturated with the solid C; the origin of the coordinates is therefore, with respect to the surface S, within the region which represents the unsaturated solutions of the solid C; this property allows of recognizing this region at once.

In general a solution formed of four independent components may precipitate various solids of definite composition, as C, C', C''. . . . By a method similar to that used for trivariant systems, we may reach the following conclusions:

The three concentrations of a solution capable of remaining in equilibrium, under atmospheric pressure and at 15° C., in contact with a *single one* of the solids C, C', C'' . . . , are represented by the three coordinates of a point M belonging to a certain surface.

This surface is formed of a certain number of curved areas S, S', S'' . . . bounded by separating lines; in other terms, it forms a polyhedral surface with curved faces.

Each of the faces S, S', S'' . . . of this polyhedral surface corresponds to one of the bodies C, C', C'' . . . and forms its *domain*.

If the representative point M belongs to the domain S of the body C, the solution which has for concentrations the three coordinates of the point M rests in equilibrium, under atmospheric pressure and at 15° C., in contact with the substance C, but not in contact with another precipitate.

If the representative point is on the boundary line of the two surfaces S, S' of the two substances C, C', the three coordinates of this point represent the three concentrations of a solution which may, under atmospheric pressure and at 15° C., remain in equilibrium in contact with a solid precipitate composed of C and C'; in such a state of equilibrium one system of four independent components is divided into three phases, so that it is no longer quadrivariant but trivariant.

Finally, if the representative point is at the cusp where the domains S, S', S'' of the three substances C, C', C'' intersect, the three coordinates of this point represent the three concentrations of a solution capable of remaining in equilibrium, under atmospheric pressure and at 15° C., in contact with a precipitate which contains the three substances C, C', C'' mixed together. In such a state the system formed of four independent components and divided into four phases has become bivariant; also, the pressure and temperature being given, the composition of each phase is determined.

114. System: Water, magnesium chloride, magnesium sulphate, chloride of potassium, potassium sulphate. Investigations of Lœwenherz, Van't Hoff, and Meyerhoffer.—An interesting example of these various considerations is found in the researches of Lœwenherz,[1] continued recently by Van't Hoff, Meyerhoffer, and Donnan.[2]

With the object of analyzing the conditions in which some of the numerous compounds are formed in the deposits of Stassfurt salt, these authors have studied, under atmospheric pressure and at 15° C., the solutions formed by mixing with water, H_2O, potassium chloride, KCl, magnesium sulphate, $MgSO_4$, and magnesium chloride, $MgCl_2$.

If desired, the system may be regarded as made up of the four independent components

$$H_2O, \quad KCl, \quad MgSO_4, \quad MgCl_2. \tag{1}$$

But we shall find it advantageous to make another choice. Within the solutions there may be formed by double decomposition, potassium sulphate, K_2SO_4, as given by the equation

$$2KCl + MgSO_4 = MgCl_2 + K_2SO_4. \tag{2}$$

There is therefore no reason for considering that the solution contains actually the substances indicated in equation (1).

Not to make any hypothesis, we shall consider the solution as being the mixture, in any state whatever, of the four following bodies:

---

[1] Lœwenherz, *Zeit. für physikalische Chemie*, v. 12, p. 459, 1894.

[2] Van't Hoff and Meyerhoffer, *Sitzungsber. d. Berliner Akad.*, 1897, p. 1019; Van't Hoff and Donnan, *ibid.*, p. 1146; Van't Hoff, Report to International Physics Congress, Paris, 1900, v. 1, p. 464.

(3) H_2O, Cl_2, K_2, SO_4, Mg.

We denote by π the molecular weight of water (18 grammes) and by w, w_2, w_3, w' the number of grammes that represent the symbols of the four other bodies.

In a molecule (or 18 grammes) of water of a given solution chemical analysis gives us

nw grammes of chlorine,
n_2w_2 grammes of potassium,
n_3w_3 grammes of SO_4,
$n'w'$ grammes of magnesium.

The composition of the solution is therefore known if we know the four numbers n, n_2, n_3, n'.

But it is not necessary to know these four numbers; if three of them, n, n_3, n', are known, the knowledge of the fourth, n_2, follows.

Thus the n_2w_2 grammes of potassium consist of p_2w_2 grammes united with p_2w grammes of chlorine, and of q_2w_2 grammes united with q_2w_3 grammes of SO_4:

$$n_2 = p_2 + q_2.$$

Also, the $n'w'$ grammes of magnesium consist of $p'w'$ grammes joined to $p'w$ grammes of chlorine, and of $q'w'$ grammes joined to $q'w_3$ grammes of SO_4:

$$n' = p' + q'.$$

Expressing the fact that the chlorine united with potassium and with magnesium gives for the total quantity of chlorine:

$$p_2w + p'w = nw, \quad \text{or} \quad n = p_2 + p'.$$

Writing that SO_4 united with potassium and SO_4 united with magnesium form the totality of SO_4:

$$q_2w_3 + q'w_3 = n_3w_3, \quad \text{or} \quad n_3 = q_2 + q'.$$

These four equations give us

$$n + n_3 = n_2 + n', \tag{4}$$

so that when n, n_3, and n' are known it is easy to calculate n_2. One may choose arbitrarily, therefore, the masses of four of the substances in group (1); the mass of the fifth is determined; of the five bodies of group (1), four only are independent; by changing the choice of independent components we have not altered their number (see Art. 87).

But in order that the system may be taken as formed by the independent components (1), the numbers n, n_2, n_3, n' must satisfy a certain condition.

All the potassium contained in 18 grammes of water will have been brought there by the chloride of potassium dissolved in it. If we denote by π_2 the number of grammes represented by 2KCl to obtain n_2w_2 grammes of potassium, it would be necessary to use $n_2\pi_2$ grammes of potassium chloride bringing n_2w grammes of chlorine.

This chlorine is not all the chlorine that 18 grammes of water contain; we must take into account that from the magnesium chloride; if π_1 is the number of grammes represented by $MgCl_2$, and if in 18 grammes of water we have dissolved $n_1\pi_1$ grammes of this substance, which would bring n_1w grammes of chlorine, we have

$$n_2w+n_1w=nw,$$

or

(5) $$n_1=n-n_2;$$

and as n_1 cannot be negative, we see that for a system to be formed of the substances (1), it is necessary that the numbers n, n_2, n_3, n' satisfy not only equation (4), but also the condition

(6) $$n-n_2 \geqq 0.$$

This is sufficient also. Let us denote by π_3 the number of grammes of $MgSO_4$; in dissolving in 18 grammes of water $n_2\pi_2$ grammes of potassium chloride, $n_3\pi_3$ grammes of magnesium sulphate, and $n_1\pi_1=(n-n_2)\pi_1$ grammes of magnesium chloride, we shall obtain the composition sought

Let us consider now the systems formed by the four following components:

(1′) H_2O, KCl, $MgSO_4$, K_2SO_4.

They may also be considered as formed of the five components (3); here again the five components will not be independent, for the condition (4) will still hold. This will not suffice in order that the system may be considered as formed by the components (1′); a condition must be joined to this.

In 18 grammes of water we put nw grammes of chlorine; all of this substance comes from potassium chloride; there must have been dissolved, therefore, $n\pi_2$ grammes of potassium chloride; but this is not all the potassium contained in 18 grammes of water; if there are π_1' grammes of K_2SO_4, and if in 18 grammes of water there are dissolved $n_1'\pi_1'$ grammes of potassium sulphate, there have been introduced $n_1'w_2$ grammes of potassium. We have, therefore,

$$nw_2+n'w_2=n_2w_2,$$

or

$$n_1'=n_2-n. \tag{5'}$$

As n_1' cannot be negative, we must have

$$n-n_2\leqq 0. \tag{6'}$$

This suffices also; for if this condition holds, it is only necessary, as is evident, to dissolve, in 18 grammes of water, $n_2\pi_2$ grammes of potassium chloride, $n_3\pi_3$ grammes of magnesium sulphate, and $n_1'\pi_1'=-(n-n_2)\pi_1'$ grammes of potassium sulphate, in order to obtain a solution of the composition indicated.

We see, therefore, that in studying the systems formed by the five components (3), only four of which are independent, in virtue of equation (4) we are studying at the same time the systems (1) and the systems (1′); we have to do with the first if $(n-n_2)$ is positive, and with the second if $(n-n_2)$ is negative.

In order to represent the composition of a solution, it suffices by equation (4) to know the value of three of the numbers n, n_2, n_3, n', or, if it is preferred, the three values

$$x=n-n_2,\quad y=n_2,\quad z=n_3. \tag{7}$$

Hence the constitution of a solution may be represented in the following manner:

Let us take (Fig. 30) a system of rectangular coordinates Ox, Oy, Oz; and prolong the x axis beyond 0 to x'. Lay off on Oy the value of n_2; on Oz the value of n_3; if $(n-n_2)$ is positive, lay off its value n_1 on Ox to the right; if $(n-n_2)$ is negative, lay off its absolute value $n_1'=-(n-n_2)$ on Ox'; these three coordinates will determine a point representing a solution of known composition.

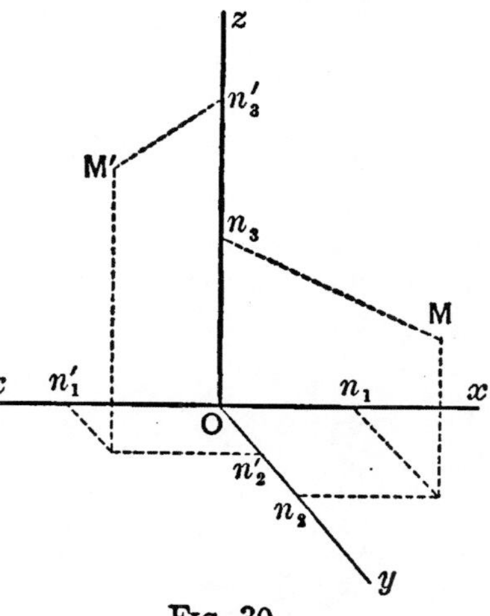

Fig. 30.

If this point M is to the right of the plane yOz, it will correspond to a positive value n_1 of (n_1-n_2); it will represent a solution obtained by mixing with 18 grammes of water $n_1\pi_1$ grammes of potassium sulphate, $n_2\pi_2$ grammes of potassium chloride, and $n_3\pi_3$ grammes of magnesium sulphate.

If this point M' is to the left of the plane yOz, it will correspond to a negative value $-n_1'$ of $(n-n_2)$; it will represent a solution obtained by mixing with 18 grammes of water $n_1'\pi_1'$ grammes of potassium sulphate, $n_2\pi_2$ grammes of potassium chloride, and $n_3\pi_3$ grammes of magnesium sulphate.

This is the method of representation devised by Van't Hoff.

It is evident that what was discussed in Art. 113 may be treated from this standpoint; under a given pressure, at a given temperature, each of the salts which may be precipitated corresponds to a surface of solubility.

The solid bodies which the preceding solution may deposit at 25° are seven in number:

Two anhydrous salts:	Potassium chloride, KCl	
	Potassium sulphate, K_2SO_4	
Three hydrated salts:	$MgSO_4 \cdot 7H_2O$	
	$MgSO_4 \cdot 6H_2O$	
	$MgCl_2 \cdot 6H_2O$	
Two double salts:	Schœnite,	$K_2Mg(SO_4)_2 \cdot 6H_2O$
	Carnallite,	$MgKCl_3 \cdot 6H_2O$

It seems, besides, that three other substances should be obtained;

Two hydrated salts: $MgSO_4 \cdot 5H_2O$
$MgSO_4 \cdot 4H_2O$
A double salt: Léonite, $K_2Mg(SO_4)_2 \cdot 4H_2O$

In any case the conditions of the formation of these substances are still little known. The authors that we have cited limited themselves to the study of the first seven; they have constructed the polyhedral surface of seven curved faces which constitute at

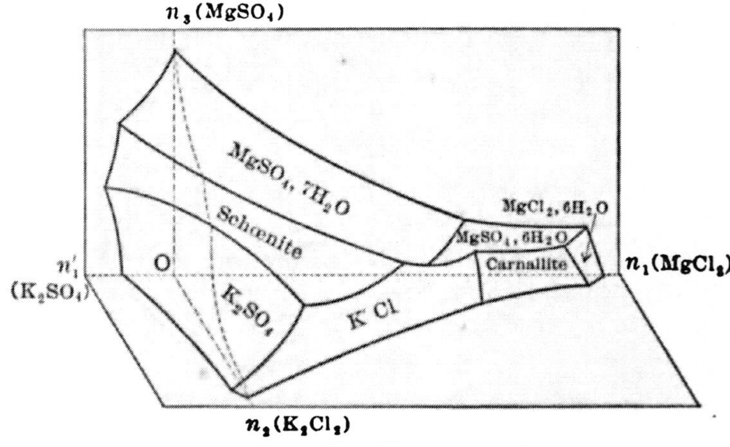

FIG. 31.

25° the domains of these salts. Fig. 31 gives a general idea of this surface.

115. System: Water, potassium chloride, sodium chloride, potassium sulphate, sodium sulphate. Studies of Meyerhoffer and Saunders.—For each temperature there corresponds a surface analogous to the preceding; for the system just studied only the surface for 25° has been constructed. A more complete study has been made by Meyerhoffer and Saunders [1] of the system

$$H_2O, \quad KCl, \quad Na_2SO_4, \quad NaCl,$$

and of the system

$$H_2O, \quad KCl, \quad Na_2SO_4, \quad K_2SO_4$$

[1] W. MEYERHOFFER and A. P. SAUNDERS, *Zeitschrift für physikalische Chemie*, v. 28, p. 453, 1899.

joined by the relation

$$2KCl + Na_2SO_4 = 2NaCl + K_2SO_4.$$

Regarding these systems almost exactly the same words may be said as in the preceding article, replacing Mg by Na_2.

Six distinct precipitates may be formed at the temperatures studied by Meyerhoffer and Saunders, as follows:

1°. Four anhydrous salts:	Sodium chloride,	$NaCl$
	Potassium chloride,	KCl
	Potassium sulphate,	K_2SO_4
	Sodium sulphate,	Na_2SO_4
2°. An hydrated salt:	Glauber's salt,	$Na_2SO_4 \cdot 10H_2O$
3°. A double salt:	Glaserite,	$K_3Na(SO_4)_2$

At each temperature and pressure the domains of the six salts form a polyhedron of six curved faces; Meyerhoffer and Saunders [1] have constructed four of these polyhedrons, those corresponding to atmospheric pressure and to temperatures 0°, 4°.4, 16°.3, and 25°.

We represent here (Figs. 32 and 33) two of these surfaces; Fig. 32 is the surface for 0°, and Fig. 33 the surface for 25°.

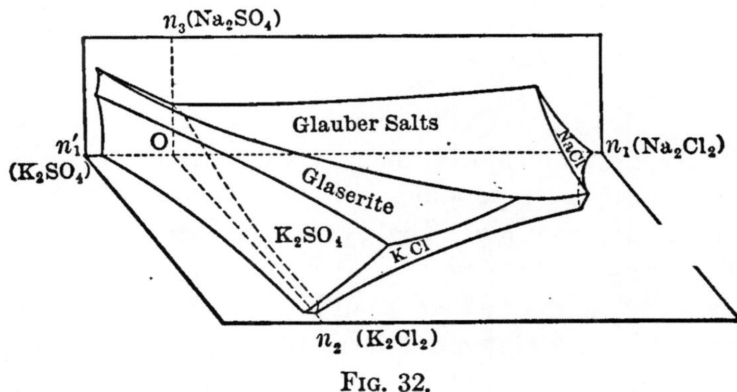

FIG. 32.

It is to be noticed that Fig. 32 has but *five* faces; at this temperature the anhydrous sodium sulphate is in no case precipitated; no domain corresponds to this salt. The same is true for the

[1] W. MEYERHOFFER and A. P. SAUNDERS, *Zeitschrift für physikalische Chemie*, v. 28, p. 453, 1899.

temperatures 4°.4 and 16°.3, which again correspond to five face surfaces. On the contrary, the surface for 25° has *six* faces; the anhydrous sodium sulphate domain figures there and is already of considerable size.

116. Four salts dissolved in water, one of them to saturation. System: Water, sodium chloride, potassium chloride, sodium sulphate, magnesium chloride.—Suppose that into the system water, potassium chloride, sodium sulphate, magnesium chloride, we introduce a new independent component, which we shall denote

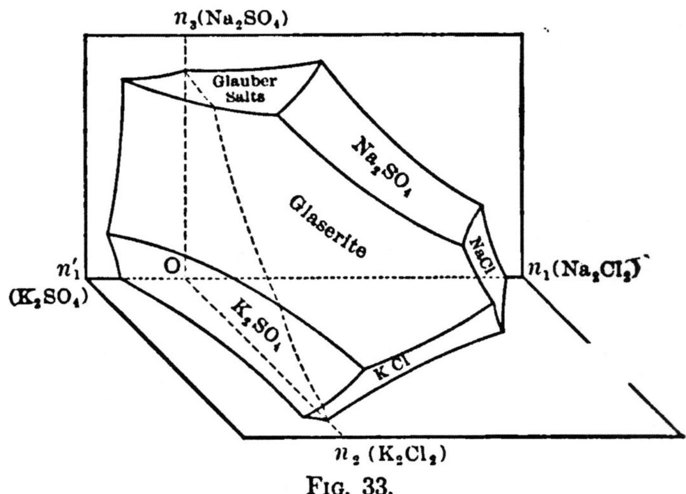

FIG. 33.

by the index 4; for example, sodium chloride, NaCl. We shall keep the index 1 for potassium chloride, 2 for sodium sulphate, 3 for magnesium chloride. The system is then formed of five independent components, $c=5$; if it existed in only two phases, $\phi=2$, the variance $V=c+2-\phi$ would be equal to 5; this *quintivariant* system would be more difficult of study than the systems of which we have just treated. But Van't Hoff and his pupils,[1]

[1] J. H. VAN'T HOFF and A. P SAUNDERS, *Sitzungsberichte der Berliner Akademie*, 1898, p. 387; VAN'T HOFF and T. ESTREICHER-ROZBIERSKI, *ibid.*, 1898, p. 487; VAN'T HOFF and MEYERHOFFER, *ibid.*, 1898, p. 590; VAN'T HOFF, *Report to Physics Congress* (Paris, 1900), v. I, p. 464; VAN'T HOFF and H. V. EULER CHELPIN, *Sitzungsber. Berl. Akad.*, 1900, p. 1018; VAN'T HOFF and MEYERHOFFER, *ibid.*, 1901.

not satisfied with fixing the pressure (1 atmosphere) and the temperature (25°), impose upon the solution the condition of being constantly saturated with sodium chloride; this condition was assuredly fulfilled in the circumstances under which were formed the salts deposits of Stassfurt. They then seek under what conditions this solution may be in equilibrium with another solid salt.

In other terms, they study this solution, formed of *five independent components*, in the presence of *two solid phases*, one of which is always sodium chloride.

Let C be the second solid phase.

Let s_4 be the ratio of the mass M_4 of sodium chloride contained in the solution to the mass M_0 of water that it contains.

The pressure π and the temperature T having invariable values, every time the two concentrations s_1, s_3 of the potassium chloride and the magnesium chloride in the solution are given arbitrarily, we shall know the concentration s_2 of the sodium sulphate and the concentration s_4 of the chloride of sodium within a solution capable of resting in equilibrium in contact with an excess of sodium chloride and a solid deposit of the salt C. In order to show completely the state of such a solution, one might make use simultaneously of two representative points: the one, M, would have for coordinates the three concentrations s_1, s_2, s_3; the other, μ, would have the coordinates s_1, s_3, s_4. When the two concentrations s_1 and s_3 were varied, the first would describe a well-determined surface S and the other an equally definite surface Σ. The simultaneous knowledge of these two surfaces would give complete information on the subject of the solutions which may remain in equilibrium, under atmospheric pressure and at the temperature of 25°, in contact with an excess of common salt and crystals of the salt C.

Each of these surfaces would possess properties similar to those possessed by the unique surface S, in the case where a solution formed of four independent components was in equilibrium with a single solid deposit.

Suppose that, besides sodium chloride, the solution may precipitate various solid salts C, C', C''. . .; to each of these salts would correspond one of the domains S, S', S'' . . . in the system of coordinates s_1, s_2, s_3, and also one of the domains Σ, Σ', Σ'' . . . in the system s_1, s_3, s_4.

In Van't Hoff's researches it was not necessary to know the concentration s_4 of the solution in sodium chloride; it sufficed to determine the composition of the solution which rested if sodium chloride was precipitated.

To determine this composition it is enough to know the respective concentrations s_1, s_2, s_3 of potassium chloride, of sodium sulphate, and of magnesium chloride; but there may be substituted for these three quantities three others which will give equally well the composition of the solution after all the sodium chloride has been precipitated.

Let w_1, w_2, w_3 be the molecular weights of potassium chloride, of sodium sulphate, and of magnesium chloride, that is to say, of the numbers of grammes represented by the formulæ

$$KCl, \quad Na_2SO_4, \quad MgCl_2.$$

For a molecule or 18 grammes of water the solution to analyze contains n_1w_1 grammes of potassium chloride, n_2w_2 grammes of sodium sulphate, and n_3w_3 grammes of magnesium chloride. Analysis gives the numbers n_1, n_2, n_3, and as evidently

$$s_1=\frac{n_1w_1}{18}, \quad s_2=\frac{n_2w_2}{18}, \quad s_3=\frac{n_3w_3}{18},$$

we see that the knowledge of the numbers n_1, n_2, n_3 is equivalent to knowing the three concentrations s_1, s_2, s_3.

Instead of having given the numbers n_1, n_2, n_3, one may, if it is preferred, determine the composition of the solution from the numbers

$$x=n_1;$$
$$y=n_1+n_2+n_3;$$
$$z=n_3.$$

These are the three quantities x, y, and z that Van't Hoff lays off on the three axes of rectangular coordinates (Fig. 34).

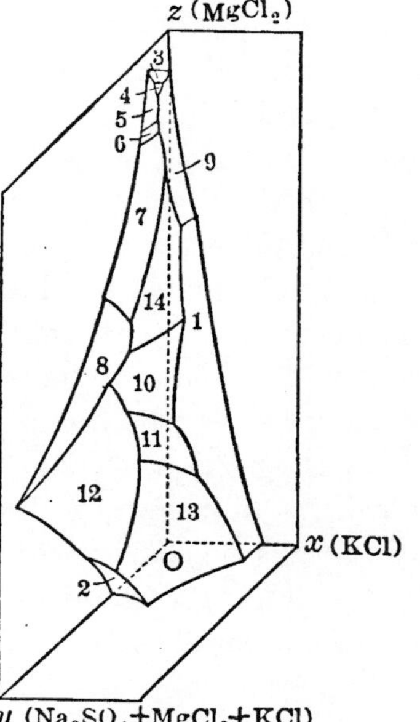

Fig. 34.

Each point represents then a solution which would have a well-determined composition after having precipitated the sodium chloride which saturates it.

At a given temperature and under a given pressure each of the precipitates, other than sodium chloride, which may be formed in the system corresponds to a domain.

There are fourteen substances, always present in excess, which may be precipitated with the sodium chloride at 25° *C.* and atmospheric pressure:

Two anhydrous salts:	(1)	KCl
	(2)	Na_2SO_4
Six hydrated salts:	(3)	$MgCl_2 \cdot 6H_2O$
	(4)	$4MgSO_4 \cdot 5H_2O$
	(5)	$MgSO_4 \cdot 4H_2O$
	(6)	$MgSO_4 \cdot 5H_2O$
	(7)	$MgSO_4 \cdot 6H_2O$
	(8)	$MgSO_4 \cdot 7H_2O$
Six double salts:	(9)	$MgKCl_3 \cdot 6H_2O$ (Carnallite)
	(10)	$MgK_2(SO_4)_2 \cdot 4H_2O$ (Léonite)
	(11)	$MgK_2(SO_4)_2 \cdot 6H_2O$ (Schœnite)
	(12)	$MgNa_2(SO_4)_2 \cdot 4H_2O$ (Astrakanite)
	(13)	$K_3Na(SO_4)_2$ (Glaserite)
	(14)	$MgSO_4 \cdot KCl \cdot 3H_2O$ (Kainite)

Fig. 34 represents the polyhedron of 13 curved faces that the domains of these salts form. The numbers preceding the salts in the column correspond in Fig. 34 to their respective domains.

117. Five salts dissolved in water, two of which to saturation; a calcium salt added to the preceding system.—A sixth component may be added to the preceding solution; provided that the number of precipitates in contact with which this solution remains in equilibrium be also increased by unity, the variance of the system will keep the same value.

Suppose that to the sodium chloride, potassium chloride, sodium sulphate, magnesium chloride we add calcium sulphate to which we give the index 5. Five concentrations, s_1, s_2, s_3, s_4, s_5, will determine the composition of the solution.

The calcium salts being very slightly soluble, one of them will be precipitated; but depending upon the conditions of composition of the solution, temperature, and pressure the calcium salt precipitated will be different.

It is question, then, of studying the solutions which may remain in equilibrium in contact with *three* precipitates one of which will always be sodium chloride and another always a calcium salt γ; the third precipitate, C, will be a salt of sodium, magnesium, or again a double salt containing two of these metals.

Suppose given the temperature, 25°, and the pressure, one atmosphere. If one wishes the solution to remain in equilibrium in contact with common salt and a couple of precipitates (C, γ), there may be given arbitrarily the values of two of the five concentrations s_1, s_2, s_3, s_4, s_5, and the values of the other three will follow. We see then that at a given pressure and temperature *each couple of precipitates* (C, γ), of which the first does not contain lime and the second is a calcium salt, will correspond to a certain *domain* S in the system of coordinates s_1, s_2, s_3; to another domain Σ in the system s_1, s_3, s_4; finally, to a third domain in the system s_1, s_3, s_5.

It is by means of the first coordinate system, or, rather, by means of the equivalent coordinate system x, y, z, defined in the preceding article, that Van't Hoff[1] and his pupils have represented the domains of various pairs of two salts which may be precipitated in the midst of the system.

This case may be simplified. The calcium salts being very slightly soluble, the concentration s_5 is always near to 0. It follows that the values which the concentrations s_1, s_2, s_3, s_4 ought

[1] VAN'T HOFF, *Reports to Physics Congress* (Paris, 1900), v. I, p. 464.

to have in order that the solution be in equilibrium in contact with common salt and with the couple (C, γ) are sensibly the same as if the concentration s_5 was equal to 0 and the calcium salt γ suppressed. For the coordinate system s_1, s_2, s_3, or what amounts to the same, for the system x, y, z, the domain of the couple (C, γ) should almost exactly coincide with the domain found for the salt C in the preceding article, or with a part of this domain.

One may therefore begin by determining the domains of the salts C, C', C'' . . . as if the system did not contain calcium sulphate, and this will give the surface represented by Fig. 34. Next it should be seen if, throughout the extent of one of these domains S, the salt C is associated with the same calcium salt γ, or if, on the contrary, it is associated with various calcium salts $\gamma_1, \gamma_2, \ldots$, case in which the domain S should be subdivided into various subdomains $S_1, S_2, \ldots$, corresponding respectively to the couples $(C, \gamma_1), (C, \gamma_2), \ldots$

The result of this discussion is represented by Fig. 35. The solid lines map the surface which has already been represented in Fig. 34. The dotted lines divide the domains of the various salts of potassium, magnesium, or sodium into subdomains relative to the various calcium salts which may be precipitated in the conditions of the experiments.

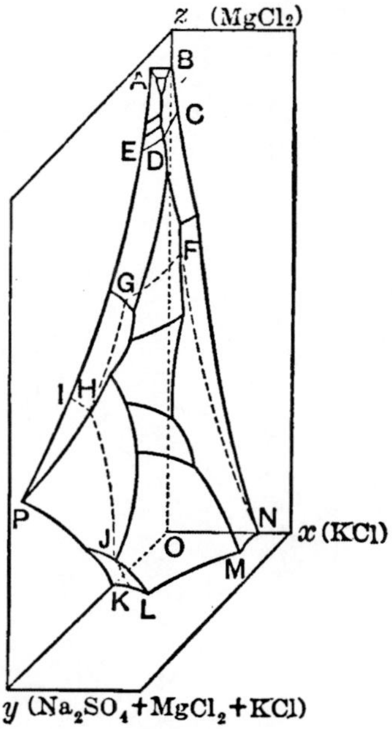

These calcium salts are four in number:

$2\,CaSO_4 \cdot H_2O$
$CaSO_4 \cdot 2H_2O$ (Gypsum)
$CaNa_2(SO_4)_2$ (Glauberite)
$CaK_2(SO_4)_2 \cdot H_2O$ (Syngénite)

Fig. 35.

The subdomains corresponding to the semi-hydrated calcium sulphate are bounded by the line *ABCDE*; the subdomains for gypsum are bounded by the line *CDEIHGFN*; glauberite precipitates in the region *IHJKP*, and the syngenite in the region *NFGHJKLM*.

We see, therefore, that the following associations in the presence of common salt in excess may be observed between a calcium salt and a non-calcium salt:

$MgCl_2 \cdot 6H_2O$	with $2CaSO_4 \cdot H_2O$;
$4MgSO_4 \cdot 5H_2O$	
$MgSO_4 \cdot 4H_2O$	
$MgSO_4 \cdot 5H_2O$	
Carnallite	with $2CaSO_4 \cdot H_2O$ or Gypsum;
$MgSO_4 \cdot 6H_2O$	with $2CaSO_4 \cdot H_2O$, Gypsum or Syngénite;
KCl	with Gypsum or Syngénite;
$MgSO_4 \cdot 7H_2O$	with Gypsum, Syngénite, or Glauberite;
Astrakanite	with Syngénite or Glauberite;
Na_2SO_5	
Kainite	with Gypsum or Syngénite;
Léonite	with Syngénite.
Schœnite	
Glaserite	

These results refer, it is understood, to the temperature of 25°.

They show how the phase rule guides the experimentalist in the analysis of systems whose complexity would otherwise defy all attacks.

CHAPTER VIII.

MONOVARIANT SYSTEMS.

118. Return to monovariant systems.—After having shown, by several examples, the services that the phase rule may render in the discussion of complicated cases occurring with trivariant or quadrivariant systems, we return to study more in detail the properties of systems whose variance has a smaller value, commencing with monovariant systems.

We call thus every system divided into ϕ phases related to the number c of components by the relation

$$\phi = c + 1.$$

For such a system the variance $V = c + 2 - \phi$ is equal to 1.

119. One component existing in two phases.—Among the monovariant systems we find in the first place all those where *a single component is divided between two phases;* among the most remarkable examples are:

1°. A solid or liquid substance is in the presence of its own vapor; the study of such systems constitutes the theory of the vaporization of liquid or solid bodies.

2°. A single substance exists simultaneously in the two states, liquid and solid, case to which applies the theory of the fusion of solids and the congelation of liquids.

3°. A single chemical substance, simple or compound, exists at once in two different solid forms, such as the yellow iodide of mercury with the red iodide.

4°. A gaseous chemical substance exists in the presence of a solid polymer, as gaseous cyanogen in the presence of solid para-cyanogen, or, again, gaseous cyanic acid with crystallized cyanuric acid.

120. Two components divided among three phases.—We find next, among the monovariant systems, those formed of *two independent components divided into three phases;* among them let us mention:

1°. The systems where a solid component and a gaseous component exist in the presence of a solid compound; such is the system formed by lime, carbonic acid gas, calcium carbonate, or, again, the system consisting of an anhydrous salt, water vapor, a definite hydrate of the same salt.

2°. The systems in which two independent components, water and an anhydrous salt, exist in three phases: a solid precipitate, anhydrous or hydrated, a solution, and water vapor.

3°. The systems where a solid hydrate of a gas is in contact with an aqueous solution of this gas and a mixture of the gas with water vapor.

4°. A mixture of two liquids, separated into two layers, beneath a simple or mixed vapor, formed from one or both of these liquids.

121. Three components with four phases.—A system for which *three independent components are divided among four phases* is also a monovariant system. As, for example:

The three independent components are water, 0, and two anhydrous salts, 1 and 2; the four phases are water vapor, an aqueous solution of the two salts, and two solid precipitates, C, C', which are bodies of definite composition formed from the three independent components 0, 1, 2 (ice, anhydrous salts, hydrated salts, double anhydrous or hydrated salts).

122. Law of equilibrium for monovariant systems. Transformation tension and transformation point.—The equilibrium states of a monovariant system obey a law which, for all these systems, has the same form, as follows:

At a given temperature the pressure for which the system is in equilibrium has an entirely definite value, which is called the Transformation Tension *at the temperature considered.* The composition and density of each phase which composes the system in equilibrium are equally definite; like the transformation tension, they do not depend upon the masses of the independent components which constitute the system. On the contrary, the masses of the various phases are not entirely determined, even when the masses of the independent components are given.

Under a given pressure the temperature for which the system is in equilibrium has a definite value, which is called the TRANSFORMATION POINT *under the pressure considered;* this equilibrium temperature does not depend upon the masses of the independent components which make up the system, and it is the same with the composition and density possessed, at the moment of equilibrium, by each of the phases into which the system is divided; nevertheless the masses of these phases are not entirely determined, even where the masses of the independent components are known.

123. Curve of transformation tensions.—Take two axes of rectangular coordinates, OT, $O\pi$ (Fig. 36); on the axis of abscissæ OT lay off a length OT measured by the temperature that we consider; through the point T draw a parallel to the axis of abscissæ $O\pi$, and on this parallel lay off a length $TM=OP$ measured by the transformation tension at the temperature considered; when the temperature assumes all possible values and the point T describes the line OT, the point M describes the curve CC', called the *curve of transformation tensions* of the monovariant system considered.

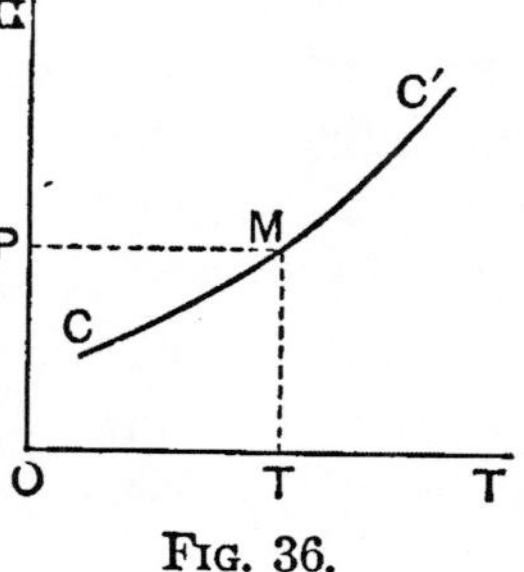

FIG. 36.

Suppose the curve of transformation tensions traced; if we know a temperature $T=OT$, a simple construction will give us the corresponding transformation tension $P=OP$; this will be the ordinate of the point M on the curve CC', which has OT as abscissa; if there is given a pressure $P=OP$, a simple construction gives us the corresponding transformation temperature $T=OT$; this is the abscissa of the point M on the curve CC' which has the ordinate OP.

124. Curve of tensions for saturated vapor.—It was in the study of the vaporization of solid and liquid bodies that physicists met, for the first time, the curve of transformation tensions of a monovariant system. The curve, in this case, is nothing else than the *curve of tensions for saturated vapor*.

125. Fusion phenomena.—The existence of a curve of transformation tensions in the phenomena of fusion was predicted theoretically in 1849 by James Thomson and verified experimentally in 1850 by William Thomson; these physicists showed that the fusing-point of ice has not an absolutely fixed value, but changes when the pressure is changed which acts upon the monovariant system composed of water and ice; other physicists have since established the same truth in studying the fusion of other substances; but, for reasons that we shall see later, it is necessary to cause the pressure to vary greatly in order to produce an appreciable change in a fusing-point; the curve of transformation tensions deviates only slightly from a straight line parallel to $O\pi$.

126. Allotropic transformations of solids.—As much may be said of the curve for transformation tensions in the case where one substance may exist in two allotropic forms, both being solid; nevertheless, if the pressure upon the system undergoes great changes, notable variations of the transformation point may be obtained; thus the red silver iodide may be transformed into the yellow silver iodide; it suffices for this, at atmospheric pressure, to increase the temperature to +146° C.; by raising the pressure to some 3000 atmospheres Mallard and Le Chatelier[1] were able to lower the transformation point to ordinary temperatures.

127. Gaseous substance and solid polymer.—The curve of transformation tensions in a system where a gaseous body exists in presence of a solid polymer has an appearance similar to that of a curve for the tension of saturated vapor; this curve rises from left to right, and that the more steeply as the temperature is higher; the existence of such a curve was discovered first by Troost and Hautefeuille[2] for the systems where gaseous cyanogen exists in presence of solid paracyanogen, then for systems where the gaseous cyanic acid exists with solid cyanuric acid.

128. Dissociation tensions.—The equilibrium of a system in which a definite solid component exists with two independent components, one of which is solid, the other gaseous, necessitates

[1] Mallard and Le Chatelier, *Journal de Physique*, 2d Series, v. 4, p. 305, 1885.

[2] Troost and Hautefeuille, *Annales de l'Ecole normale supérieure*, 2d Series, v, 2, p. 253, 1873.

that at each temperature the pressure supported by the system be equal to the transformation tension for the temperature considered; this is the fundamental law demonstrated by H. Debray, first[1] in studying the dissociation of calcium carbonate into lime and carbonic acid gas, then[2] in studying the dissociation of certain hydrated salts into anhydrous salts and water vapor.

Put with certain metallic chlorides ammonia gas is absorbed by these chlorides and forms with them definite solid compounds; the dissociation of an ammonia chloride into a metallic chloride and ammonia gas corresponds to a curve of transformation tensions which is here called *the curve of dissociation tensions.* Isambert[3] has determined a certain number of these curves and has shown their analogy to the curve for vapor tension of saturated vapors from liquids; his work has since been completed by Joannis and Croizier.[4]

Other curves of dissociation tension have been determined by chemists; we only note here the curves, determined by Joannis,[5] of the tensions of dissociation of potassammonium into potassium and ammonia gas, and of sodammonium into sodium and ammonia.

A system which encloses, simultaneously, an aqueous solution of a gas, a mixture of this gas with water vapor, and a definite solid compound formed by the union of the gas and water is in equilibrium, at each temperature, when the pressure has a definite value; the liquid mixture and the gaseous mixture has, at the same time, a definite composition; the total mass of gas and the total mass of water contained in the system do not influence either this tension nor this composition; this law was first recognized by Isambert[6] in studying the dissociation of chlorine hydrate; the curves of transformation tension of a great number

---

[1] H. DEBRAY, *Comptes Rendus*, v. 64, p. 603, 1867.

[2] *Ibid.*, v. 66, p. 194, 1868.

[3] ISAMBERT, *Comptes Rendus*, v. 66, p. 1259, 1868; *Ann. d. l'Ecole normal sup.*, v. 5, p. 129, 1868.

[4] JOANNIS and CROIZIER, *Mémoires de la Société des sciences physiques et naturelles de Bordeaux*, 4th Series, v. 5, p. 41, 1895.

[5] JOANNIS, *ibid.*, v. 5, p. 218, 1895.

[6] ISAMBERT, *Comptes Rendus*, v. 86, p. 481, 1898.

of analogous systems have been determined by H. Le Chatelier, Wroblewski, Bakhuis Roozboom, and P. Villard.[1]

129. Transformation point of double salts.—Consider a system formed by three independent components, as

Water: H_2O;
Copper acetate: $Cu(C_2H_3O_2)_2$;
Calcium acetate: $Ca(C_2H_3O_2)_2$.

This system may be divided into four phases, thus:
A liquid mixture containing the three independent components;
Crystals of hydrated copper acetate: $Cu(C_2H_3O_2)_2 \cdot H_2O$;
Crystals of hydrated calcium acetate: $Ca(C_2H_3O_2)_2 \cdot H_2O$;
Crystals of a hydrated double salt, Cupricalcium acetate:

$$CuCa(C_2H_3O_2)_4 \cdot 6H_2O.$$

Such a system is monovariant; under a given pressure there exists a well-determined temperature for which it may exist in equilibrium; this temperature is the *transformation point* of the cupricalcium acetate under the given pressure; if the temperature is less than the transformation point, the two simple salts combine and the cupricalcium acetate is formed; if, on the contrary, the temperature is higher than the transformation point, the cupricalcium acetate decomposes and there is formed copper acetate and calcium acetate; calcium acetate is uncolored, copper acetate green, and cupricalcium acetate blue, so that these transformations are accompanied by color changes which facilitate the study, as was first shown by Van't Hoff and Ch. van Deventer.[2] Reicher[3] has shown that the transformation point of cupricalcium acetate, at atmospheric pressure, is included between $+76°$ *C.* and $78°$ *C.*

This transformation point should depend upon the pressure acting upon the system; but here, as in the fusion phenomena, it is necessary to cause the pressure to vary greatly in order to produce an appreciable change in the transformation point;

---

[1] Memoir by *P.* VILLARD (*Annales de Chimie et de Physique*, 7th Series, v. 11, p. 289, 1897) contains a complete bibliography of the subject.

[2] VAN'T HOFF and VAN DEVENTER, *Recueil des Travaux chimiques des Pays Bas*, v. 6, p. 407, 1886; *Zeitschrift für physikalische Chemie*, v. 1, p. 163, 1887.

[3] REICHER, *Zeit. phys. Chem.*, v. 1, p. 221, 1887.

W. Spring and Van't Hoff[1] have s eenthis point lowered to +40° *C.* under a pressure of about 6000 atmospheres.

130. Precautions to take in the application of the preceding laws. First example: dissociation of the red oxide of mercury. Pélabon's investigations.—The application to a chemical system of the idea of the *curve of transformation tensions* may, in certain cases, necessitate certain precautions, which if neglected lead to errors.

For example, it may happen that with the same chemical substances different monovariant systems may be constituted; to each one of these monovariant systems will correspond a curve of transformation tensions, but these curves will not be superposable.

The dissociation of the oxide of mercury offers an interesting example to which this remark applies.

Suppose that the red oxide decomposes into oxygen and mercury vapor in an enclosure empty at the start. The system is divided into two phases: solid mercuric oxide and the gaseous mixture of oxygen and mercury vapor. How many independent components does it contain? In the conditions which we suppose realized it is sufficient to know the total mass of mercury, free or combined, that it contains. The system therefore does not contain two independent components, but merely one, oxide of mercury; it consists of an arbitrary mass of mercuric oxide, partly in a state of combination, partly decomposed. Summing up, we may say that we are dealing with a system formed of a *single independent component*, the substance HgO, which exists in *two phases*, a solid phase, the red oxide, and a gaseous phase, the mixture in *equivalent proportions* of oxygen and mercury vapor. Such a system is monovariant; it admits a curve *C* of transformation tensions; at each temperature *T* the curve *C* shows a corresponding transformation tension *P*.

Instead of supposing that the oxide of mercury dissociates in an enclosure empty at the start, it may be supposed to dissociate into an enclosure where oxygen or mercury vapor has been introduced; in this case the mass of oxygen, whether free or combined, and the free or combined mercury, which the system

---

[1] SPRING and VAN'T HOFF, *Zeit. phys. Chem.*, v. 1, p. 227, 1887.

includes, are no longer in equivalent proportions; these two masses are arbitrary; the system is composed of *two independent components*, oxygen and mercury.

If the system consists only of solid mercuric oxide and a mixture of oxygen and mercury vapor, it is divided into two phases only and is bivariant; for such a system there is no longer question of a curve of transformation tensions; the knowledge of the temperature is not sufficient to determine the pressure supported by the system in equilibrium.

It is no longer the same if there is introduced into the system enough mercury so that a part of this substance remains in the liquid state; the system, formed of *two independent components*, oxygen and mercury, and divided into *three phases*, red oxide of mercury, mixture of oxygen and mercury vapor, and liquid mercury, is a monovariant system; it admits of a curve of transformation tensions C'; at each temperature T the curve C' has a corresponding transformation tension P' whose value is independent of the masses of mercury and oxygen which the system contains.

The two curves have been determined by H. Pélabon;[1] they are in nowise identical; the curve C' is much higher than the curve C; for instance, at the temperature of 520° C. the transformation tension which we have denoted by P is measured by 417.6 centimetres of mercury, and the transformation tension denoted by P' by 844.0 centimetres of mercury. From this we see how unsafe it would be to speak, without further limitations, of the dissociation tension of mercuric oxide at a given temperature.

131. Second example: Dissociation of cupric oxide.—Other precautions should be taken in the application of the idea of the curve of transformation tension to a system; it may happen that a phase appears or disappears in the system; that consequently the system is monovariant or bivariant according to circumstances; that it has or has not a curve of transformation tensions.

Here is a good example taken from the investigations of Debray and Joannis.[2]

---

[1] H. Pélabon, *Mémoires de la Société des sciences physiques et naturelles de Bordeaux*, 5th Series, v. 5, 1899.

[2] H. Debray and Joannis, *Comptes Rendus*, v. 99, pp. 583 and 688, 1884.

Cupric oxide dissociates into cuprous oxide and oxygen. Two independent components, cuprous oxide and oxygen, form the system, which below a certain temperature is divided among three phases, solid cupric oxide, solid cuprous oxide, oxygen gas. The system is monovariant, admitting a curve of dissociation tensions.

Thus, at a given temperature T, equilibrium corresponds to a tension P of the oxygen atmosphere, perfectly determined by the knowledge of the single temperature T. If oxygen is introduced into the recipient so that the pressure takes on momentarily a value greater than P, the cuprous oxide absorbs oxygen and is transformed into cupric oxide until the pressure has again become equal to P. If, on the contrary, oxygen is removed, lowering the pressure below P, cupric oxide is reduced until the pressure retakes its former value.

These phenomena are very sharply produced when the temperature does not exceed a certain limit, near the fusing-point of gold; beyond this limit the two oxides exist in contact as a fused mass; instead of forming two solid phases, they form but one liquid phase; from monovariant the system becomes bivariant; at a given temperature there can be no longer any definite dissociation tension.

Suppose, for example, that at a temperature T the system is in equilibrium and the liquid mixture of cupric and cuprous oxides in the presence of an oxygen atmosphere under the pressure P; without changing the temperature, let us introduce a certain quantity of oxygen so as to increase the pressure; the liquid will absorb a part, but only a part, of the mass introduced; when equilibrium is again established the pressure will have a value P' higher than P; if, conversely, we had withdrawn oxygen so as to have lowered the pressure to less than the initial pressure P, the liquid would have liberated oxygen so as to increase the pressure, but only to a value P' less than P.

132. Dissociation of chlorine hydrate.—Analogous facts, which caught the attention of Isambert and Le Chatelier, are observed in the study of systems having for independent components water and chlorine; at low temperatures one may observe such a system divided into three phases: chlorine hydrate crystals, a liquid solution, a gaseous atmosphere of chlorine and water vapor; at each temperature equilibrium is established when the pressure has a value which depends upon the temperature alone and in nowise upon the masses of chlorine and water enclosed by the system.

If the temperature attains, then exceeds, the point of aqueous fusion of these crystals of chlorine hydrate, they disappear; the system which then contains but two phases, the liquid mixture and the gaseous mixture, is bivariant; at a given temperature the tension of the gaseous mixture which remains in equilibrium above the liquid solution may assume an infinity of values; to increase it, the addition of chlorine is sufficient; to decrease it, removal of a part of the gaseous mixture.

133. The absence of a fixed tension of dissociation distinguishes a solution from a definite compound.—These remarks suggest the means of deciding, in certain cases, some questions in dispute.

A gas, ammonia for example, is absorbed by a solid substance *C*; what is the nature of this absorption? Does the ammonia gas form with the solid a definite compound whose particles are disseminated through the solid *C* in excess, but nevertheless distinct from the solid? Or, on the contrary, does it form with the solid a *solid solution* of which each infinitely small volume contains both matter from the solid and from the ammonia?

The system is assuredly formed of two independent components, ammonia gas and the solid *C*.

According to the first hypothesis the system is divided into three phases, which are ammonia gas, the solid *C*, the solid ammoniacal compound; the system is therefore monovariant; for each temperature *T* should correspond a definite dissociation tension *P*; if the ammonia gas is removed, a certain quantity of the solid ammoniacal compound will be dissociated so as to restore the value *P* to the tension of the gaseous atmosphere, and this operation may be repeated several times up to the moment when the whole of the solid ammoniacal compound is destroyed; if ammonia gas is now added, the excess of gas will be absorbed by the solid *C* and the tension brought back to the value *P* at the moment when the whole of the substance *C* has changed over to the state of ammoniacal compound.

Quite otherwise will be the succession of phenomena according to the second hypothesis; the system, formed only of two phases, ammonia gas and the solid solution, is bivariant; at a single temperature T the system may be observed in equilibrium for an infinity of different values for the tension of the ammonia gas; the pressure exerted by this gas at the moment of equilibrium will increase if ammonia gas is added to the system, and will diminish if gas is withdrawn.

As long ago as 1867, Isambert made use of this criterion to demonstrate that ammonia gas when absorbed by metallic chlorides forms with these bodies definite compounds, while when absorbed by carbon it forms with this substance a solid solution.

134. The zeolites are solid solutions.—In the last few years the criterion suggested by Isambert has been used anew and has furthered the establishment of interesting consequences.

Nature furnishes us with a certain number of hydrated silicates which mineralogists call *zeolites;* the dehydration of certain hydrates offers curious peculiarities; *analcime*, for instance, may be completely dehydrated without any sudden variation in form or optical properties of the crystals being observed; Georges Friedel[1] has shown that analcime had not, at a given temperature, an invariable dissociation tension; let us suppose the temperature constant; in a first equilibrium state the tension of the water vapor which exists in equilibrium above the crystals has the value P; remove a portion of this water vapor; the analcime will undergo a certain dehydration and the tension of the water vapor will increase, but only to a value P', less than P; and so on; analcime is therefore not a definite hydrate, but only a *solid solution* in which water is mixed with an anhydrous silicate.

Tammann[2] has extended Friedel's observation to a great number of hydrated silicates studied by mineralogists, and also to hydrated platinocyanide of magnesium:

$$MgPt(CN)_4+Aq.$$

---

[1] G. Friedel, *Bulletin de la Société de Minéralogie*, v. 19, p. 363, 1896; v. 21, p. 5, 1898.

[2] Tammann, *Wiedemann's Annalen*, v. 73, p. 16, 1897; *Zeit. phys. Chem.*, v. 27, p. 323, 1898.

135. The existence of a dissociation tension does not always prove the existence of a definite compound. Dissociation of palladium hydride.—The absence of a fixed dissociation allows also, in certain cases, to demonstrate that a body is not a definite compound; prudence must be exercised in concluding from the existence of a fixed dissociation tension the existence of a definite compound; the study of the absorption of hydrogen by palladium will show us that such a conclusion must sometimes be accepted with caution.

Following Troost and Hautefeuille,[1] let us take palladium, kept at a constant temperature, and introduce hydrogen into the enclosure about this metal; the hydrogen is in part absorbed and the tension attains a certain fixed value P; introduce more hydrogen, some of this is also absorbed and the pressure reassumes the value P; this may go on until the palladium has absorbed a quantity of hydrogen proportional to its mass; beyond this, each time that hydrogen is introduced it is found that, after equilibrium is established, the tension of the gas attains a value higher than in the preceding operations.

Troost and Hautefeuille interpreted these observations by admitting that there is first formed a hydride of palladium, of definite composition, to which they assign the formula Pd_2H; it is only when all the palladium would have passed over to the hydride that the hydride in its turn would absorb hydrogen, forming a solid hydrogen solution; to the first form of reaction corresponded a fixed dissociation tension, while the second would be characterized by the absence of such a tension.

There may be some doubt as to the validity of this conclusion.

From the existence of a fixed dissociation tension, the only consequence which we may deduce with certainty is that the system is monovariant; and as the system is certainly formed of two independent components, palladium and hydrogen, this conclusion is equivalent to the following: the system is divided into three phases.

To suppose, with Troost and Hautefeuille, that the system is composed of gaseous hydrogen, solid palladium, and hydride of

---

[1] Troost and Hautefeuille, *Ann. de Chimie et de Physique*, 5th Series, v. 2, p. 279, 1874.

solid palladium, is to make an hypothesis which accords with this certain consequence, but which is not equivalent to it; other hypotheses, in fact, accord equally well with this consequence; such, for example, as the following made by Roozboom and Hoitsema:[1] the solid mass that the system includes would be formed by the juxtaposition of two solid solutions, containing each a different quantity of hydrogen, comparable consequently to the unequally concentrated layers into which a liquid mixture of water and ether is divided.

May one decide between these hypotheses?

If we examine the question more closely, we come, as is easy to show, to results which differ according to the hypothesis made and which may be compared with experimental data.

Let us take, at a given temperature T, a given mass of palladium, a gramme for example; also take two rectangular coor-

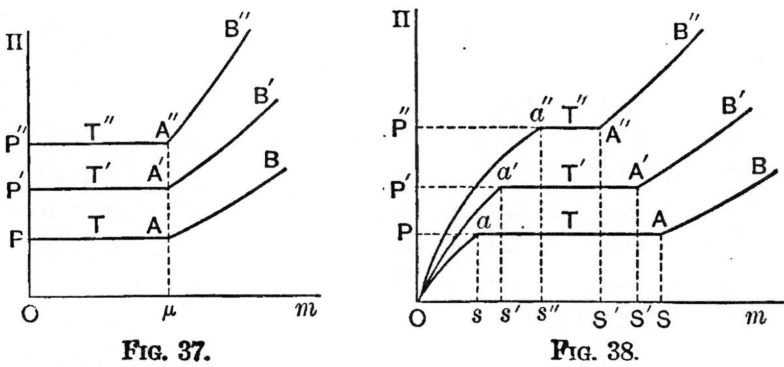

FIG. 37. FIG. 38.

dinate axes (Figs. 37 and 38), and along the axis of abscissæ lay off the mass m of hydrogen absorbed by a gramme of palladium, while along the axis of ordinates lay off the pressure π of the hydrogen in the enclosure.

Let us first trace the consequences of the hypothesis of Troost and Hautefeuille.

As long as the pressure π is less than the dissociation tension P of the hydrogenized palladium at the temperature T, the palladium does not absorb hydrogen; the representative point describes a segment OP of the line $O\pi$.

---

[1] HOITSEMA, *Archives néerlandaises d. sciences exactes et naturelles*, v. 30, p. 44, 1895; *Zeit. f. phys. Chem.*, v. 17, p. 1, 1895.

From the moment the pressure exceeds the value P, the hydride of palladium commences to be formed; as an increasingly greater fraction of our mass of palladium passes over to the state of hydride, just so fast the mass m of hydrogen absorbed increases, while the tension of hydrogen, once equilibrium is established, remains constant and equal to P; the representative point describes a segment of the straight line PA, parallel to OM.

This goes on until the instant that the gramme of palladium employed has passed entirely over to the state of hydride; at this moment the mass m of hydrogen absorbed has a definite value μ, which is the mass that must be combined with a gramme of palladium to obtain the substance Pd_2H, about $\frac{1}{212}$ gramme.

From this moment the palladium hydride as a solid solution absorbs hydrogen; the pressure of the hydrogen in the system in equilibrium does not guard a fixed value; it increases at the same time as the mass of hydrogen absorbed by the palladium hydride; the representative point describes a straight line AB which rises from left to right.

If we repeat the same operations at other temperatures, T', T'', more and more elevated, we shall obtain the lines $OP'A'B'$, $OP''A''B''$, analogous to $OPAB$; the lines PA, $P'A'$, $P''A''$, parallel to the axis Om, will be higher and higher, because to the more and more elevated temperatures, T, T', T'', correspond the increasing dissociation tensions P, P', P'' of palladium hydride; but the points A, A', A'' lie all on the same parallel to the axis $O\pi$, for their abscissa has a value μ which in nowise depends upon the temperature.

Let us now reconsider the same operations and represent the results according to the hypothesis of Roozboom and Hoitsema (Fig. 38).

At a given temperature T the hydrogenized palladium may exist in two solid solutions of different composition; when the hydrogenized palladium thus separated exists in the presence of an atmosphere of hydrogen, one has to do with a monovariant system; where such a system is in equilibrium at a temperature T, not only has the tension of the hydrogen a value P which depends on the single temperature T, but also the two solid solutions have compositions entirely fixed by the knowledge of the one temperature T; in the first, containing the less hydrogen, a gramme of palladium is united with s grammes of hydrogen; in the second, a gramme of palladium unites with S grammes of hydrogen; constant for a given temperature, the numbers s and S vary with the temperature.

At a given temperature T let us take a gramme of palladium and increase the mass m of hydrogen absorbed by this palladium.

While m is less than s, the hydrogenized palladium is composed of a single solid solution; the system, which includes only two phases, is bivariant; it does not have a fixed dissociation tension; the tension of the gaseous hydrogen for the system in equilibrium increases with the amount of hydrogen held in the solid solutions; the representative point describes a curve Oa which rises from left to right.

When the mass m of hydrogen absorbed by a gramme of palladium attains, then exceeds s, the hydrogenized palladium divides into two solid solutions, the one of s, the other of S grammes of hydrogen; when the mass m increases, the two quantities s and S remain constant, but the mass of the first solution decreases and the second increases; the system is monovariant; the tension of hydrogen keeps a constant value P and the representative point describes a segment aA of a straight line parallel to Om; the two extremities a and A of this segment have for abscissæ s and S respectively.

When the mass m of hydrogen absorbed by a gramme of palladium attains, then exceeds S, the hydrogenized palladium forms but a single solid solution; the system again becomes bivariant; the tension of the hydrogen increases with the richness of the solid solution; the representative point describes a curve AB which rises from left to right.

Let us repeat the same operations at the temperatures T, T', T'', higher and higher; we shall obtain a series of analogous lines $OaAB$, $Oa'A'B'$, $Oa''A''B''$, about which we may make the following remarks:

The segments aA, $a'A'$, $a''A''$, parallel to the axis Om, are more and more elevated, because to the increasingly higher temperatures T, T', T'' correspond greater and greater transformation tensions P, P', P''.

The origins of these segments a, a', a'' have abscissæ s, s', s'' different from zero and differing among themselves.

The extremities A, A', A'' of these segments have abscissæ S, S', S'' which differ among each other.

It suffices now to compare Figs. 37 and 38 to see that experiment will allow deciding between the hypothesis of Troost and Hautefeuille and that of Roozboom and Hoitsema. The experiment has been tried by the latter; the curves that they have obtained, absolutely irreconcilable with the disposition of Fig. 37, possess, on the contrary, differences from Fig. 38 which are easily explicable.

136. Robin's law.—In order that a monovariant system taken at a certain temperature and under a certain pressure may be in equilibrium, it is necessary that this pressure be equal to the transformation tension relative to this temperature; in other words, it is necessary that the representative point which has for abscissa this temperature and for ordinate this pressure be located on the curve of transformation tensions.

What happens if the representative point having temperature and pressure as ordinate is not on the curve of transformation tensions? A very simple rule answers this question; it is due to G. Robin,[1] and is thus stated:

If the representative point is above the curve of transformation tensions, every transformation taking place at the temperature and under the pressure considered is accompanied by a diminution in volume of the system; the opposite takes place if the representative point is below the curve of transformation tensions.

Take, for example, a m novariant system formed by a liquid in contact with its vapor; at the temperature T the tension of saturated vapor has the value P; if the pressure π has a value greater than P, the temperature remaining at T, the system will be the seat of a modification accompanied by a lessening in volume, that is, by a condensation of vapor; if, on the contrary, at the same temperature T, the pressure is less than the tension P of saturated vapor, the system will be the seat of a modification

[1] G. Robin, *Bulletin de la Société philomathique*, 7th Series, v. 4, p. 24, 1879.

accompanied by an increase in volume, that is to say, of vaporization of the liquid.

Let us take as a second example a monovariant system formed of calcium carbonate, carbonic acid gas, and lime; denote by P the dissociation tension of calcium carbonate for the temperature T; if, at the same temperature, the pressure is greater than P, the system will be the seat of a reaction accompanied by a decrease in volume, that is, of a combination of carbonic acid gas with lime; if, on the contrary, the pressure is less than P, the system will be the seat of a reaction accompanied by an increase in volume, or of dissociation of calcium carbonate.

The accordance of these conclusions with the facts is beyond criticism.

137. Moutier's Law.—Some years before Robin stated the rule which permits predicting the nature of the modification which is produced in a monovariant system when the representative point is above or below the curve of transformation tensions, J. Moutier had stated an analogous rule; the latter predicts the nature of the modification of which the system is the seat according as the representative point is to the left or to the right of the curve of transformation tensions.

This rule is as follows:

Let π be a pressure arbitrarily given and θ the transformation point at this pressure. If, at the pressure π, the temperature T has a value less than θ, the representative point is to the left of the curve of transformation tensions; in these conditions the system is the seat of a certain modification; taking place at the same pressure π and at the temperature θ, this modification will liberate heat.

If, at the pressure π, the temperature T has a value greater than θ, the representative point is to the right of the curve of transformation tensions; in these conditions the system is the seat of a modification; accomplished at the same pressure π and temperature θ, this modification would absorb heat.

For example, at a given pressure and at the boiling-point corresponding to this pressure, vaporization of the liquid absorbs heat; the condensation of the vapor liberates heat; therefore, at this same pressure, for temperatures less than the boiling-point the vapor condenses, while at temperatures higher than the boiling-point the liquid vaporizes.

At a given pressure and at the fusing-point corresponding to this pressure, the fusion of a solid absorbs heat, while the freezing of the liquid sets heat free; therefore, at this same pressure, for temperatures less than the fusing-point the liquid freezes, while at temperatures higher than the fusing-point the solid melts.

138. False equilibria in monovariant systems.—These examples show how easy it is, in most cases, to apply Robin's or Moutier's rule to a given monovariant system; nevertheless, when predictions of the modifications of a given monovariant system are sought by aid of these rules, one must not lose sight of what has been said at the end of Chapter VI (Arts. 98 and 99) concerning the phenomena of false equilibrium.

When thermodynamics indicates that a modification is impossible, such a modification cannot take place; but when it announces that a change should occur, it may happen that no change is produced, and the system remains in equilibrium.

Striking examples of this general remark may be found.

Moutier's rule teaches that at a temperature less than the fusing-point the solid cannot melt, while the liquid should freeze; and in fact at a temperature below the fusing-point the solid never does pass into the liquid state; but it may very well happen that the liquid does not freeze and yet remains in equilibrium; it is in general sufficient, in order to observe this phenomenon of *surfusion,* to avoid shocks and especially the introduction of a particle of the so id which the liquid would give in freezing.

Robin's rule teaches that under a pressure higher than the tension of saturated vapor corresponding to the temperature of the experiment, the vapor should pass into the liquid state; as a fact, if compression is carefully brought about and if the vapor is free from all dust and liquid drops, it is easy to obtain a vapor under a higher pressure than the tension of saturated vapor without condensation; this was first observed by Coulier and afterwards studied by Wüllner and Grotian.

It suffices for the moment to mention these phenomena, which will be studied farther on (see Chap. XVII, Arts. 274–278).

139. Another form of Moutier's law.—As simple as Robin's rule, and also antedating it, Moutier's rule surpasses the former by the originality of views that it introduces into chemical statics.

Under a pressure π let θ be the value of the transformation point of a monovariant system; let us consider a reaction taking place in this system under the same pressure π and at the temperature T, different from θ; this reaction calls into play a certain quantity of heat, which depends upon the temperature T, as we have seen (Chap. III, Art. 41); if, therefore, without changing the pressure, we cause the temperature T to vary, making it approach θ, the value of the quantity of heat set free by the reaction will vary, and even its sign may change; this last circumstance will certainly not be produced if the temperatures T are not too far distant from the transformation point θ; let us suppose that this is not produced for the monovariant systems which we shall study and in the conditions under which we study them; Moutier's rule may then be stated:

At a given pressure, every change produced in a monovariant system at a temperature less than the transformation point is accompanied by a liberation of heat; every modification produced at a temperature higher than the transformation point is accompanied by an absorption of heat.

140. Corollary to this law.—Moutier deduced a consequence which is almost identical with this proposition, but which has the advantage of better emphasizing its importance:

Imagine that in the same monovariant system, under the same pressure, but at two different temperatures, two reactions, the inverse of each other, are observed; the reaction taking place at the lower temperature is exothermic, the one produced at the higher temperature is endothermic.

Thus, in the same monovariant system formed of cupric oxide, cuprous oxide, and oxygen, and under the same pressure, two inverse reactions may be observed; at a certain temperature, oxidation of curpous oxide; at a higher temperature, the dissociation of cupric oxide; the dissociation of cupric oxide absorbs heat, the oxidation of cuprous oxide liberates heat.

141. Consequence relative to very low temperatures; the principle of maximum work is exact at these temperatures.—Let us suppose for an instant that we have never to consider other than monovariant systems, and let us see what follows from the preceding proposition.

If, under a given pressure, as that of the atmosphere, the temperature was sufficiently lowered so that it became less than all the transformation points of the systems considered, it would be impossible to observe reactions other than those accompanied by a liberation of heat; all the spontaneous decompositions would be decompositions of endothermic compounds, all the spontaneous syntheses would be syntheses of exothermic compounds.

On the contrary, if the temperature were raised sufficiently to become higher than all the transformation points of the systems studied, all the possible reactions would be accompanied by an absorption of heat; all spontaneous decompositions would be decompositions of exothermic compounds, all spontaneous syntheses would be syntheses of endothermic compounds.

In other terms, at a sufficiently low temperature the principle of maximum work would rigorously apply.

142. Consequence for high temperatures.—On the other hand, at sufficiently high temperatures this principle would be turned upside down and replaced by its opposite; it is therefore clear that, in creating the chemistry of high temperatures, H. Sainte-Claire Deville had to meet a host of facts irreconcilable with the principle set forth by J. Thomsen; we see why he could obtain the dissociation of a great number of exothermic compounds, such as water or carbonic acid gas (Arts. 49, 50, 51); it is as easily understood how his disciples could, at very high temperatures, reproduce endothermic substances which are spontaneously destroyed at much lower temperatures, as we shall see in the following chapter.

Evidently these conclusions are not logically established by what precedes except on the condition to suppose monovariant all the chemical systems found in nature; but their generality, which will be demonstrated in the following chapter, may already be anticipated; as long ago as 1877, Moutier affirmed it with great clearness.[1]

[1] J. MOUTIER, *Bulletin de la Société philomathique*, 3d Series, v. 1, p. 96, 1877. This important contribution to the history of chemical mechanics

These results transformed profoundly the ideas previously held concerning the opposition which exists between exothermic and endothermic reactions.

For the chemists of the beginning of the nineteenth century every combination was exothermic, every decomposition endothermic.

For the thermochemists who accepted without restriction the principle of maximum work an exothermic reaction was one susceptible of producing itself; an endothermic reaction could not take place without the aid of external energy.

In modern chemical mechanics an exothermic reaction is one susceptible of producing itself at low temperature; an endothermic reaction is one that may produce itself at high temperature.

143. Similarity of Moutier's and Robin's laws. Form of the curve of transformation tensions.—Comparing the two laws stated by J. Moutier and by G. Robin, we are immediately led to important conclusi ns.

Consider a monovariant system; let M (Fig. 39) be a point of the transformation curve of this system; θ, P are abscissa and ordinate of this point.

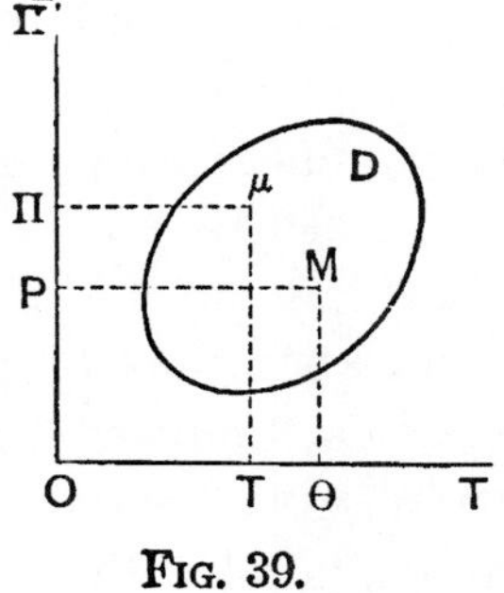

Fig. 39.

Let us suppose that the system studied can present only two kinds of modifications, the inverse of each other; for example, the fusion of a solid and the freezing of a liquid, or the combination of carbonic acid gas with lime and the dissociati n of calcium carbonate.

Among the modifications which this system may have, there are some which, taking place at the temperature θ and under the pressure P, would absorb heat; there are others, the inverse of the former, which liberate heat. Let us choose one of these latter.

Taking place at the temperature θ and pressure P, it would determine a certain variation of the volume of the system; this

is reproduced in Duhem's *Introduction à la Mécanique Chimie*, p. 147 (Gand, 1893).

variation may be a *decrease*, and we shall call this the *first case;* it may be an *increase*, and we shall call this the *second case.*

When the temperature and pressure change gradually, the quantity of heat developed by a definite change varies in a continuous manner, and this is also true for the change in volume accompanying this modification; one may therefore always suppose this change of temperature and pressure small enough so that it does not necessitate any change of sign either in the quantity of heat involved in the modification or in the variations of the volume which it causes.

Applied to the modification which we have chosen, this remark may be put into the following form:

About the point M it is always possible to draw a region D small enough so that it has the following property: at the temperature T and pressure Π which serve as coordinates for any point μ, in this region, the modification considered would cause a liberation of heat; there will be a diminution of volume for the first case, and an increase of volume for the second case.

From now on, let us consider only the points of this region and apply Moutier's rule to them; first, this shows us that at the temperature T and pressure Π which serve as coordinates to one of these points μ, the modification considered cannot be produced, unless the point μ is to the *left* of the curve of transformation tensions.

Let us now make use of Robin's rule; for the *first case* the modification may be produced at the temperature T and pressure Π only if the point μ is *above* the curve of transformation tensions; in the *second case*, on the contrary, it can be produced only if the point μ is *below* the curve of transformation tensions.

In order that these conclusions may be in accord, it is necessary that the portion of the region D which is *to the left* of the curve of transformation tensions be, at the same time, in the *first case above* this curve, and in the *second case below*, which leads to the theorem:

Let θ and P be the coordinates of a point M on the curve of transformation tensions of a monovariant system; choose a modification of this system which liberates heat when it is supposed to take place at the temperature θ and the pressure P; if this modification *is accompanied by a decrease in volume of the system, the curve of transformation tensions rises from left to right in the neighborhood of the point M* (Fig. 40); *if this modification is accompanied by an increase in volume of the system, the curve of transformation tensions descends from left to right near the point M* (Fig. 41). Let us apply this theorem to some simple cases.

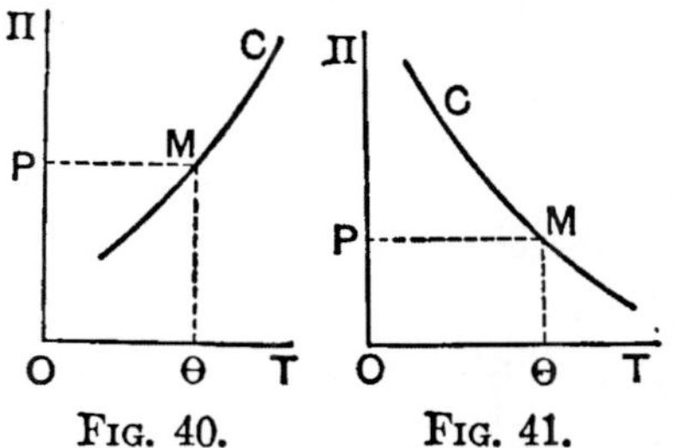

FIG. 40. FIG. 41.

At a given temperature and under the tension of vapor saturated at this temperature, the condensation of vapor to the liquid state is accompanied by a liberation of heat and a decrease in volume; the curve of tensions of saturated vapor should rise from left to right; the tension of saturated vapor is the greater the higher the temperature.

At a given temperature and under a pressure equal to the dissociation tension for this temperature, the combination of carbonic acid gas with lime is accompanied by a decrease in volume of the system and by a liberation of heat; the curve of dissociation tensions of calcium carbonate should rise from left to right.

Under a given pressure and at a temperature equal to the fusing-point for this pressure, the freezing of a liquid sets heat free. Most liquids decrease in volume on freezing; the curve of transformation tensions should therefore rise from left to right; some liquids, such as water, increase in volume in freezing; for these substances the curve of transformation tensions should descend from left to right; whence the following proposition, which has been verified by a great number of observers:

For almost all solids whose fusion is accompanied by expansion, the fusing-point rises at the same time as the pressure supported by the system increases; for solids, such as ice, whose fusion is accompanied by contraction, the fusing-point is lowered when the pressure increases.

We shall go farther and find the value of the tangent to the curve of transformation tensions; but in order to be able to state the important equation which determines the value of this quantity, we must first speak of the modifications that a monovariant system may be supposed to assume.

144. In every monovariant system there may be two modifications, the inverse of each other, which change the masses of the phases without changing their composition.—In certain monovariant systems only two modifications can be observed, the inverse of each other; thus, in a system which encloses a liquid and its vapor, only the vaporization of the liquid or the condensation of the vapor; in a system containing a solid and the liquid resulting from its fusion, only the fusion of the solid or the freezing of the liquid can be observed; in a system containing carbonate of calcium, lime, and carbonic acid gas, one may observe only the combination of carbonic acid with lime or the dissociation of the carbonate of calcium.

In the various systems we have mentioned each phase is a substance of definite composition; it is therefore evident that the two opposite kinds of modification, which may be produced, change the masses of the various phases without changing their composition.

Some other monovariant systems are more complicated.

Take, for example, a system whose independent components are water and sodium chloride and which is divided into three phases: solid chloride of sodium, an aqueous solution, and water vapor; it may be supposed that sodium chloride dissolves without condensation of water vapor, that water vapor condenses without dissolving sodium chloride, that when a mass of sodium chloride is dissolved and a mass of water vapor condensed, the ratio of these two masses may have any value whatever; it may similarly be imagined that the system undergoes any modifications the opposite of the preceding.

It is quite clear that in general these modifications change the composition of the solution of sodium chloride; if, for example, sodium chloride is dissolved without condensing water vapor, the solution becomes more concentrated; if one condenses water vapor without dissolving sodium chloride, the solution becomes more dilute.

It is always possible to imagine two kinds of modification, the opposite of each other, which do not change the concentration of the sodium chloride solution; the first of these two kinds of modification consists in dissolving a certain mass of sodium chloride and, at the same time, in condensing a certain mass of water, the ratio of the first mass to the second being exactly equal to the concentration of the solution; the second modification is to take from the solution a mass of sodium chloride and a mass of water vapor, the ratio of the first mass to the second being again equal to the concentration of the solution.

This remark may be generalized into the following proposition:

In any monovariant system, two kinds of modification the opposite of each other may be imagined which change gradually the mass of each of the phases without altering the composition of any of them.

145. The equilibrium of a monovariant system is indifferent.—This proposition gives rise to a consequence which should be noted.

Let T be any temperature whatever; in order that a monovariant system be in equilibrium at this temperature T, it is necessary and sufficient that the pressure have a definite value P, which is the transformation tension for this temperature, and that each of the phases into which the system is divided have a definite composition; then it is easy to see that if *the temperature T and the pressure P are kept constant, the monovariant system is in indifferent equilibrium.*

One may, in fact, without changing either the temperature T or the pressure P, impose on the system a modification which changes the masses of the various phases without changing the composition of any of them; such a modification will not, therefore, disturb the equilibrium of the system.

Let us take, for instance, a system which encloses a liquid and its vapor; at a given temperature this system is in equilibrium if the pressure is equal to the tension of saturated vapor; without changing temperature or pressure, condense a certain mass of vapor or vaporize a certain mass of liquid; the system is modified, but all the states through which it passes are equilibrium states; the equilibrium of a system formed by a liquid and its vapor is therefore a state of indifferent equilibrium.

146. Law of Clapeyron and Clausius.—The conclusion that we have just drawn is, besides, as we shall see, not the only one that may be deduced from the existence of modifications which leave unaltered the composition of each of the phases of a monovariant system; it is from the consideration of such modifications that we are going to determine the tangent to the curve of transformation tensions.

Take, in a monovariant system, a point M (Fig. 42) on the curve C of transformation tensions; let θ and P be the coordinates of this point; suppose, what has not been necessary to this time, that θ is not the temperature on any thermometer, but the *absolute temperature;* draw through M a tangent Mt to the curve C; this tangent makes with the line OT or with its parallel, $M\tau$, an angle α; it is the trigonometrical tangent of this angle which we wish to know.

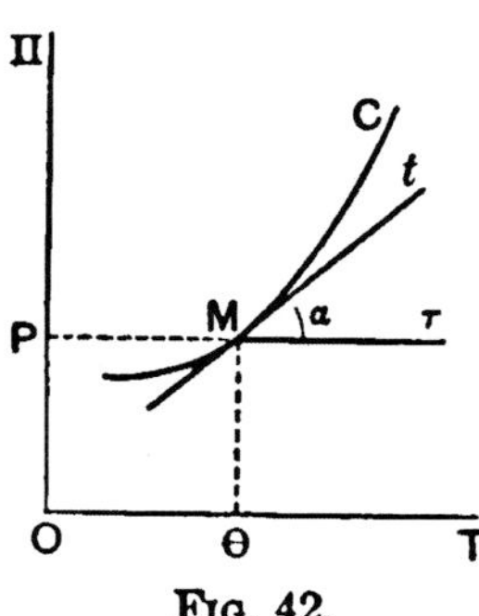

Fig. 42.

Among the modifications which change the masses of the phases into which a monovariant system is divided without changing the composition of any of these phases, there are necessarily some which at the temperature θ and pressure P liberate heat. Take one of these modifications and let Q be the quantity of heat, positive by definition, which it liberates. Denote by V the increase in volume of the system due to this modification; if this change is accompanied by a contraction, V is negative.

Thermodynamics gives us the following equation:

$$\tan \alpha = -\frac{E}{\theta} \cdot \frac{Q}{V}, \tag{1}$$

E being the mechanical equivalent of heat.

This formula agrees, as is easily seen, with the propositions which we have obtained by combining Moutier's and Robin's rules.

If the modification considered is accompanied by a decrease in volume, V is negative, $\tan \alpha$ is positive, and the curve of transformation tensions rises from left to right.

If the modification considered is accompanied by an increase in volume, V is positive, $\tan \alpha$ negative, and the curve of transformation tensions descends from left to right.

This relation is one of the oldest known to thermodynamics;

it already existed, although in an incomplete form, in the commentary given by Clapeyron on the work of Sadi Carnot on the motive power of heat; it was later completed and demonstrated by R. Clausius among his earliest researches in thermodynamics.

Clausius applied it in the first place to the best known of monovariant systems, that formed by a liquid and its vapor. Let us see what form this relation takes for such a system.

147. Application to vaporization.—As modification leaving invariable the compo ition of the system and producing a liberation of heat, we may take the condensation of a gramme of vapor; at the temperature θ, under the pressure P which is the tension of saturated vapor at this temperature, the quantity of heat Q liberated by this modification is exactly the *heat of vaporization* of the liquid considered, at the temperature θ. At the temperature θ and pressure P let v be the volume of a gramme of vapor and v' the volume of a gramme of liquid; v is called the *specific volume of the saturated vapor* at the temperature T, and v' the *specific volume of the liquid* at the same temperature. The change considered produces an increase in volume,

$$V=v'-v;$$

but v' is less than v, so that V is negative and corresponds to a decrease in volume.

The formula (1) therefore becomes

$$\tan \alpha=\frac{E}{\theta}\cdot\frac{L}{v-v'}. \tag{2}$$

This relation has in the theory of saturated vapors a very great importance, which we only mention here, without entering into detailed explanations which interest especially the physicist.

148. Application to fusion. Variation of fusing-point with pressure.—J. Thomson was the first to remark, in 1849, that a relation similar in all respects to equation (2) should apply to the curve which represents the variations of the fusing-point of a substance with the pressure. To pass from the preceding to the present case it suffices to substitute the word *liquid* for *vapor* and the word *solid* for *liquid;* by this substitution L becomes the heat of fusion, v the specific volume of the liquid at the point of fusion θ under the pressure P, v' the specific volume of the solid under the same conditions.

There is a preliminary remark to be made:

Whether it is a question of a heat of vaporization or a heat of fusion, the order of magnitude of L remains the same, at least under ordinary conditions; this cannot be said of the expression $(v-v')$ by which L is divided; the volume occupied by a gramme of vapor is in general incomparably greater than the volume occupied by a gramme of the same substance in the liquid or solid state; it is then evident that, in the case of fusion, tan α will have a value very much greater than that assumed by this quantity for most cases of vaporization.

If the specific volume of the liquid exceeds that of the solid $(v-v'>0)$, the curve of fusion rises from left to right very steeply; the fusing-point is the higher as the pressure on the system increases; but in order to attain an appreciable rise of the fusing-point, a very considerable increase must be given to the pressure.

If the specific volume of the solid is greater than that of the liquid $(v'-v>0)$, the curve of fusion rises from right to left with great rapidity; the fusing-point is lowered with increase of pressure, but the lowering of the fusing-point is noticeable only with a great increase of pressure.

This latter case is the same as that of the fusing-point of ice.

Equation (2) not only allows us to affirm that a great change of pressure produces a slight variation of the fusing-point; it also permits calculating, approximately but quite sufficiently, what variation of the fusing-point is produced by a given increase in pressure; one may, in fact, within certain limits, replace the curve of fusion by its tangent.

Denote by θ_0 (Fig. 43) the fusing-point under the pressure π_0 for atmospheric pressure; by θ_1 the fusing-point at the pressure π_1; let M_0 be the point whose coordinates are θ_0, π_0, and M_1 the point of coordinates θ_1, π_1; in the right triangle $M_0\mu M_1$ we have

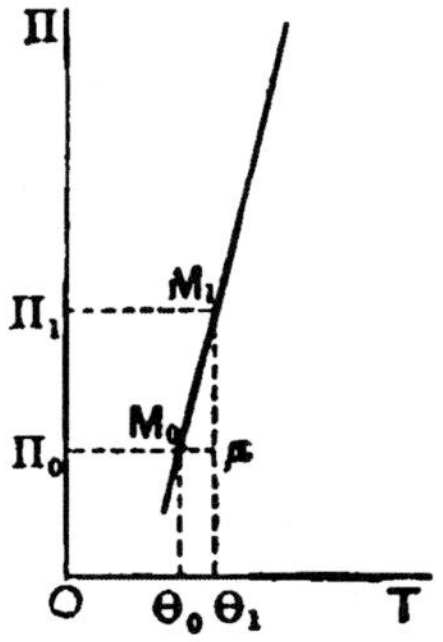

$$\tan\alpha = \frac{M_1\mu}{M_0\mu} = \frac{\pi_1-\pi_0}{\theta_1-\theta_0}.$$

Fig. 43.

Comparing this with equation (2), we get

$$\theta_1-\theta_0 = \frac{\theta_0}{E}\cdot\frac{v-v'}{L}(\pi_1-\pi_0). \qquad (3)$$

By means of this equation James Thomson showed that for an increase in pressure from 1 to 8.1 atmospheres the fusing-point of ice would be lowered by $\frac{1}{107}$ degrees Fahrenheit; very precise experiments made by William Thomson gave a lowering of $\frac{1}{106}$ degrees.

Almost all substances other than ice increase in volume upon melting; for these substances the fusing-point rises with increase of pressure; by a very ingenious method Bunsen verified qualitatively this prediction of theory for spermaceti and paraffine; various physicists have extended this verification to a great number of substances.

Others have gone farther and sought a quantitative confirmation for formula (2); taking this formula as basis, they have calculated the increase that the fusing-point should have for certain substances for an increase of one atmosphere in pressure, and they have compared the number calculated with the number observed; the following table shows such a series of comparisons:

Substance studied.	Rise of Fusing-point.		Observer.
	Observed.	Calculated.	
Acetic acid.	0°.02425	0°.02421	De Visser
Benzine.	.0294	.02936	Demerliac
*P*aratoluidine.	.0187	.0188	"
Naphtylamine α.	.0170	.0170	"

We therefore have very exact verifications of the relation of Clapeyron and Clausius.

149. Application to the allotropic transformation of a solid into another solid.—Formula (2) evidently applies almost without modification to the case in which a substance may exist in two distinct solid forms a and b, resulting from an allotropic or isomeric transformation; suppose that the form b passes into the form a with liberation of heat; under the pressure P there exists a transformation point θ; at temperatures lower than θ the form b passes over to a; above θ, a changes to b When, at the temperature θ and pressure P, a gramme of the form b passes into the form

a, L is the quantity of heat set free; if v_a, v_b are the volumes occupied by a gramme of forms a and b respectively, we may write

$$\text{(4)} \qquad \tan \alpha = \frac{E}{\theta} \cdot \frac{L}{v_b - v_a}.$$

The heat of transformation L and the difference $(v_b - v_a)$ of the specific volumes are, in general, quantities of the same order of magnitude as a heat of fusion and the difference between the specific volume of the liquid and that of the solid taking part in this fusion; one might therefore develop here considerations analogous to those we mentioned for the variation of the fusing-point, and conclude likewise that a considerable increase in pressure will produce only a slight change in the transformation point.

A rough verification of formula (4) has been tried by Mallard and Le Chatelier by studying the allotropic changes of silver iodide. An investigation of the modifications of ammonium nitrate has given Silvio Lussana a more precise verification of formula (4), or rather of the formula

$$\text{(5)} \qquad \theta_1 - \theta_0 = \frac{\theta_0}{E} \cdot \frac{v_b - v_a}{L} (\pi_1 - \pi_0),$$

which is analogous to formula (3).

Crystallized solid ammonium nitrate occurs in four distinct allotropic forms which we shall denote by a, b, c, d; each of these forms may be derived from the preceding with liberation of heat.

Under atmospheric pressure each of these forms passes over into the following when the temperature reaches a certain transformation point; the three transformation points have about the following values:

$$\theta_{ab} = 273° + 34°;$$
$$\theta_{bc} = 273° + 86°;$$
$$\theta_{cd} = 273° + 125°.$$

Corresponding to these transformation points are the following heats of transformation:

$$L_{ab} = 5.02 \text{ cal.}$$
$$L_{bc} = 5.33 \text{ “}$$
$$L_{cd} = 11.86 \text{ “}$$

The first change is accompanied by a variation in volume (in cubic centimetres per gramme):

$$v_b - v_a = 0.01964.$$

The second modification is accompanied by a volume change of opposite sign:

$$v_c - v_b = -0.00854.$$

Finally, for the third change, the sign of $v_d - v_c$ is known, but not its magnitude,

$$v_d - v_c > 0.$$

For any one of these modifications let θ_0 be the transformation point under atmospheric pressure, and θ this point under a pressure π; these transformation points may be observed either by heating or cooling the substance; from which we have two distinct values, both deduced from observation, for the difference $(\theta - \theta_0)$; we shall denote them by $(\theta - \theta_0)_{\text{hot}}$ and $(\theta - \theta_0)_{\text{cold}}$; these values are given in the following table together with $(\theta - \theta_0)_{\text{calc.}}$, the approximate values of $(\theta - \theta_0)$ which are deduced from equation (5).

It is evident that the study of the transformations $a-b$ and $b-c$ furnishes remarkable confirmations of formula (5); concerning the transformation $c-d$ for which a quantitative comparison is impossible, it agrees qualitatively with equation (5).

Transformation.	Pressure Π in Atmospheres.	$(\theta - \theta_0)$hot	$(\theta - \theta_0)$cold	$(\theta - \theta\ )$calc.
$a-b$	50	+1°.60	+1°.43	+1°.67
	100	3 .14	2 .82	2 .94
	150	4 .32	4 .65	4 .41
	200	6 .02	5 .89	5 .88
	250	7 .31	7 .40	7 .35
$b-c$	50	−0°.70		−0°.70
	100	1 .47		1 .40
	150	2 .12		2 .10
	200	2 .82		2 .80
	250	3 .56		3 .55
$c-d$	50	+0°.58	+0°.65	
	100	1 .20	1 .15	
	150	1 .88	1 .74	>0
	200	2 .31	2 .36	
	250	3 .15	3 .00	

150. Application to dissociation.—Equation (1) is very easily applied to the case of dissociation of which the dissociation of calcium carbonate is the type.

Consider w grammes of the compound, calcium carbonate in this case, equal to its molecular weight; in order to form w grammes of the compound it is necessary to combine p grammes of the gaseous component, here carbonic acid, with p' grammes of the solid component, lime.

Let us suppose the compound exothermic; taking place at the temperature θ, under a pressure equal to the dissociation tension P, the formation of a gramme of the compound sets free L calories; the formation of w grammes of the compound liberates $Q=wL$ calories.

Let u be the volume occupied by a gramme of the compound, at the pressure P and temperature θ, v the volume of a gramme of the gaseous component, v' the volume of a gramme of the solid component; the formation of w grammes of the compound is accompanied by an increase in volume

$$V=uw-pv-p'v',$$

and equation (1) may be written

$$\tan\alpha=\frac{E}{\theta}\frac{wL}{pv+p'v'-uw}. \tag{6}$$

Further, in many cases one may with sufficient exactness replace this formula by a simpler approximate formula. The volume u occupied by a gramme of the compound, and the volume v' occupied by a gramme of the solid component, are negligible compared with the volume v occupied by a gramme of the gaseous component, so that for equation (6) may be substituted

$$\tan\alpha=\frac{E}{\theta}\cdot\frac{wL}{pv}. \tag{7}$$

Making use of this formula and of the results of Debray's investigations, giving an approximate value for $\tan\alpha$, Peslin was able as early as 1871 to calculate the quantity of heat L set free by the formation of a gramme of calcium carbonate at a temperature included between the cadmium and zinc boiling-points; he found this quantity of heat to have the value

$$L=293.3 \text{ cal.}$$

Favre and Silbermann have determined the heat of formation of calcium carbonate at a temperature which they have not stated; they found

$$L=308.1 \text{ cal.}$$

Much more satisfactory applications of Clapeyron's formula to dissociation phenomena have been made by Bonnefoi.[1] They treat of the combinations of lithium chloride and lithium bromide with ammonia and the amines; the following reactions liberate a quantity of heat Q which has been both measured by thermochemical methods ($Q_{obs.}$) and calculated by Clapeyron's equation from the dissociation tensions ($Q_{calc.}$).

The concordance is as satisfactory as could be wished.

Reaction.	$Q_{obs.}$	$Q_{calc.}$
$LiCl+NH_3=LiCl, NH_3$	11.857	11.9
$LiCl, NH_3+NH_3=LiCl, 2NH_3$	11.502	11.6
$LiCl, 2NH_3+NH_3=LiCl, 3NH_3$	11.096	11.1
$LiCl, 3NH_3+NH_3=LiCl, 4NH_3$	8.927	8.9
$LiBr+NH_3=LiBr, NH_3$	13.293	13.37
$LiBr, NH_3+NH_3=LiBr, 2NH_3$	12.644	12.74
$LiBr, 2NH_3+NH_3=LiBr, 3NH_3$	11.526	11.51
$LiBr, 3NH_3+NH_3=LiBr, 4NH_3$	10.635	10.51
$LiCl+NH_2CH_3=LiCl, NH_2CH_3$	13.815	13.83
$LiCl, NH_2CH_3+NH_2CH_3=LiCl, 2NH_2CH_3$	12.007	12.10
$LiCl, 2NH_2CH_3+NH_2CH_3=LiCl, 3NH_2CH_3$	10.870	10.93
$LiCl+NH_2C_2H_5=LiCl, NH_2C_2H_5$	13.834	13.71
$LiCl, NH_2C_2H_5+NH_2C_2H_5=LiCl, 2NH_2C_2H_5$	10.983	11.09
$LiCl, 2NH_2C_2H_5+NH_2C_2H_5=LiCl, 3NH_2C_2H_5$	10.570	10.50

[1] J. BONNEFOI, *Combinaisons des sels haloïdes du lithium avec l'amoniac et les amines* (Thèse de Montpellier, 1901).

CHAPTER IX.

MULTIPLE OR TRANSITION POINTS.

151. The same substance in the three states: solid, liquid, gaseous. Triple point.—A great number of substances of definite composition may be observed, according to circumstances, in the three states of solid, liquid, and gas; may these three forms coexist in equilibrium? The system formed of a single independent component, divided into three phases, would be an invariant system; as we know, there can be but one temperature θ and one pressure $\mathfrak{P}$ for which the system is in equilibrium. If, as in the preceding chapter, we lay off on a set of coordinates axes the temperatures T as abscissæ and the pressures π as ordinates (Fig. 44), a definite point, $\mathfrak{J}$, of coordinates θ, $\mathfrak{P}$, will represent the conditions in which one may observe the substance in equilibrium simultaneously in the three forms solid, liquid, vapor.

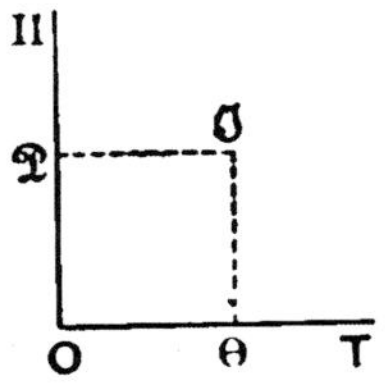

Fig. 44.

At the temperature θ and pressure $\mathfrak{P}$ the solid and the liquid may remain in equilibrium in contact with each other, so that the point $\mathfrak{J}$ belongs to the curve F of fusion tensions; the liquid may remain in equilibrium in contact with the vapor, so that the point $\mathfrak{J}$ belongs to the curve V of tensions of saturated vapor from the liquid; the solid may stay in contact and in equilibrium with the vapor, so that the point $\mathfrak{J}$ belongs to the curve V' of tensions of saturated vapor from the solid. Therefore *the representative point of the equilibrium conditions of the invariant system formed of a given substance in the three forms solid, liquid, vapor, is a point where the three following curves meet:*

1°. *The curve of fusion tensions;*

2°. *The curve of saturated vapor tensions from the liquid;*

3°. *The curve of saturated vapor tensions from the solid.*

From this comes the name *triple point* given to the point $\mathfrak{J}$.

It may be shown also that any two of the three curves of which we have just spoken can never meet outside of the triple point $\mathfrak{J}$.

152. The transformation curves in the neighborhood of the triple point.—It is evidently important to know how these three curves are arranged.

The curve of fusion tensions divides the plane into two regions; every point in the region situated to the right of this curve represents conditions in which the solid melts, while the liquid cannot freeze; every point to the left of this curve represents conditions in which the solid cannot melt, while the liquid freezes; if, in this region, the liquid is observed in equilibrium, it is because of a phenomenon of false equilibrium; the liquid is in *surfusion.*

At the triple point, the two curves V and V' of the tensions of saturated vapor cut the curve F of fusion tensions, so that to the left of this curve there exists certainly a branch of the curve V and a branch of the curve V'; let us see where these branches lie.

Can the curve V' exist above the curve V?

Suppose this to be so and take (Fig. 45) a point M of abscissa T and ordinate π, situated between these two curves.

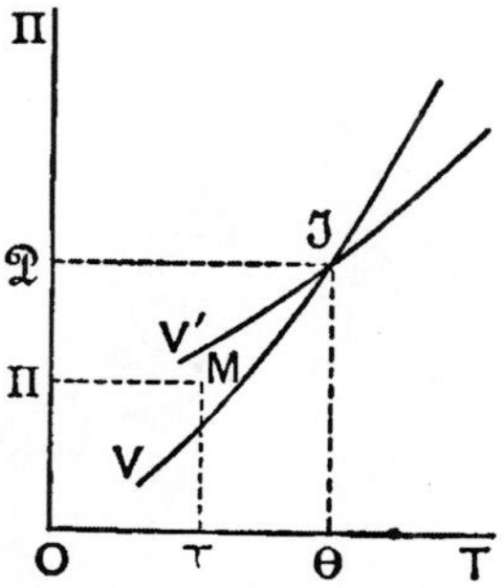

At the pressure π and temperature T the liquid freezes, for the point M is to the left of the curve F of fusion tensions; the solid vaporizes, for the point M is below the curve V' of tensions of saturated vapor from the solid; the vapor condenses to the liquid state, for the point M is below the curve V of the tensions of saturated vapor from the liquid. One may therefore, at the constant temperature T and constant pressure π, put the system through a real closed cycle, which causes it to pass from the liquid to the solid state, from the solid to the vapor, and from the vapor to the liquid state.

Fig. 45.

But, as the external forces reduce to a constant pressure, they admit a potential (Chap. I, Art. 12), so that the work which they perform about a closed cycle is zero. Our hypothesis leads us to regard as realizable a closed cycle, described at constant temperature, with kinetic energy constantly zero and total work accomplished equal to zero; we have seen (Chap. V, Art. 66) that such a cycle can be realized only by the aid of positive external work; the hypothesis assumed is therefore inadmissible, and we may state the following theorem:

To the left of the curve F of the fusion tensions, and consequently in the conditions where the liquid is in surfusion, the curve V of the tensions of saturated vapor from the liquid is above the curve V′ of the tensions of saturated vapor from the solid.

This theorem gives a correct idea of the positions of the three curves F, V, V'.

There are two cases to distinguish:

FIRST CASE.—*Fusion is accompanied by an increase in volume.* This is the case occurring the more frequently; acetic acid gives us an often-cited example. In this case the curve of fusion F rises steeply from left to right; the two curves V and V' rise gradually from left to right; the three curves appear as shown in Fig. 46.

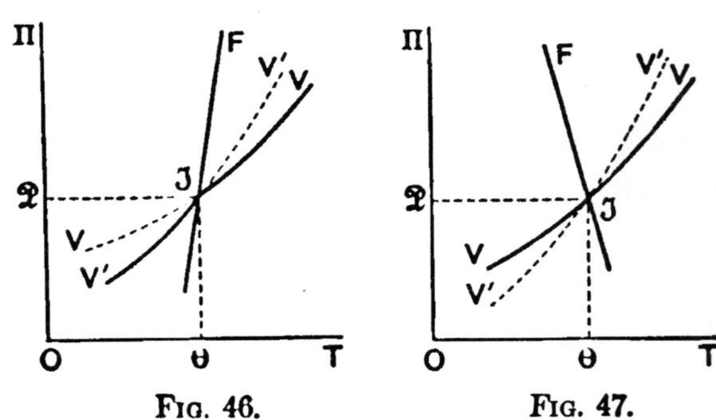

FIG. 46. FIG. 47.

SECOND CASE.—*Fusion is accompanied by a decrease in volume.* This is the case with the fusion of ice. Here the fusion curve F rises abruptly from right to left; the two curves V and V' behave as in the preceding example; Fig. 47 shows their arrangement.

153. Historical.—Most careful observations were for a long time powerless to detect a difference between the tension of saturated vapor from an undercooled liquid and the tension of saturated vapor from the solid formed by its freezing.

We owe to Regnault a series of *Researches undertaken to decide if the solid or liquid state of substances exercises an influence on the elastic force of the vapors that they give off in vacuo at the same temperature.* The study of water, acetic acid, hydrocarbide of bromine and benzene, led the eminent physicist to the following conclusion: "We must therefore admit that the molecular forces which determine the solidification of a substance exercise no sensible influence upon its vapor tension in vacuo; or, more exactly, if an influence of this nature exist, the variations produced are so small that they cannot be detected in a certain manner from my experiments."

"To sum up, my experiments prove that the passage of a body from the solid to the liquid state produces no appreciable change in the curves of the elastic forces of its vapor; this curve conserves a perfect regularity before and after the transformation."

As early as 1858 this conclusion was attacked, in the name of thermodynamics, by G. Kirchoff, and it is to the investigations of theorists such as Kirchoff, James Thomson, and J. Moutier that we are indebted to-day for our knowledge of the properties of vapors emitted by a given substance in the two states solid and liquid.

154. Experimental verifications.—But if experiment did not precede theory, the latter once established has received precise confirmations from experiment.

Here is an experiment due to Gernez which shows that the tension of saturated vapor from an undercooled liquid exceeds the tension of saturated vapor of the same substance, taken at the same temperature, but in the solid state.

An inverted U tube, whose ends are sealed off (Fig. 48) contains in the branch *A* liquid acetic acid, and in the tube *B* crystallized acetic acid; the whole is kept at a very nearly constant temperature by packing in sawdust; in these conditions the liquid acetic acid is in a state of surfusion; the tension of saturated vapor from the liquid is greater than that from the crystallized acetic acid; therefore, little by little, a distillation is established from the branch *A* to the branch *B*; after several months the liquid mass has considerably diminished in the branch *A*, while in the branch *B* the sides of the tube are carpeted with crystals.

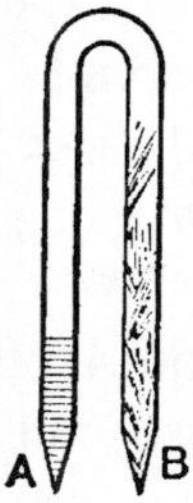

A B

FIG. 48.

But we may go farther and give a quantitative verification to the theory; by methods of great accuracy, Lord Ramsay and Young on the one hand and S. V. Fischer on the other, have measured, below 0°, the tension of water vapor from ice and that from undercooled water; the second exceeds the first, but by a very small quantity, 0.155 millimetres of mercury at $-5°$ and 0.22 mm. at $-10°$. Nevertheless these measurements agree very exactly with the theorems stated by Moutier and G. Kirchoff; the same is true for the measurements of vapor tensions of benzene in the liquid and solid states, as studied by Fischer.

From the computations of Robert von Helmholtz, the temperature of the triple point for water would barely exceed the temperature of fusion under atmospheric pressure; it would be $+0°.0076$ C.; the tension of saturated water vapor above the liquid would surpass that above ice by 0.000332 mm.; on account of their smallness these numbers escape any exact experimental verification.

155. Modifications of phosphorus. Researches of Troost and Hautefeuille.—We have just studied substances taken at a temperature near to the triple point; the tension of saturated vapor from the solid and the tension of saturated vapor from the undercooled liquid, equal to each other at the triple point alone, differ only very slightly from each other; only by the most skilfully arranged experiments could one detect this difference.

If we could study the tensions of saturated vapor from the undercooled liquid and from the solid which arises from freezing at a temperature widely separated from the triple point, it is to be presumed that the two tensions of saturated vapor will be extremely different and that even rough experiments would be sufficient to distinguish them. The vaporization of liquid white phosphorus and of solid red phosphorus will give us an example of the vaporization of a single substance in two different states, at temperatures very distant from the triple point.

White solid phosphorus at ordinary temperatures and white liquid phosphorus above its fusing-point are transformed slowly into red solid phosphorus under the action of light; at higher temperatures the transformation becomes more and more rapid and takes place even in the dark; the inverse transformation of red phosphorus into white does not occur at any temperature attainable in the laboratory; the curve of transformation tensions of red into white phosphorus, if it exists, is beyond the field of observable temperatures.

The transformation of liquid white phosphorus into red takes place with liberation of heat; thus, according to Hittorf, when at 280° liquid white phosphorus is rapidly transformed into solid red phosphorus, the heat set free is so great that the temperature is suddenly raised from 280° to 370° *C.*

The transformation of liquid white into solid red phosphorus takes place also with decrease in volume.

At temperatures attainable in the laboratory we may observe, as has been said, the exothermic transformation of liquid white phosphorus into the red solid form, but not the inverse transformation; the region where the observations are made is, therefore, according to Moutier's rule, to the left of the curve of tensions of transformation from red phosphorus into liquid white phosphorus; if the transformation of red into liquid white phosphorus is possible, it is at temperatures much higher than those realizable in our laboratories.

All these properties allow us to regard liquid white phosphorus as a liquid which may freeze in various forms; certain of these forms, as amorphous or crystallized white phosphorus, cannot exist at the relatively high temperatures which we consider; as for the form red phosphorus, its point of fusion into the liquid white form, if it exists, is very much higher than the temperatures which we can reach; at these temperatures *the liquid white phosphorus should be considered as a liquid in surfusion with respect to the solid red phosphorus.*

At any given temperature the tension of saturated vapor of liquid white phosphorus should exceed the tension of saturated vapor of red phosphorus.

Thus if liquid white phosphorus is heated to 440°, for example, as was done by Troost and Hautefeuille,[1] the vapor pressure attains quickly 7.75 atmos., which is, at this temperature, the

---

[1] Troost and Hautefeuille, *Annales de l'Ecole supérieure*, 2d Series, v. 2, p. 253, 1873.

tension of saturated vapor from liquid white phosphorus; but the white phosphorus is gradually changed into red, and when this transformation is complete the tension of the phosphorus vapor is lowered to 1.75 atmospheres.

If red phosphorus is heated to 440°, the vapor pressure increases slowly up to 1.75 at. and remains stationary when this value has been reached.

Troost and Hautefeuille denoted the pressure of 1.75 at. as being at 440° the *transformation tension of phosphorus vapor;* from what precedes we see that this pressure should be considered simply as the tension of saturated vapor from red phosphorus.

Troost and Hautefeuille determined at various temperatures the tensions of saturated vapor from white liquid phosphorus and from red phosphorus; at the same temperature the first tension is always very considerably higher than the second. Below are the results of these measurements:

Temperature.	Tension of Saturated Vapor.	
	From White Phosphorus.	From Red Phosphorus.
360° C.	3.20 at.	0.12 at.
440	7.75	1.75
487	——	6.80
494	18.00	——
503	21.90	——
510	——	10.8
521	26.20	——
531	——	16.0
550	——	31.0
577	——	56.0

Above 510° the transformation of liquid white phosphorus into red phosphorus is so rapid that the tension of saturated vapor from white phosphorus has not the time to become established and can no longer be measured.

156. Researches of G. Lemoine.—When phosphorus vapor is brought to a high temperature and then suddenly cooled to room temperature, it condenses in the form of white phosphorus; this fact may be of use in explaining Lemoine's observations.[1]

If we enclose in a flask a mass of white or red phosphorus, and if we keep this flask at a fixed high temperature T long enough for equilibrium to be established, we shall have, at the end of the experiment, the flask filled with phosphorus vapor, whose tension will be equal to the tension of saturated vapor from red phosphorus at the temperature of the experiment, and the mass of solid in excess will be in the form of red phosphorus. If, for instance, the temperature is 440° C., we shall have a mass of phosphorus vapor which will fill the flask at a pressure of 1.75 at.; this mass will be equal to as many times 3.6 gr. as there are litres in the volume of the flask; the rest of the phosphorus will be in the red form. By suddenly cooling the flask we shall find a mass of phosphorus, soluble in carbon bisulphide, equal to 3.6 gr. per litre; the rest will be insoluble in carbon bisulphide.

Lemoine was not content to study the state of equilibrium which is set up in such a flask after a continued heating; by submitting such flasks to a sudden cooling after variable times of heating, he studied the way in which this equilibrium is established when starting either with white or with red phosphorus.

In the first case the mass of white phosphorus which the flask holds should constantly decrease up to the limiting value, which is 3.6 gr. per litre; in the second case this quantity should increase, reaching, without ever exceeding, the same limit.

157. Anomaly observed by Lemoine.—The results of Lemoine's observations always accord with prediction in the first case, but not always in the second; if a considerable quantity of red phosphorus is heated in a closed vessel, it is noted that the mass of white phosphorus coming from the condensation of the phosphorus vapor increases at first so as to exceed 3.6 gr. per litre, passes through a maximum, then decreases to 3.6 gr.

For example, here are the results of experiments where Lemoine heated 30 gr. of red phosphorus to 440° in flasks of 1 litre;

---

[1] G. Lemoine, *Annales de Chimie et de Physique*, 4th Series, v. 24, p. 129, 1871.

the masses of white phosphorus obtained after variable times of heating are the following:

Time in Hours.	Masses of White Phosphorus.	Time in Hours.	Masses of White Phosphorus.
0 h. 30 min.	4.54 gr.	23 h.	3.90 gr.
2 h.	4.75	32	3.74
8 h.	4.40	47	3.72

158. Explanation of this anomaly by Troost and Hautefeuille. —The explanation of this anomaly has been given by Troost and Hautefeuille.[1]

There exists not merely one, but a great number of varieties of red phosphorus; the properties of red phosphorus depend upon the temperature at which it is produced; prepared at 265° it is a bright red having brilliant vitreous breaks, suggesting by its brightness realgar; red phosphorus obtained at 440° is orange, its breaks are dull and granular; obtained above 500° it is compact, has a bright violet hue, a conchoid break, and in its cavities are formed geodes of red phosphorus crystals.

These various kinds differ in density and heat of combustion.

Let us take for standard of comparison crystallized red phosphorus, whose density is 2.34.

Commercial red phosphorus has a heat of combustion which exceeds by 568 calories per gramme that of crystallized red phosphorus.

Red phosphorus kept at 265° for 650 hours has a density of 2.148; its heat of combustion exceeds by 320 calories per gramme that of crystallized red phosphorus.

Kept 540 hours at 360° the density of red phosphorus is 2.19, and its heat of combustion exceeds by 298 calories that of red crystallized phosphorus.

Phosphorus prepared at 580° has a heat of combustion of 50 calories less than that of crystallized red phosphorus.

---

[1] Troost and Hautefeuille, *Annales de Chimie et de Physique*, 5th S., v. 2, p. 155, 1874.

"Red phosphorus does not assume immediately the peculiar appearance that we have pointed out; this is acquired slowly if the experiment is performed at a moderately high temperature, and very rapidly if above 500°."

"The varieties which cease to be modified by a new heating for a great number of hours at the same temperature, pass from one form to the other by insensible changes when they are brought to a higher temperature kept a long time constant."

We see by what precedes that the varieties obtained at low temperatures should be regarded as existing in the state of false equilibrium, when they are brought to a higher temperature; they change into the variety corresponding to this temperature; the reasoning given in Art. 152 may therefore be applied here and shows that the first varieties have a tension of saturated vapor higher than the last; Troost and Hautefeuille have in fact proved that at a given temperature the various forms of red phosphorus have a tension which is the higher in proportion as they are prepared at a lower temperature; thus red phosphorus obtained at 265° should behave at 440°, with respect to the red phosphorus obtained at this latter temperature, as white phosphorus acts with respect to red phosphorus; this is what Troost and Hautefeuille have shown.

Lemoine having worked at 440° with commercial red phosphorus prepared between 250° and 300°, the vapor tension should have, as in Troost and Hautefeuille's experiments, increased rapidly up to the tension of saturated vapor from commercial red phosphorus, then decrease to the tension of saturated vapor of red phosphorus prepared at 440°.

159. The triple point considered as transition point.—Consider a system that may contain water in the three forms liquid, ice, and vapor; at a temperature differing from the temperature θ of the triple point $\mathfrak{J}$ the system cannot be in equilibrium unless at least one of the three phases has disappeared.

Three kinds of systems formed of two phases may be observed: the system made of ice and water, that of ice and vapor, and the one formed of water and vapor.

For the first system to be in equilibrium, the point representing this state must be on the line FF_1 (Fig. 49); in order that the second system may be in equilibrium it is necessary that the point representing it lie on the line $V'V_1'$; and for the third to be in equilibrium its representative point must lie on VV_1.

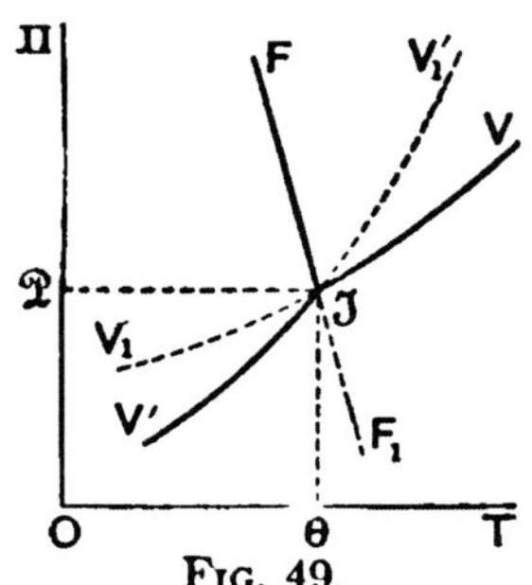

Fig. 49.

These conditions are not in general sufficient to assure equilibrium.

Consider, at a temperature less than θ, a system formed of liquid water and vapor and suppose that the representative point is on the curve $V_1\mathfrak{J}$ of tensions of saturated vapor from liquid water; will the system be in equilibrium? No, for the vapor, whose tension is greater than that of saturated vapor from ice at the same temperature, may condense into the form ice; the liquid, whose temperature is less than the fusing-point for the same pressure, may freeze. The system formed of liquid water and vapor is therefore not in equilibrium in the conditions which we have described, or at least, if it may be observed in equilibrium, it is one of those *false equilibria* to which we have called attention; it is through these false equilibria that we can trace, in certain cases, the line $V_1\mathfrak{J}$ of tensions of saturated vapor from water in surfusion.

Take next, at a temperature above θ, a system formed of ice and vapor and suppose that the representative point is on the curve $\mathfrak{J}V_1'$ of tensions of saturated vapor from ice; the representative point being above the curve $\mathfrak{J}V$ of tensions of saturated vapor from liquid water, the vapor may condense as liquid; the representative point being to the right of the curve of fusion points FF_1, the ice may melt; therefore the system is not in equilibrium.

Take, finally, a system formed of ice and liquid at a temperature higher than θ, and suppose that the representative point is on the curve $\mathfrak{J}F_1$ of the points of fusion; the representative point being below the two curves of tensions of saturated vapor, the liquid water and the ice may evaporate and the system is not in equilibrium.

In short, at temperatures below θ two kinds of true equilibrium may be observed:

1°. Systems consisting of ice and water vapor; it is necessary and sufficient that the representative point be on the branch $V'\mathfrak{I}$ (indicated by a full line) of the curve of tensions of saturated vapor from ice.

2°. Systems formed of ice and water; it is necessary and sufficient that the representative point be on the branch $F\mathfrak{I}$ (indicated by a full line) of the curve of fusing-points.

At temperatures above θ there may be had a single kind of system in the state of true equilibrium, namely, systems formed of liquid water and vapor; it is necessary and sufficient that the representative point be on the branch $\mathfrak{I}V$ (indicated by a full line) of the curve of tensions of saturated vapor from liquid water.

A similar treatment may be given to a system where a single substance may exist in the three forms solid, liquid, vapor, and when the fusion is accompanied by an increase in volume. In this case the curve of fusing-points F_1F (Fig. 50) rises from left to right.

The results attained are the following:

At temperatures less than the temperature θ of the triple point there may be observed in equilibrium only a single kind of monovariant system, that formed of solid and vapor; in order for this equilibrium to take place it is necessary and sufficient that the representative point be on the branch $V'\mathfrak{I}$ of the curve of tensions of saturated vapor from the solid.

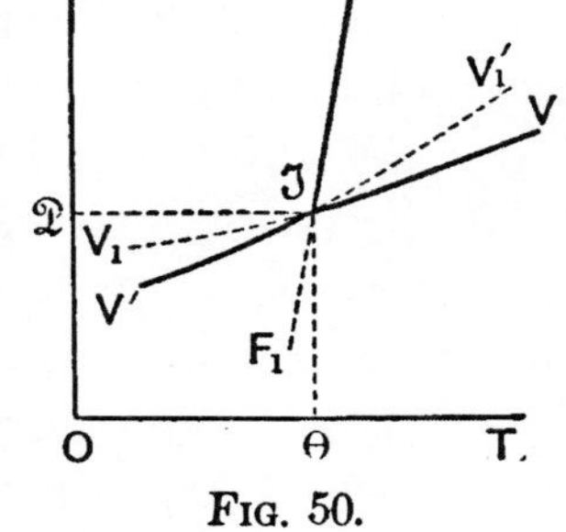

Fig. 50.

At temperatures higher than the temperature θ of the triple point two kinds of monovariant systems in equilibrium may be observed:

1°. Systems formed of liquid and vapor; it is necessary and sufficient for this that the representative point be on the branch $\mathfrak{I}V$ of the curve of tensions of saturated vapor from the liquid.

2°. Systems formed of solid and liquid; it is necessary and sufficient that the representative point lie on the branch $\mathfrak{I}F$ of the curve of fusing-points.

Let us generalize the observations which we have made for the two cases we have treated:

When a single substance may exist in three different states, these three states united form an invariant system; this system cannot be in equilibrium except at a single temperature θ and at a single pressure $\mathfrak{P}$, temperature and pressure of the triple point.

By grouping these states two by two, three monovariant systems are obtained; each of these systems can be in equilibrium only if the temperature is to a well determined side of the temperature θ of the triple point; some can be in equilibrium only at temperatures less than θ; others only at temperatures higher than θ.

If, therefore, a monovariant system is taken in equilibrium and the temperature is so varied that it reaches the temperature θ of the triple point, at the instant this value is passed, the monovariant system considered can no longer be kept in equilibrium; it *passes* perforce into another kind of system, whence the names *transition temperature, transition point,* given by Bakhuis Roozboom to the temperature θ and the triple point $\mathfrak{J}$.

160. Generalization of the preceding ideas. Quadruple points.—These ideas may be generalized.

Take a system formed of two independent components and suppose that it may be divided into four phases. If the four phases coexist in the system, it is invariant; there is a single temperature θ and a single pressure $\mathfrak{P}$, coordinates of a definite point $\mathfrak{J}$, at which this invariant system may be observed in equilibrium.

If one of the four phases is supposed to be excluded, a monovariant system is obtained; according as the excluded phase is one or another of the four possible phases, there may be formed four distinct monovariant systems.

In order that one of these four monovariant systems may be in equilibrium it is necessary that the representative point be on the curve of transformations corresponding to this system; to the four possible monovariant systems correspond four curves of transformation tensions.

It is clear that each of these four curves must pass through the point $\mathfrak{J}$, which is therefore a *quadruple point.*

To have one of our four systems in equilibrium it is necessary for the point which represents the temperature and pressure to be on the curve of transformation tensions of this system; but this does not suffice; the curve of transformation tensions does not represent, in its entirety, states of equilibrium; only one branch of this curve should be kept, and this branch has the point $\mathfrak{J}$ for an end, so that *the quadruple point is a transition point.*

161. Quadruple points in the study of hydrates from a gas.—Bakhuis Roozboom and, after him, other chemists have studied a certain number of quadruple points; the first quadruple points which have been discussed are found in the study of systems whose two independent components are water and a gas and which may have four phases:

Ice;
Solid hydrate;
Liquid mixture;
Gaseous mixture.

162. Quintuple points.—We may go farther; in a system formed of three independent components, susceptible of having five phases, a *quintuple point* will be met, where the five curves of transformation will cut which correspond to the five monovariant systems obtained by excluding successively one of the five phases; this quintuple point is a transition point.

Here are two examples of quintuple points analyzed by Bakhuis Roozboom:

Take a system formed of three independent components:

Water: H_2O;
Sodium sulphate: Na_2SO_4;
Magnesium sulphate: $MgSO_4$.

At a temperature of 22° the following five phases may co-exist:

1°. Water vapor, *V*;

2°. A liquid mixture L, formed of water, sodium sulphate, and magnesium sulphate;

3°. Crystals N of hydrated sodium sulphate: $Na_2SO_4 \cdot 10H_2O$;

4°. Crystals *M* of hydrated magnesium sulphate: $MgSO_4 \cdot 10H_2O$;

5°. Crystals *A* of astrakanite, whose formula is

$$Na_2MgS_2O_8 \cdot 4H_2O.$$

The temperature, $\theta = 22°$, is that of the quintuple point $\mathfrak{J}$; in the neighborhood of this point the five curves of transformation are arranged as is shown in Fig. 51.

Take a system whose independent components are:

Water;

Cupric acetate: $Cu(C_2H_3O_2)_2$;

Calcic acetate: $Ca(C_2H_3O_2)_2$.

At a temperature of 76° there may coexist;

1°. Water vapor, V;

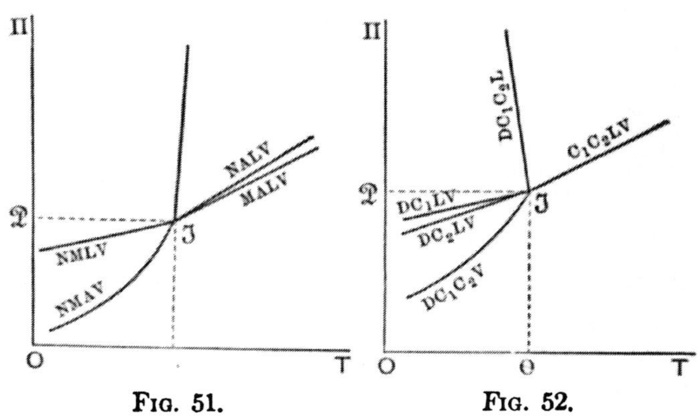

Fig. 51. Fig. 52.

2°. A liquid mixture L, formed of water, cupric acetate, and calcium acetate;

3°. Crystals C_1 of hydrated cupric acetate: $Cu(C_2H_3O_2) \cdot H_2O$;

4°. Crystals C_2 of hydrated calcic acetate: $Ca(C_2H_3O_2) \cdot H_2O$;

5°. Crystals D of cupri-calcic acetate $CuCa(C_2H_3O_2)_2 \cdot 6H_2O$.

The temperature 76° is the temperature of a quintuple point $\mathfrak{J}$; in the neighborhood of this point the five curves of transformation tensions are arranged as shown in Fig. 52.

CHAPTER X.

THE DISPLACEMENT OF EQUILIBRIUM.

163. In general a modification which leaves invariable the composition of each phase cannot be imposed upon a system whose variance exceeds 1.—We have seen (Chap. VIII, Art. 145) that we could always impose upon a monovariant system a modification which, while changing the mass of the different phases, left invariable the composition of each; we concluded from this that if we took a monovariant system in equilibrium at a given temperature and under the transformation tension corresponding to this temperature, and if, moreover, the temperature and pressure were kept constant, the equilibrium of the monovariant system was indifferent.

These properties no longer hold when we study systems whose variance exceeds unity or, at least, are only met with in certain particular cases; thus in the next chapter we shall see, in studying bivariant systems, that such a system may sometimes have a particular equilibrium state where are found all the properties which the equilibrium states of a monovariant system have manifested; but even in the case where such a state of equilibrium will be possible it will represent an exception among the cases of equilibrium of the system considered.

In general, for a bivariant system, every modification which alters the mass of the phases alters at the same time the composition of some among them.

Take, for instance, a bivariant system consisting of two independent components, water and sodium chloride, divided into two phases, solid sodium chloride and an aqueous solution of this salt; if we seek to increase the mass of the liquid phase and decrease the mass of the solid phase, we shall necessarily decrease the concentration of the solution.

164. In general the equilibrium of a system whose variance exceeds 1 is stable.—Suppose given a pressure and temperature and let us keep them constant. The bivariant system studied can be put into equilibrium at this temperature and pressure; for this it is necessary that each phase composing the system has a definite composition; take such a state of equilibrium, and, starting with this state, subject the system to a slight modification, which causes the masses of one or of some of the phases to vary; this modification changes the composition of at least one of the phases into which the system is divided; the system which was in equilibrium before the modification can no longer be so after this change; the original equilibrium state was therefore not a state of indifferent equilibrium.

If the temperature and pressure are kept constant, this state of equilibrium is *stable;* when it has been disturbed by a small modification similar to that we have spoken of, the system undergoes spontaneously a modification in the opposite direction.

Take, for example, one system formed of solid sodium chloride and an aqueous solution of this salt; suppose it in equilibrium at a given pressure and temperature; under these conditions the solution is *saturated* with sodium chloride; without changing the temperature or the pressure, imagine that any cause brings into the solution a small quantity of salt; immediately the solution, become supersaturated, will deposit sodium chloride and return to its original concentration; imagine, on the contrary, that the saturated solution deposits a small quantity of sodium chloride; it will cease to be saturated and will dissolve salt until it has returned to its original composition.

What we have just said regarding a bivariant system may be repeated for any multivariant system whatever: *If certain exceptional equilibrium states which are indifferent are excluded, every state of equilibrium of a bivariant or multivariant system is stable when the temperature and pressure are kept constant.*

165. Displacement of equilibrium by variation of the pressure.—From the fact that the equilibrium of a bivariant or multivariant system is stable when the pressure and temperature are constant there follow two laws of great importance: the *law of the displacement of equilibrium by variation of the pressure* and the *law of the displacement of equilibrium by variation of the temperature.*

Consider the first of these laws, which may be stated in the following way:

Take a system in stable equilibrium at a given temperature and pressure; without changing the temperature, INCREASE THE PRESSURE *by a small amount; in general the equilibrium will be disturbed; the system will be the seat of a small reaction which will bring it into a new state of equilibrium; if the same reaction is supposed to take place, starting from the initial equilibrium state, without change of pressure or temperature, it would be accompanied by a* DECREASE IN VOLUME *of the system.*

If we had disturbed the initial equilibrium by DECREASING *slightly* THE PRESSURE, *we should have produced a small reaction in the system; taking place in the initial state of equilibrium, without change of temperature or pressure, this reaction would have been accompanied by an* INCREASE IN VOLUME *of the system.*

166. Various applications.—Here is an example showing how easy this law is to apply:

A vessel, brought to 1000° at a pressure π, contains oxygen, hydrogen, and water vapor; let x be the ratio between the mass m of the water vapor which the system encloses and the mass M of this vapor which it would enclose if the combination of oxygen and hydrogen were carried to the disappearance of one of the two gases. For the system in equilibrium at the pressure π this ratio x has a certain value X.

Leaving the temperature equal to 1000°, cause the pressure to take on a value π' a little greater than π; a reaction is produced in the system; the ratio x passes from the value X to the value X' which assures the equilibrium at the temperature of 1000° and under the pressure π'; accomplished at the constant temperature 1000° and under the constant pressure π, this reaction will be accompanied by a decrease in the volume of the system.

Now, at constant temperature and pressure, the formation of a certain quantity of water vapor, reaction which increases x, is accompanied by a decrease in volume; the dissociation of a certain quantity of water vapor, reaction decreasing x, is accompanied by an increase in volume; it follows, therefore, that X' is greater than X.

Whence the following conclusion:

At a given temperature, 1000° for example, a system is taken which encloses a given mass of oxygen (free or combined) and a given mass of hydrogen (free or combined); the ratio of the mass of water vapor contained by the system in equilibrium to the mass of water vapor which it would contain if the combination was complete is the greater as the pressure is higher.

In other terms, at a given temperature a decrease in pressure favors the dissociation of water, an increase in pressure aids the combination of oxygen and hydrogen.

When a mixture of chlorine and water vapor passes over partially at constant temperature into oxygen and hydrochloric acid the reaction is accompanied by an increase in volume; this reaction is helped, therefore, by decreasing the pressure; by increasing the pressure the inverse reaction is aided.

167. Case of combinations without contraction.—When iodine vapor and hydrogen combine at constant temperature and pressure to form hydriodic acid the reaction is accompanied by no change in volume, at least in the conditions where the iodine vapor, hydrogen, and hydriodic acid may be treated as perfect gases. In reasoning as we have for the system oxygen, hydrogen, water vapor, we may draw from the law of the displacement of equilibrium by pressure the following conclusions:

A change in pressure without variation of temperature cannot, in such a system, help either the formation or the dissociation of hydriodic acid; if a system of definite elementary composition is taken and heated to a definite temperature, the ratio Y between the mass of free hydrogen and the total hydrogen, free or combined, which it encloses will have a value independent of the pressure supported by the system.

168. Experimental verifications: hydriodic acid.—This proposition has been verified by G. Lemoine.[1]

A system containing hydrogen and iodine vapor in equivalent

---

[1] G. Lemoine, *Annales de Chimie et de Physique*, 5th Series, t. 12, p. 145, 1877.

proportions is brought to the sulphur boiling-point and subjected to various pressures; to each value π of the pressure corresponds a value of Y as follows:

π	Y	π	Y
4.5 at. 2.3	0.24 0.25	0.9 at. 0.2	0.26 0.29

In this series of experiments, while the pressure has passed from one value to another twenty-two times smaller, the ratio Y has changed by only $\frac{1}{5}$ of its value; if account is taken of the fact that the substances are not perfect gases, of the difficulty of the experiments and the numerous sources of error, such a conclusion appears justified.

169. **Selenic acid.**—The formation of gaseous selenic acid from liquid selenium and hydrogen is also a reaction which, taking place at constant temperature and pressure, produces but a very slight change in volume. Take now a system where hydrogen and gaseous selenic acid exist in presence of liquid selenium; the system is in equilibrium at a given temperature at the pressure π; it contains a mass m of selenic acid; if the whole of the hydrogen passed into the form of selenic acid, it would enclose a mass M; the ratio $\frac{m}{M}$ has a certain value X; if, without changing the temperature, the value π of the pressure is changed, the value of X should undergo only small variations.

Ditte[1] tried to verify this proposition experimentally; Pélabon[2] made a more careful verification; here are the results of his observations:

TEMPERATURE 620°. The total pressure of the gaseous mixture reduced to 23° was 520 mm. of mercury:

$$X = 0.4067.$$

[1] DITTE, *Annales de l'Ecole normale supérieure*, 2d Series, v. I, p. 293, 1873.

[2] H. PÉLABON, *Mém. de la Société des Sciences physiques et naturelles de Bordeaux*, 5th Series, v. 3, p. 11·41; *Sur la dissociation de l'acide selenhydrique*, Paris, A. Hermann, 1898.

The total pressure reduced to 23° was 1270 mm.:

$$X=0.4112.$$

The total pressure reduced to 25° was 1520 mm.:

$$X=0.4200.$$

The total pressure reduced to 22° was 3016 mm.:

$$X=0.4230.$$

TEMPERATURE 475°. The total pressure of the gaseous mixture reduced to 22° was 1450 mm. of mercury:

$$X=0.3840.$$

The total pressure reduced to 23° was 2556 mm.:

$$X=0.3917.$$

TEMPERATURE 325°. The total pressure of the gaseous mixture reduced to 17° was 825 mm. of mercury:

$$X=0.182.$$

The total pressure reduced to 15° was 3240 mm.:

$$X=0.206.$$

170. Variation of the solubility of a salt with pressure.—We owe to F. Braun[1] other interesting verifications of the principle of the displacement of equilib ium by varying the pressure; they are drawn from the study of saturated solutions.

Consider at a gi en pressure and temperature a bivariant system whose two components are a salt and water and whose two phases are the solid salt and an aqueous solution; when the system is in equilibrium, the solution has a definite concentration S; it is *saturated* at the given pressure and temperature.

Imagine that a very small mass of salt passes from the solid phase into the almost saturated solution; the volume of the solid phase decreases by a quantity which is known when the density of the salt is known; the volume of the liquid phase undergoes an

[1] F. BRAUN, *Wiedemann's Annalen*, v. 30, p. 250, 1887.

increase which may be calculated when we know the law of the variation of the density of the solution for concentrations near to S; the volume of the system undergoes a modification which may be either an increase or a decrease.

At ordinary temperature and pressure, the solution of alum or of sodium sulphate with the molecules of water in an aqueous solution almost saturated with the same salt is accompanied by a *contraction* of the system; in the same conditions, the solution of ammonium chloride is accompanied by an *expansion.*

At a given temperature take a saturated solution of a definite salt, in the presence of an excess of the salt, at the pressure π; S is the concentration of the solution. Give to the pressure a value π', slightly greater than π, keeping the temperature constant; the equilibrium will be disturbed and the composition of the solution will vary until its concentrations has taken the value S' which corresponds to saturation at the pressure π'.

The modification undergone by the system while the solution changes from the concentration S to the concentration S' should be a modification which, at constant pressure and temper ture, would involve a decrease in volume; if the dissolving of the salt in a nearly saturated solution takes place with contraction, this modification consists in the dissolving of a certain mass of salt, and S' is greater than S; if the dissolving of the salt in a nearly saturated solution takes place with dilatation, this modification consists in the precipitation of a certain quantity of salt, and S' is less than S. The following propositions may therefore be stated:

If, at a given temperature, the dissolving of a salt in a nearly saturated solution is accompanied by contraction, the solubility of the salt increases with pressure; if, on the contrary, the dissolving of the salt in the nearly saturated solution is accompanied by expansion, the solubility of the salt decreases as the pressure increases.

The first case is illustrated, as we have said, by alum and sodium sulphate with ten molecules of water; if we compress very slowly, in a piezometer, a saturated solution of alum or of sodium sulphate with ten molecules of water, in the presence of an excess of the same salt, the solution will dissolve a new quantity of salt; it will remain clear during the compression; relieved cautiously and brought back to ordinary pressure, it will possess the properties of a supersaturated solution; the crystals still remaining in excess will have eaten faces.

An illustration of the second case is ammonium chloride; if a saturated solution of ammonium chloride in the presence of crystals of this salt becomes supersaturated, the solution will deposit on the crystals a part of the salt which it contains.

The solution of sodium chloride offers interesting peculiarities.

Let us study at a constant temperature, as 15°, the dissolving of a small mass of sodium chloride in a nearly saturated solution of this salt; this phenomenon is produced with contraction of the system if it takes place under a constant pressure less than 1530 atmospheres; on the contrary, it is accompanied by dilatation if it takes place under a pressure greater than 1530 atmospheres; when the pressure exceeds 1530 atm. and continues to increase, the solubility diminishes; at the constant temperature of 15° the pressure of 1530 atm. corresponds to a maximum of solubility of sodium chloride in water.

The existence of this maximum of solubility has been shown by F. Braun; one compresses very slowly, to a pressure much higher than 1530 atmospheres, a saturated solution of sodium chloride in the presence of crystals of sea-salt; after return to the ordinary pressure, the crystals placed in the piezometer are examined; their faces are eaten and carry little cubical crystals of sodium chloride; the sodium chloride must therefore dissolve during a part of the time of compression and precipitate during the rest of this time.

171. Displacement of equilibrium by variation of the temperature.—The very simple and most fruitful law of the displacement of equilibrium by the variation of the pressure was stated by H. Le Chatelier [1] in 1884; a short time previously, J. H. Van't Hoff [2] had announced the still more important law of the displacement of equilibrium by variation of the temperature.

There are actually two laws of the displacement of equilibrium by variation of the temperature; the one supposes the system kept at constant pressure, the other at constant volume; these two

[1] H. Le Chatelier, *Comptes Rendus*, v. 99, p. 786, 1884.

[2] J. H. Van't Hoff, *Etudes de dynamique chimique*, Amsterdam, 1884.

laws having exactly the same form, we shall be content to state the first; it will suffice in our statement to replace the words *constant pressure* by the words *constant volume* to give the second law.

A chemical system is in stable equilibrium at a given pressure and at a temperature T; keeping the pressure constant, INCREASE *slightly the temperature to T′; the equilibrium is disturbed; in order to reach the new equilibrium state corresponding to the given pressure and to the temperature T′, the system must undergo a certain change of state; if this change of state was produced under the given constant pressure and at the invariable temperature T′, it would be accompanied by an* ABSORPTION OF HEAT.

172. Lowering of freezing-points of solutions.—Let us see, from an example, the importance of this law.

Under a given pressure, such as the atmospheric pressure, and at the temperature T, there is stable equilibrium in a bivariant system formed by ice in contact with a salt solution; s is the concentration of the solution.

Without changing the pressure, give the pressure a new value T', slightly higher than T; the equilibrium is disturbed; in order to reestablish it, the solution must assume a concentration s' different from s.

Now the modification undergone by the system while the concentration of the solution changes from s to s' would absorb heat if it took place at constant temperature and pressure; of the two changes of state of which the system is capable, fusion of a part of the ice, freezing of a part of the solvent, the first only fulfils the conditions which we have indicated; therefore the passage of the solution of concentration s to the concentration s' has necessitated the fusion of a certain mass of ice, so that s' is inferior to s.

Thus the concentration of a salt solution, which can remain under a given constant pressure in equilibrium with ice, decreases as the temperature increases. This may be stated in the following way:

Under a given pressure, the freezing-point of the solvent in a solution of given nature is lowered as the solution becomes more concentrated.

This lowering of the freezing-point of a liquid by mixing with it a foreign substance has been known a long time. Berthollet attributes the discovery to Blagden, who in 1788 observed it in dissolving salts in water. It has since been the object of numerous investigations. Let us note in particular those of Raoult,[1] who studied the freezing of solutions of water, benzene, nitro-benzene, ethylene bromide, formic acid, acetic acid, etc. The laws to which these experiments led Raoult have become the foundation of an important branch of physical chemistry, *cryoscopy*.

173. Lowering of the tension of the saturated vapor of solutions.—Consider again the preceding reasonings, but replacing the word *ice* by the word *vapor*; of the two modifications which may be produced in the system, condensation of a certain mass of vapor, vaporization of a certain quantity of solvent, the first, at constant temperature and pressure, liberates heat, the second absorbs it; therefore the passage of the solution from concentration s to the concentration s' has necessitated the vaporization of a part of the solvent, so that the concentration s' is greater than the concentration s. From this is derived the following proposition:

Under a given pressure, the boiling-point of a solution of given nature is increased as the solution becomes more concentrated.

If we consider a solution of given concentration s, at each temperature it will have a well-defined pressure of saturated vapor; if, on the two rectangular coordinate axes sOT, $O\pi$ (Fig. 53) we lay off the temperatures T as abscissæ and the pressures π as ordinates, to the concentration s there will correspond a curve C of tensions of the saturated vapor; to a concentration s' will correspond another analogous curve, C'.

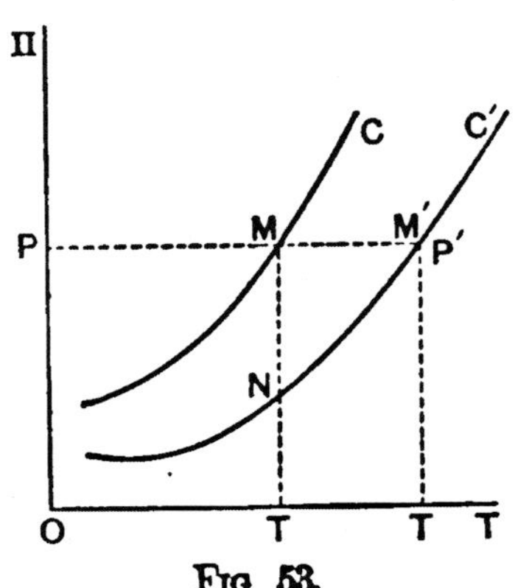

FIG 53.

All the curves C, C', . . . rise from left to right.

Take a particular value P of the pressure π and trace the line PP' parallel to OT, for which all the points have for ordinate this value $\pi = P$ of the pressure. This

[1] F. M. RAOULT, *Comptes Rendus*, v. 95 to 99, 1880 to 1884.

line PP' cuts the curve $C, C', \ldots$ at the points $M, M' \ldots$ which have for respective abscissæ $T, T', \ldots$ These temperatures T, $T', \ldots$ are, under the pressure P, the respective boiling-points of the solutions of concentrations $s, s', \ldots$ According to the preceding theorem, if s' is greater than s, T' is greater than T, and the point M' is to the right of the point M.

Furthermore, the line TM certainly cuts the curve C' in a point N located below the point M; but TM is at the temperature T, the tension of saturated vapor from the solution at concentration s; TN is, at the same temperature, the tension of saturated vapor from the solution at concentration s'; the following theorem may therefore be stated:

At a given temperature the tension of saturated vapor from a solution is diminished as the concentration of the solution is increased.

174. Dissociation of exothermic compounds and formation of endothermic compounds by rise in temperature.—Some purely chemical applications will indicate more clearly the importance of the law stated by Van' Hoff.

Under constant pressure, let us study a bivariant or multivariant system in which a certain chemical compound may be formed or destroyed; in a given state, the system encloses a mass m of this compound; the elementary composition of the system is such that the mass of this compound would have the value M if the combination were carried as far as possible; let x be the ratio $\frac{m}{M}$.

For the system in stable equilibrium at the temperature T x has a value X; if the temperature passes from the value T to a slightly higher value T', x assumes a new value X' near to X; is X' greater or less than X? Such is the question which the law of displacement of equilibrium by variation of the temperature permits us to answer.

When x passes from the value X to the value X', the system is the seat of a certain reaction; accomplished at constant pressure and temperature, this reaction would absorb heat; therefore if the formation, under constant pressure, of the compound considered, liberates heat, this reaction is a decomposition and X' is less than X; if the formation of this compound under constant pressure absorbs heat, the reaction is a combination and X' is greater than X.

We have then the double proposition:

If, without changing the pressure, the temperature of a system containing an exothermic compound (at constant pressure) is gradually raised, the proportion of the non-dissociated compound is more and more diminished.

If, without varying the pressure, the temperature of a system containing an endothermic compound (at constant pressure) and the elements whose combination may form this compound is gradually raised, the proportion of the compound in the system is increased.

Water vapor and carbonic acid are bodies which are formed, under constant pressure, with liberation of heat; if, therefore, at a constant pressure, as atmospheric pressure, for example, the temperature of a system containing one of these compounds is raised, this compound will dissociate more and more completely, as has been verified by the memorable researches of H. Sainte-Claire Deville.

175. Actions produced by a series of electric sparks; interpretation given by H. Sainte-Claire Deville. Apparatus with cold and with hot tubes.—When a series of electric sparks are passed for a sufficient time through a gas formed with liberation of heat, it often happens that this gas is more or less completely decomposed; ammonia gas, for example, is almost completely decomposed into nitrogen and hydrogen; hydrochloric acid, on the contrary, undergoes only a trace of decomposition.

Perrot,[1] causing considerable quantities of water vapor to pass between the multiple sparks of an induction coil, obtained a partial decomposition of water vapor into its elements. H. Sainte-Claire Deville[2] did not hesitate to see in this experiment the ana logue of Grove's experiment. The spark, a line of fire at very high temperature, dissociates the water vapor as does the mass of incandescent platinum; the oxygen and hydrogen liberated

[1] PERROT, *Comptes Rendus*, v. 47, p. 351, 1858; *Recherches sur l'action chimique de l'étincelle d'induction de l'appareil de Ruhmkorff.* Thèse, Paris, 1861.

[2] H. SAINTE-CLAIRE DEVILLE, *Bibliothèque universelle, Archives*, Nouvelle période, v. 6, p. 267, 1859.

are abruptly cooled by contact with the cool gases encountered a few millimetres away from the spark; brought to a temperature at which their direct combination is no longer produced, they may elude observation.

If this explanation is correct, if the actions which determine a series of sparks are merely actions which are produced at a very high temperature and which, thanks to the quickness of cooling, have not the time to be completely reversed, it should be possible to reproduce these actions without making any use of electricity; for this it would suffice to circulate the gases to be studied in a space where a very cold region is in immediate contact with a very hot region.

Here is how H. Sainte-Claire Deville realized these conditions:

Take a porcelain tube and place it in a furnace, which may be heated to a very high temperature; the ends of the tube are closed by corks each pierced with two holes. The gas may be let in and out through one hole of each end; through the other two holes is passed a tube of silvered brass, 8 mm. in diameter, running the length of the porcelain tube and through which a rapid current of cold water is passed. Finally, two small screens of unglazed porcelain separate the cool portions of the porcelain tube outside the furnace from the hot part inside.

"This brass tube even in its hottest parts is cooled to about 10° by the current of cold water. The velocity of this water is such that in passing through the incandescent tube the latter does not sensibly heat it.

"There is thus in a small space a cylindrical porcelain surface very strongly heated and a concentric brass surface very cold.

" . . . In order to give an idea of the strange manner in which this apparatus acts, I shall say that one may with impunity cover the metallic tube with the most alterable organic substances, plunge them into a bright brazier, with which I operate, and note in this way certain decompositions. If the layer of alterable substance is sufficiently thin, it will always be protected against the action of the fire by the current of fresh water passing through the metallic tube. It suffices that the latter has thin walls and that they be good heat conductors. The mass of very hot gas being quite insensible in comparison with the mass of the cooling apparatus, and the conductivity of the gas being nearly zero, the cooling of the matter experimented upon will always be sudden, and one will have quite the same conditions realized with the electric spark."

176. Dissociation of carbonous oxide, of sulphurous and hydrochloric acid gases. Synthesis of ozone.—If a current of carbonous acid gas is passed through this apparatus, the gases coming from the tube contain a certain quantity of carbonic acid, while the cold metallic tube becomes covered with a carbon deposit; the carbonous oxide is therefore partially decomposed into carbonic oxide and carbon according to the equation

$$2CO = CO_2 + C.$$

This reaction is the same one taking place when a series of electric sparks are passed through a eudiometer containing carbonous oxide.

A current of sulphurous anhydride may be passed through the apparatus with cold and hot tubes after having plated the brass tube with a thick layer of pure silver; the silver has no sensible action on the sulphurous acid at the temperature of 300° and, *a fortiori*, at the temperature of 10° at which it is kept during these experiments; at the end of a certain time the silver is found to be considerably blackened by its transformation into silver sulphide and covered with a layer of sulphuri anhydride, which absorbs moisture readily from the air and produces, in a solution of barium chloride, an abundant precipitate. The sulphurous anhydride has therefore been decomposed into sulphuric anhydride and sulphur, according to the equation

$$3SO_2 = 2SO_3 + S.$$

By various experiments H. Sainte-Claire Deville showed that this is also the equation for the partial decomposition undergone by sulphurous anhydride in a eudiometer in which passes a series of electric sparks.

When a train of electric sparks are made to pass through a eudiometer containing hydrochloric acid, a small quantity of this acid is decomposed into hydrogen and oxygen.

This same decomposition takes place at the hightest temperatures which may be produced by laboratory furnaces.

To demonstrate this, pass a current of pure, dry hydrochloric acid through the apparatus with cold and hot tubes after having covered the cold tube with a layer of silver amalgam, unattackable by the hydrochloric acid at the low temperature at which it is kept. After several hours the mercury and even the silver have some chloride on the surface, for upon wetting the tube with ammonia the tube is blackened and the ammonia takes up a small amount of silver chloride.

These various experiments place beyond doubt the hypothesis formulated by H. Sainte-Claire Deville: the eudothermic decompositions which are produced by the passage of a long series of sparks within a gas are due to the high temperature produced by the spark; they are likewise confirmations of the principle of the displacement of equilibrium by variation of temperature.

The passage of a series of electric sparks through a gaseous system is not merely susceptible of producing certain decompositions; it may also give rise to certain syntheses. We do not wish to speak here of sudden combinations and explosives, such as the combination of oxygen and hydrogen which a single spark suffices to provoke, but slow combinations determined by the passage of frequent electric sparks, prolonged over several hours; the type of these syntheses is the partial transformation of oxygen into ozone:

$$3O_2 = 2O_3.$$

If H. Sainte-Claire Deville's way of veiwing this is correct, these syntheses should not be regarded as indirect reactions, rendered possible by a certain electrical action, but as reactions which are produced directly at high temperatures; the apparatus with cold and hot tubes should suffice to produce them without making any use of electricity.

Troost and Hautefeuille have in fact shown that if a current of oxygen is passed through a hot tube brought to 1300° or 1400°, while the cold tube is covered with a layer of pure silver, there collects on this tube, after a certain time, silver dixoide, certain indication of a transformation of oxygen into ozone by contact with the extremely hot porcelain.

But, according to the determinations of Berthelot, the reaction

$$3O_2=2O_3$$

absorbs 61.4 cals.; hence the direct formation of ozone at a higher temperature is a remarkable confirmation of the law of the displacement of equilibrium with variation of temperature.

177. Synthesis of acetylene.—We see from this experiment that the formation of ozone within the oxygen traversed by a series of electric sparks should be regarded as a reaction which is produced of itself at a high temperature; the same interpretation should be accepted for a great number of syntheses produced by a series of sparks or by the electric arc.

Thus when a current of hydrogen is passed between the two carbon electrodes of an electric arc, acetylene gas is formed, as was shown by Berthelot;[1] the formation of acetylene in these circumstances should be regarded as a reaction which produces itself at the extremely high temperature of the electric arc.

According to Berthelot, the reaction

$$2C+2H=C_2H_2,$$

which represents the formation of acetylene, absorbs 58.1 cals. The formation of acetylene at the temperature of the electric arc should also be regarded as a consequence of the law of the displacement of equilibrium.

Analogous examples exist in great numbers; we shall limit ourselves to those cited above.

178. Case of reactions which neither absorb nor liberate heat.—A particularly interesting case is the one for which *the compound contained in the system is formed, under constant pressure, without liberation or absorption of heat;* in this case a reasoning similar in all points to that which we have developed above shows us that X' can neither be greater nor less than X; *the proportion of the compound contained by the system, when it is in equilibrium under a given pressure, is independent of the temperature.*

179. Phenomena of etherification.—The studies of Berthelot on etherification furnish an application of this law.

[1] BERTHELOT, *Comptes Rendus*, v. 54, pp. 640 and 1042, 1862.

The etherification of alcohol by acetic acid does not give rise to any appreciable quantity of heat. If these substances are mixed in equivalent proportions and left long enough for equilibrium to be established, it is found that the proportions of etherized acid are the following:

At room temperature, after 16 years..........	0.652
At 100°, after a very long time..............	0.656
At 170°, after 42 hours......................	0.665
At 200°, after 24 hours......................	0.673
At 220°, after 38 hours......................	0.665

These numbers should be regarded as identical.

180. Minimum dissociation of hydrogen selenide.—One may, in all the preceding statements, replace the words *constant pressure* by the words *constant volume* without changing the accuracy of these statements, so that the following considerations are justified:

At constant pressure raise the temperature of a system which contains liquid selenium, hydrogen, and gaseous hydrogen selenide; for the system in equilibrium the ratio X between the mass of sel nide formed and he possible mass of selenide varies as the temperature is raised; this ra io increases at first, passes through a maximum, then diminishes while the temperature continues to rise.

Ditte [1] was the first to announce a maximum for the ratio X; unfortunately his observations were erroneous, due to the partial absorption of hydrogen selenide by the liquid selenium; H. Pélabon,[2] eliminating this source of error, was able to study the variations of the ratio X with the temperature and established beyond doubt the existence of a maximum for this ratio; this maximum corresponds to a temperature of 575°, and its value is about 0.41.

It should be concluded that there is *absorption of heat* when, at constant volume and at a constant temperature less than 575°,

---

[1] Ditte, *Annales de l'Ecole normale supérieure*, 2d S., v. 1, p. 293, 1873.

[2] H. Pélabon, *Mém. de la Soc. des Sciences physiques et naturelles de Bordeaux*, 5th S., v. 3, p. 241; *Sur la dissociation de l'acide sélenhydrique*, Paris, A. Hermann, 1898.

liquid selenium and hydrogen combine to form hydrogen selenide; on the contrary, when this reaction takes place at a temperature higher than 575° it must *liberate heat.*

Hautefeuille had already shown that, at constant pressure at ordinary temperature, the formation of hydrogen selenide from hydrogen and liquid selenium was an endothermic reaction; Fabre [1] has recently given an exact determination of the heat of formation of hydrogen selenide in these conditions. If it is observed, also, that the combination, under constant pressure, of hydrogen and liquid selenium is accompanied by almost no change in volume, it is seen that the heat of formation under constant pressure is sensibly equal to that under constant volume. Thus is verified the first part of the preceding statement, a consequence of the principle of displacement of equilibrium by variation of the temperature.

181. Similarity of the preceding principle and Moutier's Law. At very low temperatures the principle of maximum work is exact.—This principle leads, for bivariant and multivariant systems, to conclusions similar in all respects to those drawn (Art. 141 and 142) for monovariant systems from Moutier's Law; greatly dissociated at a high temperature, an exothermic compound remains less altered, in a system in equilibrium, as the temperature is lowered; an endothermic compound, on the contrary, is formed in very small quantities at low temperature; as the temperature rises its stability increases. At a very low temperature, for a system in equilibrium, one may regard the dissociation of exothermic compounds as almost null, the dissociation of endothermic compounds as almost complete; every endothermic compound is spontaneously resolved into its elements; every exothermic compound is formed spontaneously at the expense of its elements; otherwise expressed, *at a very low temperature the principle of maximum work applies to all reactions without exception.*

As the temperature is raised higher and higher, we see increase the number of reactions, decompositions of exothermic compounds, or syntheses of endothermic compounds which are exceptions to the principle of maximum work. According to the happy ex-

---

[1] FABRE, *Annales de Chimie et de Physique*, 6th S., v. 10, p. 482.

pression of Van't Hoff, *this principle would be rigorously exact only at 0° absolute.*

However, if we desire to understand fully the sense and importance of this proposition, we must not forget the existence of states of false equilibrium of which the preceding theory takes no account; no reaction which contradicts this theory is ever observed, but, on the other hand, a great number of reactions predicted as necessary by this theory do not take place; the system which should show them remains in equilibrium.

CHAPTER XI.

BIVARIANT SYSTEMS. THE INDIFFERENT POINT.

182. Various types of bivariant systems: Solutions and double mixtures.—A bivariant system is one divided into a number of phases equal to the number of independent components which form it; a system of two independent components and divided into two phases is the most generally studied type.

This type may be divided into two classes.

It may happen that one of the two phases of the system consists of a definite compound, containing one of the two components or both, while the other phase is a mixture of the two components in varying proportions; a system which contains sodium chloride crystals in the presence of an aqueous solution of sodium chloride, a system containing ice in the presence of an aqueous solution of potassium nitrate, one with crystals of hydrated sodium sulphate ($Na_2SO_4 \cdot 10H_2O$) in the presence of an aqueous solution of sodium nitrate give us three cha acteri tic examples belonging to this class which we shall call the class of *solutions*.

It may happen, on the contrary, that each of the two phases into which the system is divided is a mixture in variable proportion of the two independent components; this takes place when a liquid mixture of water and alcohol is in the presence of a mixed vapor which contains these two substances at once; this also is true when a liquid mixture of ether and water divides into two layers which have different compositions; such systems form the category of *double mixtures*.

183. Law of equilibrium for bivariant systems. This equilibrium is stable in general.—These two categories of bivariant systems obey the same law which we have stated in studying the phase rule (Art. 96). If there is given arbitrarily a temperature and pressure, it is possible, in general, to observe this system in equilibrium at this temperature and pressure; the composition of each of the two phases into which the system in equilibrium is divided is determined by the knowledge of this temperature and pressure.

In all cases, by the words *is determined* is not to be understood a determination which excludes all ambiguity; it may happen, and does in certain cases which we shall meet in this chapter, that at a given pressure and temperature a bivariant system formed of the same independent components presents *two* distinct states of equilibrium corresponding to different compositions of the several phases.

This ambiguity disappears when there is given not only the nature of the two independent components, the temperature and pressure, but also the mass of each of the independent components; in this case, not only is the composition of each of the phases composing the system in equilibrium known, but also, save for a case which we shall treat at length in this chapter, the mass of each of the phases is determined.

Consequently, when a bivariant system is thus given, it is impossible, except for the particular case we have just mentioned, to vary the masses of the different phases which are held in equilibrium without causing their composition to vary, so that at constant pressure and temperature the system in equilibrium could not undergo any modification without destroying the equilibrium; the exceptional case aside, the equilibrium of a bivariant system is not an indifferent equilibrium; this distinguishes the bivariant systems sharply from the monovariant systems.

It may be shown that the state of equilibrium of a bivariant system is stable except for the special case, when it is indifferent; this proposition is of considerable importance, for it shows that in general one may apply to bivariant systems the two laws of displacement of equilibrium by variation of pressure and of the displacement of equilibrium by change of temperature; in fact, in the preceding chapter we have borrowed several examples of these laws from the study of bivariant systems.

184. Solutions. Saturation. Solubility curve.—Consider first solutions.

Two independent components, water, which we shall denote by the index 0, and an anhydrous salt, which we shall denote by the index 1, form the system, which is divided into two phases; the one is a solid salt, anhydrous or hydrated, of definite composition; the other is a mixture of variable composition; this mixture encloses a mass of water M_0 and a mass of anhydrous salt M_1; the ratio $\frac{M_1}{M_0}=s$ is the *concentration* of the solution.

Take a pressure π, which we shall suppose always the same; it may be, for example, atmospheric pressure; take besides a temperature T; suppose that, at this pressure π and temperature T, the solution is in equilibrium with an excess of solid salt, case in which it is said to be saturated with this salt; the concentration of this saturated solution will have a well-determined value S. Take two rectangular coordinate axes (Fig. 54); on the axis of abscissæ OT lay off the values of the temperature; on the axis of ordinates Os lay off the concentrations; the concentration S of the saturated solution at the temperature T is represented by a point M having the coordinates T, S; when, without changing the pressure, the temperature T is varied, the point M describes a curve whi h is the *solubility curve* of the salt studied at the pressure considered.

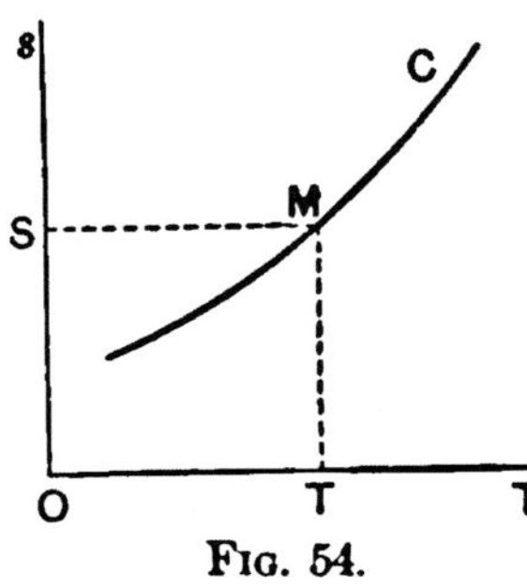

FIG. 54.

185. For a hydrated salt two saturated solutions correspond to each temperature. The solubility curve has two branches.—What we have just said supposes that a single point M corresponds to the temperature T or, in other terms, that *the concentration of the solution, saturated at the temperature T, has a value determined without ambiguity.* If the solid precipitate enclosed by the system is an *anhydrous* salt, the above is certainly true; but it may be otherwise *if this precipitate is a hydrated salt; it may happen,* in this case, *that to the same temperature T correspond two distinct saturated solutions, one of concentration S_1, the other of* *greater concentration, S_2; the first richer in water than the hydrated salt, the other less rich in water than the hydrated salt.*

These two solutions are represented by the two points M_1 and M_2 (Fig. 55) which have the same abscissa T and the ordinates S_1 and S_2; when the temperature T varies, these two points M_1, M_2 describe two curves C_1, C_2, which together compose the *solubility curve* for the hydrate; the *lower branch*, C_1, represents the saturated solutions richer in water than the hydrate; the *upper branch*, C_2, represents the saturated solutions less rich in water than the hydrate. It may be said that the lower branch alone exists for the case in which the precipitate is an anhydride salt; it exists also alone for a great number of cases in which the precipitate is an hydrated salt.

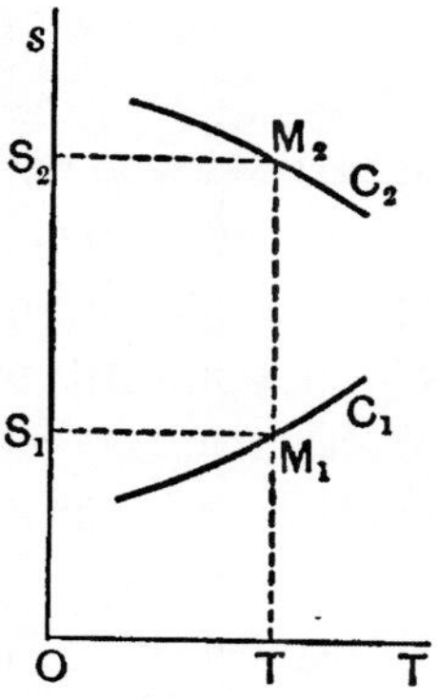

Fig. 55.

186. Non-saturated and supersaturated solutions.—The various equilibrium states of which we have just spoken are all stable. If a small quantity of solid salt precipitates from a saturated solution, the solution is brought in o a state in which it can no longer abandon solid salt, but where it dissolves that which is thrown into it, a fact we express by saying it is *unsaturated*. If a small quantity of solid salt is dissolved in a satuiated solution, the solution is brought at once into a state where it is impossible for it to dissolve the least solid particle; according to the predictions of thermodynamics, it ought to give up the salt it contains in excess and return to the concentration corresponding to saturation; we know that this modification is not always produced, and that the solution may remain in the state of false equilibrium; it is then said to be *supersaturated*.

A saturated solution therefore becomes non-saturated by the subtraction of a small quantity of solid salt, and supersaturated by the addition of a small quantity of the same salt.

If a solution is richer in water than the solid salt precipitated, which is always the case for anhydrous salts, the addition of a small quantity of salt to the solution increases its concentration; if, on the contrary, the solution is less rich in water than the solid salt, the addition of a small quantity of this salt to the solution decreases its concentration.

At this point we may evidently state the following propositions:

If a saturated solution is represented by a point of the lower branch C_1 of the solubility curve, this solution becomes non-saturated when the concentration is diminished and supersaturated when the concentration is increased; if, on the contrary, a saturated solution is represented by a point on the upper branch C_2 of the solubility curve, this solution becomes non-saturated when the concentration is increased and supersaturated when the concentration is diminished.

Otherwise expressed, *the unsaturated solutions are represented by the points of the plane TOs* (Fig. 56) *which are situated below the lower branch C_1 or above the upper branch C_2 of the solubility curve; the supersaturated solutions are represented by the points situated between the two branches.*

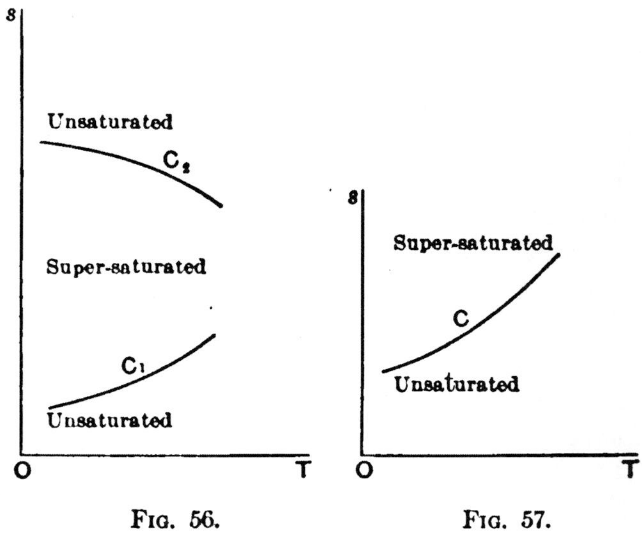

Fig. 56. Fig. 57.

For the case in which only the lower branch exists, as for the saturated solutions of anhydride salts and for the saturated solutions of a great number of hydrated salts, *the non-saturated solutions are represented by the points of the plane TOs* (Fig. 57) *which are above the solubility curve C, and the supersaturated solutions by the points which are below this curve.*

187. Heat of solution in saturated solutions.—When a very small mass of salt m passes, at the temperature T, from the precipitate state into a solution nearly saturated at this temperature T, the phenomenon is accompanied by a certain absorption of heat; the quantity of heat absorbed which, other things being equal, is proportional to the small mass m, depends on the temperature T at which the phenomenon is produced; this quantity of heat absorbed may be represented by the product mL, L being a fixed coefficient for a definite temperature, but variable with the temperature; L is what is called the *heat of solution in saturated solution* of the salt considered, at the temperature T.

The quantity of heat absorbed of which we have just spoken is, in certain cases, negative in other words, the dissolving of a salt in an almost saturated solution may be accompanied by liberation of heat; in this case the heat of solution L is negative.

It is evident that if, at a definite temperature, there exist two distinct saturated solutions of concentrations S_1, S_2, there correspond to these two solutions two distinct heats of solution L_1, L_2.

188. Displacement of equilibrium by variation of temperature.—The equilibrium states which we have just studied being all stable, we may apply to them the law of displacement of equilibrium by variation of the temperature.

A system, containing the salt spoken of in contact with the solution, is in equilibrium at the temperature T; the saturated solution has the concentration S; without changing the pressure, bring the temperature to a value T' a little higher than T; the equilibrium is broken and there is produced in the system a change of state which brings the concentration to the value S', characterizing the saturated solution at the new temperature T'.

If this same change of state is produced without variation of temperature, it should absorb heat; this change of state consists therefore in the solution of a small quantity of salt if the heat of solution in saturated solution is positive; it consists in the precipitation of a small quantity of salt if the heat of solution in saturated solution is negative.

We recall that the mixture of a small quantity of precipitate with the solution increases the concentration of this solution if it is richer in water than the precipitate, and decreases the concentration of the solution if it is less rich in water than the precipitate; we may state the following propositions:

If the heat of solution in saturated solution is positive, the lower branch of the solubility curve rises from left to right, the upper branch of the solubility curve descends from left to right. If the heat of solution in saturated solution is negative, the lower branch of the solubility curve descends from left to right; the upper branch of the solubility curve rises from left to right.

Let us make some applications of this proposition to the lower branch of the solubility curve, the only one existing for the case in which the precipitate is anhydrous, and for a great number of cases where it is hydrated.

Most salts dissolve in water with absorption of heat; also most solubility curves rise from left to right; the salt is more soluble as the temperature is higher.

Sulphate of sodium gives us an example, however, of a salt less soluble with rise of temperature.

At temperatures below 23° a solution of sodium sulphate remains in equilibrium in contact with a precipitate of hydrated sodium sulphate, $Na_2SO_4 \cdot 10H_2O$; this salt dissolves with absorption of heat; its solubility increases when the temperature rises. At temperatures higher than 23° sodium sulphate with ten molecules of water cannot be absorbed in equilibrium in contact with a solution of sodium sulphate; on the other hand, the latter may rest in equilibrium in contact with a precipitate of anhydrous sodium sulphate; the solubility of anhydrous sodium sulphate diminishes with increase of temperature; the heat of solution of anhydrous sodium sulphate is negative, as has been shown by Pauchon.[1]

Calcium hydrate and cerium sulphate behave like anhydrous sodium sulphate.

Calcic orthobutyrate with one molecule of water has a solubility which decreases as the temperature rises to 60°; at 60° this solu-

---

[1] PAUCHON, *Comptes Rendus*, v. 97, p. 1555, 1883.

bility passes through a minimum; it then increases with the temperature; the law of the displacement of equilibrium by variation of the temperature gives us then the following information:

Above 60° calcic orthobutyrate dissolves, in nearly saturated solution, with liberation of heat; at 60° the heat of solution in saturated solution is equal to 0; beyond 60° this heat becomes positive.

Chancel and Parmentier[1] have verified experimentally the first part of this statement.

189. Precautions regarding the use of the preceding law.— The law of the displacement of equilibrium by variation of the temperature is an exact theorem, which leads surely to just conclusions, provided that, in applying it, the exact conditions indicated in its statement are fulfilled; without this precaution one may, by an unjustified application of this principle, draw false conclusions, as for example:

Calcic isobutyrate with five molecules of water is the more soluble the higher the temperature is raised; consequently the heat of solution of this salt *in saturated solution* is positive; Chancel and Parmentier,[2] having measured the heat of solution of calcic isobutyrate, found it negative and concluded, therefore, that the law of displacement of equilibrium by variation in temperature was not always exact; Le Chatelier[3] pointed out very justly that these physicists had measured not the heat of solution *in saturated solution*, but the heat of solution *in very dilute solution*, a quantity which may be very different from the first, which may even have another sign; by direct experiment he proved that the heat of solution of hydrated calcium isobutyrate, *in saturated solution*, is positive as required by the law of the displacement of equilibrium by variation of the temperature.

190. The two branches of the solubility curve of a hydrate join each other at the indifferent point where the saturated solution has the same composition as the hydrate.—Let us take a

---

[1] CHANCEL and PARMENTIER, *Comptes Rendus*, v. 104, pp. 474 and 881, 1887.

[2] *Ibid.*

[3] H. LE CHATELIER, *Comptes Rendus*, v. 104, p. 679, 1887.

hydrated salt whose solubility curve consists of two branches; suppose that the heat of solution in saturated solution be positive for both branches. The lower branch, C_1 (Fig. 58), rises from left to right; the upper branch, C_2, descends from left to right.

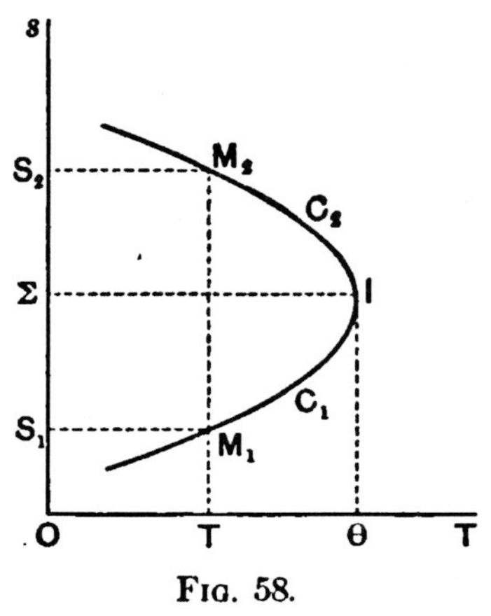

Fig. 58.

To the same temperature T correspond a point M_1, of ordinate S_1, on the branch C_1, and a point M_2, of ordinate S_2, on the branch C_2; as the temperature rises, the two points M_1, M_2 approach each other and the two concentrations S_1, S_2 approach each other.

May it happen that at some temperature θ the two points M_1, M_2 unite in a single point I, that the two concentrations S_1, S_2 assume a common value Σ?

The concentration S_1 is the concentration of a solution richer in water than the hydrated salt with which it is saturated; the concentration S_2 is the concentration of a solution less rich in water than the same hydrated salt; if these two concentrations S_1, S_2 approach a common limit Σ, Σ is certainly the concentration of a solution having exactly the same composition as the hydrated salt with which it is saturated.

Thus *the two branches of the solubility curve of a hydrate may, for a certain value θ of the temperature, unite in a common point I, a point where the saturated solution has the same composition as the hydrate in contact with which it remains in equilibrium.*

In what manner is the junction of the two branches of the solubility curve made? One might be tempted, in replying to this question, to apply again to each of these two branches the law of the displacement of equilibrium with change of temperature; this would be an unwarranted application of this principle; in fact, the equilibrium state of the saturated solution at the temperature θ is no longer a state of stable equilibrium; the saturated solution having, at this temperature, the same composition as the precipitate, one may, without varying the composition of the two phases and, moreover, not disturbing the equilibrium, suppose that a certain mass of hydrated salt dissolves or precipitates; it is therefore clear that the solution saturated at θ degrees is in *indifferent equilibrium* with the hydrated solid salt; also, we shall give the name *indifferent point* to the point I, of coordinates θ, Σ, which represents this solution.

The law of the displacement of equilibrium by variation of temperature being unable to instruct us on the behavior of the solubility curve in the neighborhood of the point I, we should seek this information from a special theorem; this special theorem has been indicated by J. Willard Gibbs, and teaches us that—

The two branches C_1, C_2 of the solubility curve of the hydrate join each other at the point I, so as to form a curve without cusp, giving at the point I a tangent parallel to Os.

191. The temperature of junction is the aqueous fusing-point of the hydrate.—The solubility curve C_1IC_2 divides the plane into two regions; one of these regions, cross-hatched in Fig. 59, is in

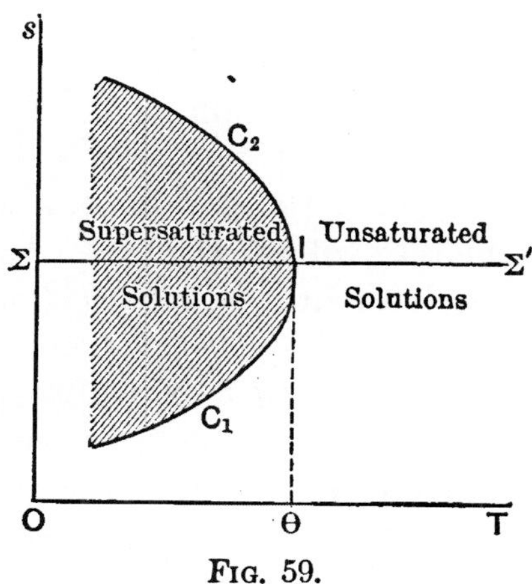

FIG. 59.

the concavity of this curve; every point in this region represents a supersaturated solution of the hydrate, the other region represents, by its various points, all the non-saturated solutions of the hydrate.

Parallel to OT trace the line $\Sigma\Sigma'$, whose various points have for constant ordinate the concentration of a solution of the same composition as the hydrate; this line passes through the point I;

the points in this line having abscissæ less than θ are to the left of I and represent supersaturated solutions; the points having abscissæ greater than θ, to the right of the point I, represent non-saturated solutions.

Take a solution, of concentration Σ, separated from any solid precipitate; at a temperature higher than θ this solution will be in equilibrium; but if the temperature falls below θ, this solution, supersaturated, can no longer remain in equilibrium, except for a phenomenon of false equilibrium; it may precipitate the hydrate, and as this precipitation does not alter its composition, the modification will continue until all the liquid is used; *θ is therefore the temperature at which a solution of the same composition as the hydrate solidifies.*

Take also a certain mass of hydrate in the solid state and free from all trace of solution; at a temperature less than θ this hydrate cannot undergo aqueous fusion, for the solution engendered, having a concentration Σ, would be supersaturated and would resolidify; on the contrary, at a temperature higher than θ, if this hydrate could be observed in equilibrium, this state of equilibrium would be unstable; if the hydrate undergoes a trace of aqueous fusion, the solution resulting, of concentration Σ, would be non-saturated; it would commence to dissolve a new mass of hydrate; this dissolving not changing the composition of the solution, the dissolving will continue until all the hydrate is fused; *the temperature θ is therefore the temperature at which the solid hydrate undergoes complete aqueous fusion.*

192. Experimental investigations of Guthrie, Roozboom, and other observers.—The ideas which we have just set forth exist in germ in the theoretical works of J. Willard Gibbs, but they have been brought to light largely by the theoretical and experimental researches of Roozboom and Guthrie.

In 1884 Guthrie [1] described the indifferent point of ethylamine hydrate, indifferent point which corresponds to the temperature −8°; in 1885 Roozboom [2] studied the indifferent points of hydrochloric and hydrobromic hydrates; in 1889, in a work of capital

---

[1] GUTHRIE, *Philosophical Magazine*, 5th S., v. 18, p. 22, 1884.

[2] ROOZBOOM, *Recueil des Travaux chimiques des Pays Bas*, v. 3, p. 84, 1884; v. 4, p. 102, 1885.

importance,[1] he found $+30^\circ.2$ *C.* the temperature of the indifferent point of the hydrate $CaCl_2 \cdot 6H_2O$.

Pickering [2] recognized for the sulphuric hydrates $SO_3 \cdot 5H_2O$ and $SO_3 \cdot 2H_2O$ the two branches C_1, C_2 of the solubility curve, and he could follow each of these branches over a considerable temperature range; for the hydrate $SO_3 \cdot H_2O$ he found an indication of the existence of the upper branch for solutions more concentrated than the hydrate.

Pickering [3] likewise undertook the study of the combinations which the amines form with water, a subject which had already furnished Guthrie examples of indifferent points; Pickering found in his turn the existence of these points.

In a very important investigation on the hydrates of ferric chloride, Roozboom [4] showed the existence of an indifferent point for each of the four hydrates which ferric chloride may form. These indifferent points correspond to the following temperatures:

For $Fe_2Cl_6 \cdot 12H_2O$	$\theta = +37^\circ$ *C.* approx.
$Fe_2Cl_6 \cdot 7H_2O$	$\theta = +32.5$
$Fe_2Cl_6 \cdot 5H_2O$	$\theta = +56$
$Fe_2Cl_6 \cdot 4H_2O$	$\theta = +73.5$

Van't Hoff and Meyerhoffer [5] recognized the existence of the two branches of the solubility curve and of the indifferent point for the hydrate $MgCl \cdot 12H_2O$; this indifferent point corresponds to the temperature $-16^\circ.3$ *C.*

Finally, Le Chatelier [6] has studied with great care the solubility of lithium borate in water; lithium borate furnishes the hydrate $Li_2B_2O_4 \cdot 16H_2O$; the solubility curve of this hydrate is composed of two branches; the lower branch, C_1, corresponding

---

[1] Roozboom, *Recueil des Travaux chimiques des Pays Bas*, v. 8, p. 1, 1889; *Archives néerlandaises des Sciences exactes et naturelles*, v. 23, p. 199, 1889; *Zeitschrift für physikalische Chemie*, v. 4, p. 31, 1889.

[2] Pickering, *Journal of Chemical Society*, v. 57, p. 338, 1890.

[3] *Ibid.*, v. 63, pp. 141 and 890, 1893.

[4] Roozboom, *Archives néerlandaises*, etc., v. 28, 1892; *Zeitschrift für physikalische Chemie*, v. 10, p. 447, 1892.

[5] Van't Hoff and Meyerhoffer, *Sitzungsberichte der Berliner Akad.*, Feb. 4 and 18, 1897.

[6] H. Le Chatelier, *Comptes Rendus*, v. 124, p. 1091, 1897.

to solutions less concentrated than the hydrate, could be followed from the temperature −60° C.; the upper branch, C_2, corresponding to solutions more concentrated than the hydrate, could be followed from a point whose abscissa corresponds to the temperature +34° C.; these two curves unite in an indifferent point *I*, whose abscissa corresponds to the temperature +47° C.; the behavior of the two curves in the neighborhood of the point *I* indicates clearly that they meet at this point and that their common tangent is parallel to *Os*.

The hydrates are not the only substances which give us such phenomena; every time that one may dissolve in variable proportion in a liquid 0 a substance 1 susceptible of forming with this liquid a solid compound 2 of definite composition, one may repeat for these three substances 0, 1, 2 all that we have just said about water, an anhydrous salt, and the hydrate formed by their union.

Iodine, dissolved in liquid chlorine, may give iodine chloride, ICl, capable of being deposited in the solid state; this solid chloride may exist in two allotropic forms denoted by the symbols ICl_α and ICl_β; the first form has the fusing-point +27°.2 C., and the second the fusing-point +13°.9 C.; Stortenbeker [1] has shown that each of these two temperatures corresponds to an indifferent point, the one for the solubility curve of ICl_α in liquid chlorine, the one for the solubility curve of ICl_β in the same solvent.

The substance 0 may be a melted anhydrous salt, the substance 1 another anhydrous salt, the substance 2 a double salt formed by the combination of the first two in definite proportions; Le Chatelier [2] has studied several systems of this kind.

The solution of lithium carbonate in melted potassium carbonate gives a solid double salt whose formula is $KLiCO_3$; the temperature of the indifferent point is 515° *C*. The melted mixture of sodium borate and sodium pyrophosphate gives a double salt formed by the union of one molecule of each of the simple salts; the temperature of the indifferent point is about 960° *C*. Besides these examples furnished by melted salts, we may

[1] W. Stortenbeker, *Recueil des Travaux chimiques des Pays Bas*, v. 6, 1888; *Zeitschrift für physikalische Chemie*, v. 3, p. 11, 1888.

[2] H. Le Chatelier, *Comptes Rendus*, v. 118, p. 801, 1894.

note the case studied by Kuriloff,[1] the mixture of picric acid, $C_6H_2\cdot(NO_2)_3\cdot OH$, and β-naphtol, $C_{10}H_7OH$, giving the substance $C_6H_2(NO_2)_3OHC_{10}H_7OH$, which in the presence of a liquid mixture of picric acid and β-naphtol has an indifferent point very sharply marked at the temperature $+157°$ *C*.

193. Indifferent point of a double mixture.—A solution supporting the pressure π and brought to the temperature T is in indifferent equilibrium in contact with a hydrated salt if, at this temperature and under this pressure, the saturated solution has the same composition as the hydrate; when a double mixture is in equilibrium at the pressure π and temperature T the composition of each of the two phases into which it is divided is determined; if these two phases have the same composition, this state of equilibrium is indifferent.

Suppose, for instance, that a mixture of volatile liquids is in the presence of the mixed vapor which it emits; at the temperature T and pressure π the liquid mixture and the mixed vapor which remain in equilibrium have definite compositions; if the liquid mixture and the mixed vapor happen to have the same composition at a certain temperature and pressure, the equilibrium of the system for this temperature and pressure is evidently indifferent; it is in fact clear that, without changing the composition of any of the phases, consequently without disturbing the equilibrium of the system, we may either vaporize a part of the liquid mixture or condense a part of the mixed vapor.

194. Two theorems of Gibbs and of Konovalow.—Under what circumstances shall we observe such a state of indifferent equilibrium? Two important theorems, discovered by J. Willard Gibbs, found anew by D. Konovalow, give us this information. Here are these two theorems:

First Theorem of Gibbs and Konovalow.—*Under a constant pressure cause the composition of the liquid mixture to vary in a well-defined way; the boiling-point of this mixture changes; if, for a certain composition of the liquid mixture, the boiling-point passes through a maximum or minimum, this liquid mixture gives off a saturated vapor of the same composition, and reciprocally.*

---

[1] Kuriloff, *Zeitschrift f. physikalische Chemie*, v. 23, pp. 90 and 673, 1897.

SECOND THEOREM OF GIBBS AND KONOVALOW.—*At constant temperature cause the composition of the liquid mixture to vary in a definite way; if, for a certain composition of the liquid mixture, the tension of the saturated vapor passes through a maximum or a minimum, this liquid mixture gives off a saturated vapor of the same composition, and reciprocally.*

195. Application of the first theorem to mixtures of volatile liquids.—Let us consider in detail the consequences of these two important theorems beginning with the first.

Take a liquid mixture containing two substances 1 and 2; a gramme of this mixture contains X grammes of the substance 2 and $(1-X)$ grammes of substance 1; in just the proportion that the substance 2 increases in the mixture, X will increase; starting from the value 0 at the instant the liquid contains only the substance 1 in a state of purity, X approaches 1 as the mixture approaches the substance 2 taken in the state of purity.

Consider this liquid mixture at the constant pressure π; for each value of X will correspond a boiling-point T; if we take X (Fig. 60) for abscissa and T for ordinate of a certain point M, the locus of the points M will be a curve C; this curve will start from the point M_1, which has for abscissa O and for ordinate the temperature T, the boiling-point, under the constant pressure π, of the liquid 1 taken in the state of purity; it ends at the point M_2, which has 1 for abscissa and for ordinate the temperature T_2, boiling-point, at the same pressure π, of the liquid 2 taken in the state of purity.

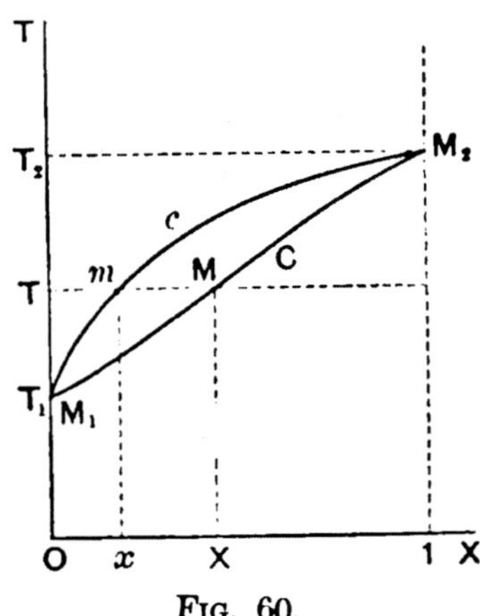

FIG. 60.

At the temperature T and under the pressure π a gramme of the saturated vapor in equilibrium with the liquid of concentration X contains x grammes of the substance 2 and $(1-x)$ grammes of substance 1; take a point m, in the plane XOT, having x as abscissa and T for ordinate; this point m *will correspond* to the point M; the ensemble of two corresponding points M, m, having a common abscissa, will inform us as to the composition of the liquid mixture and of the mixed vapor which may coexist in equilibrium under the pressure π at the temperature T. While the point M describes the curve C from M_1 to M_2, the point m describes another curve, c, which likewise joins the point M_1 to the point M_2.

Let us suppose, to speak definitely, that the substance 2 is, under the pressure π, less volatile than the substance 1; its boiling-point T_2 at this pressure will be higher than the boiling-point T_1 of the liquid 1. Three special cases may then arise, concerning which the principles of thermodynamics give us the following information:

FIRST CASE: THE CURVE C RISES CONSTANTLY FROM THE POINT M_1 TO THE POINT M_2. *In this case the curve c rises also constantly from M_1 to M_2, and except at the points M_1 and M_2 it is always above the curve C.*

This case is represented in Fig. 60.

SECOND CASE: BETWEEN THE POINTS M_1 AND M_2 THE CURVE C HAS A POINT I (FIG. 61), OF ABSCISSA ξ AND ORDINATE θ, HIGHER THAN ALL THE OTHERS.

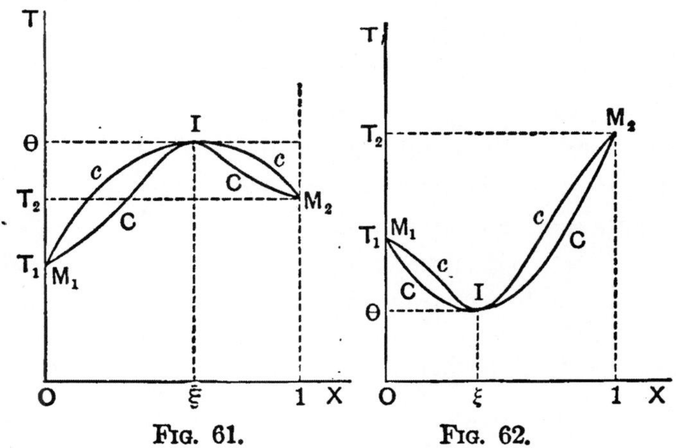

FIG. 61. FIG. 62.

According to the first theorem of Gibbs and of Konovalow, this point I is an *indifferent point;* under the pressure π, at the temperature θ, the liquid mixture and the saturated mixed vapor have the same composition, $X=x=\xi$. *The curve c also passes through the point I which is, on this curve, a point above all the others; outside of the points M_1, I, M_2 the curve c is always above the curve C.*

THIRD CASE: BETWEEN THE POINTS M_1 and M_2 THE CURVE C (FIG. 62) HAS A POINT I, OF ABSCISSA ξ AND ORDINATE θ, LOWER THAN ALL THE OTHERS. *In this case the curve* c *also passes through the point I, which is, for this curve, a point lower than all the others; outside of the points* M_1, I, M_2 *the curve c is always above the curve C.*

Experiment gives us numerous examples of each of these three cases.

The first case is by far the most frequent; it occurs with the following mixtures:

Water–methyl alcohol;
Water–ethyl alcohol;
Water–acetic acid;
Water–butyric acid.

Here are some examples of the second case:

Water–propyl alcohol;
Water–butyryl alcohol;
Carbon sulphide–ethyl alcohol;
Carbon sulphide–ethyl acetate;
Carbon tetrachloride–methyl alcohol.

The first two mixtures have been studied by Konovalow,[1] the next two by Brown,[2] and the last by Thorpe.[3]

According to Konovalow, the mixture water–formic acid furnishes us an example of the third kind.

196. Distillation of a mixture of two volatile liquids under constant pressure.—These various principles are going to permit us to study the phenomena which accompany the distillation of a mixture of two liquids under constant pressure. In the still the liquid mixture has a mixed vapor above it; one may regard this vapor as having sensibly the composition of the saturated vapor in equilibrium with the mixed liquid for the conditions of temperature and pressure which reign in the still. At every instant a portion of this vapor condenses outside of the still and a new mass of liquid vaporizes.

---

[1] D. KONOVALOW, *Wiedemann's Annalen*, v. 14, pp. 34 and 219, 1881.
[2] BROWN, *Quarterly Journ. of the Chem. Soc. of London*, v. 80, p. 529, 1881.
[3] THORPE, *Quarterly Journ. of the Chem. Soc. of London*, v. 85, p. 544, 1879.

The following proposition, which we shall take as starting-point, may be demonstrated:

If the saturated vapor enclosed by the still has not the same composition as the liquid beneath it, the boiling-point of the liquid rises as the distillation proceeds.

Take, to start with, a liquid mixture which comes under the first of our three cases.

At a given moment the liquid contained in the still has a certain composition X, abscissa of a certain point M (Fig. 63) of the curve C; the temperature within the still is the boiling-point T of the liquid of composition X, that is to say, the ordinate of the point M; on the curve c there is a point m, having the same ordinate as the point M; the abscissa x of this point m gives us the composition of the vapor which fills the still at the instant considered.

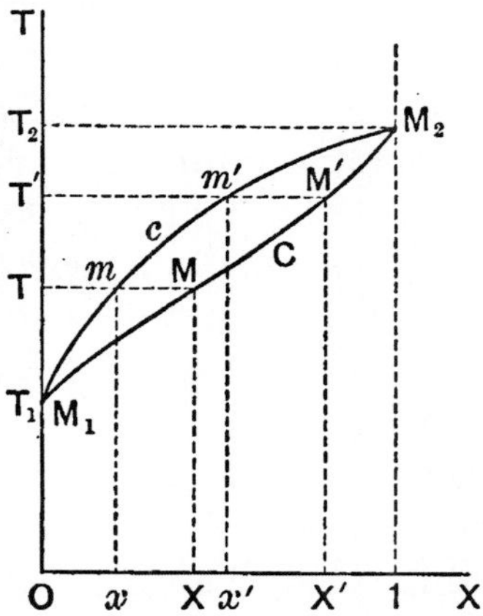

Fig. 63.

The composition of the vapor differs from the composition of the liquid; the boiling-point of the liquid contained in the still rises therefore, with the effect of distillation. After a certain time this boiling-point has assumed a value T' higher than T; if we draw a line, parallel to X, all of whose points have the ordinate T', this line cuts the lines C, c at the points M', m', which have as abscissæ X' and x' respectively; X' is the composition of the liquid in the still, x' the composition of the enclosed vapor, at the moment when the boiling-point becomes T'.

We may therefore state the following proposition:

When we distil under constant pressure a liquid mixture which is among the first of our three cases. the composition of the liquid remaining in the still and the composition of the vapor which distils vary always in the same way and tend to contain only the less volatile of the two mixed substances.

Every one knows that the above is true for the distillation of a mixture of water and alcohol.

Quite otherwise is the action for a mixture which is of the *second kind.*

197. **Mixtures which pass over entirely by distillation, without variation of the boiling-point.**—Denote always by ξ and θ the coordinates of the indifferent point I.

Reasoning as in the preceding case, we may establish without difficulty the following propositions: *When a liquid mixture is distilled whose initial composition corresponds to a value of X less or greater than ξ, the boiling-point constantly rises and approaches θ; the composition of the liquid in the still and the composition of the vapor in it vary always in the same way, so as to approach the common composition ξ.*

What will happen at the moment when, the liquid and vapor having assumed the common composition ξ, the boiling-point will have reached the value θ? Our principle, according to which the boiling-point should constantly rise during the distillation, is no longer applicable; on the contrary, as the vapor within the still passes over, a vapor of the same composition may replace it without either the composition of the liquid or the value of the boiling-point changing. *When the composition of a liquid has reached the value ξ and the boiling-point the value θ, there is established a permanent régime of distillation in which the boiling-point keeps the value θ, while the vapor which distils and the liquid contained in the still keep a constant composition ξ.*

This régime of distillation is stable. Thus, if any cause whatever disturbs it in one direction or the other, the action of the distillation itself will tend, as we have seen, to reestablish it. A mixture which, as the mixture of formic acid and water, comes under our *third case*, may have a *permanent régime* of distillation; if the liquid mixture has the composition ξ which accords with the indifferent point, the vapor has the same composition; the distillation may then be produced without change of composition of the liquid or of the vapor, consequently without variation of the boiling-point, which remains equal to θ; but this permanent régime is *unstable;* if any circumstance whatever disturbs it, however slightly, the distillation will deviate more and more from this régime. In fact, reasoning as we have done for the first case, we may establish the following proposition:

If the value of X which denotes the initial composition of the liquid is less than ξ, the distillation has for effect to increase constantly the proportion of the fluid 1 in the vapor and in the liquid, both of which tend to contain this substance alone; if, on the contrary, the value of X which indicates the initial composition of the liquid is greater than ξ, the distillation has for effect to increase constantly the proportion of the fluid 2 in the vapor and in the liquid, both of which tend to contain this substance only.

198. These mixtures are not definite compounds. Researches of Roscoe and Dittmar.—Let us return to the permanent and stable régime which characterizes our second case.

Acted upon by a constant pressure π, the liquid of composition ξ distils at a constant temperature θ, furnishing a vapor which has an identical composition; it behaves, therefore, like a liquid body of definite composition which vaporizes and whose boiling-point is θ under the pressure π. Nevertheless, if one were tempted to take it for a definite compound, there is a property which would permit us to distinguish it from one. The composition of a definite compound does not change with the pressure to which it is submitted; on the contrary, if, instead of distilling one liquid mixture under a pressure π, we distil it under a different pressure, π', the liquid mixture capable of passing entirely over in the distillation without change of composition and without variation of boiling-point will correspond to a value ξ' of X which will not, in general, be equal to ξ.

A solution of hydrochloric acid, subjected to atmospheric pressure, begins to boil at a temperature which rises gradually by distillation to 110° C.; it then distils in constant proportion of water and hydrochloric acid; this mixture had been regarded by Bineau as a definite chemical compound represented by the formula $HCl \cdot 8H_2O$. Roscoe and Dittmar[1] did not consider the matter in this way, and they showed its inaccuracy by causing hydrochloric acid solution to boil under various pressures. The boiling attains in every case a permanent régime; but instead of reproducing constantly the supposed hydrate $HCl \cdot 8H_2O$,

---

[1] Roscoe and Dittmar, *Liebig's Annalen*, v. 113, p. 327, 1859; *Annales de Chimie et de Physique*, 3d S., v. 58, p. 492, 1860.

the mixture which distilled during the permanent régime had a composition variable with the actual pressure, and the richer in acid as the pressure rose.

One may see this from the following table, where π denotes the pressure in centimetres of mercury, and ξ the number of grammes of hydrochloric acid contained in 1 gramme of the solution which has a constant boiling-point under the pressure π.

π	ξ	π	ξ	π	ξ
5	0.232	80	0.202	170	0.188
10	0.229	90	0.199	180	0.187
20	0.223	100	0.197	190	0.186
30	0.218	110	0.195	200	0.185
40	0.214	120	0.194	210	0.184
50	0.211	130	0.193	220	0.183
60	0.207	140	0.191	230	0.182
70	0.204	150	0.190	240	0.181
76	0.2024	160	0.189	250	0.180

When one distils any aqueous solution whatever of nitric acid under atmospheric pressure, there always comes a moment when the temperature becomes stationary at 123° *C.* at which the mixture passes over unchanged by the distillation; 1 gramme of this mixture contains 0.68 grammes of acid, NO_3H; if a mixture richer in acid is distilled, very concentrated acid first passes over, a portion of this acid even decomposes, and when the temperature reaches 123° the liquid which passes over and that remaining have the same concentration; when a weaker acid is distilled, water with more or less acid passes over until the temperature reaches 123°.

This mixture, which has a fixed boiling-point and passes over as a whole by distillation, is not a definite hydrate; H. Roscoe[1] has shown that its composition varies with the pressure under which the distillation takes place; 1 gramme of this mixture contains 0.68 gr. of the acid NO_3H if the distillation takes place under atmospheric pressure; if the distillation takes place under the pressure of 7 centimetres of mercury, this gramme of mixture

[1] Roscoe, *Liebig's Annalen*, v. 116, p. 203, 1860.

contains but 0.667 gr. of acid; it contains 0.686 gr. if the distillation is under a pressure of 122 cm. of mercury.

199. Application of the second theorem of Gibbs and of Konovalow to mixtures of volatile liquids.—The study of the tensions of saturated vapors of a mixture whose composition X is varied at a constant temperature T offers remarkable similarities in all points to those we have made on the subject of boiling-points under a given pressure.

At the temperature T, let P_1 and P_2 be the pressure of saturated vapors of the liquids 1 and 2 taken in the state of purity; suppose also that the liquid 1 be more volatile than the liquid 2, so that P_1 exceeds P_2.

The mixture of composition X has, at the temperature considered, a tension of saturated vapor Π. Take a point N (Fig. 64) of abscissa X and ordinate Π; when X varies from 0 to 1 the point N will describe a curve D joining the point N_1, of coordinates O, P_1, to the point N_2, of coordinates 1, P_2.

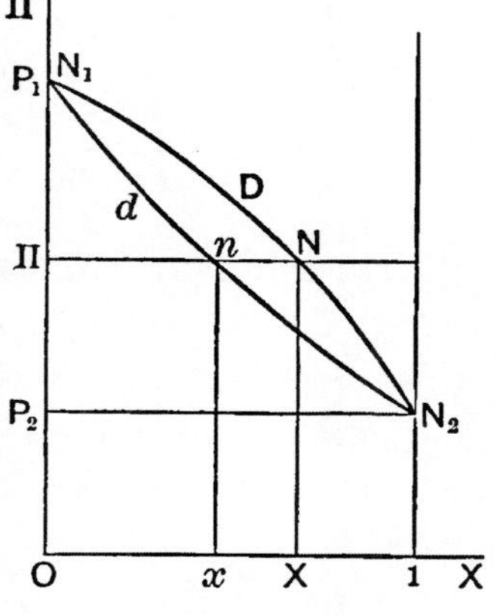

Fig. 64.

The liquid mixture whose composition is X and whose vapor tension Π is in the presence of a saturated vapor of composition x; the point n, of abscissa x and ordinate Π, associated with the point N of the same ordinate, concludes the representation of an equilibrium state of the system. When X varies from 0 to 1, x varies likewise from 0 to 1 and the point n describes a curve d joining N_1 to N_2.

There are three principal cases to distinguish:

First Case: The curve D descends constantly from the point N_1 to the point N_2.—*In this case the curve d descends likewise constantly from the point N_1 to the point N_2; the curve d is, throughout its length, below the curve D.*

Fig. 64 shows this case.

Second Case: Between the points N_1 and N_2 (Fig. 65) the curve D has a point I, of ordnate P smaller than all the others.—*According to the second theorem of Gibbs and of Konovalow, this point is an indifferent point when the liquid and the saturated* *vapor have the same composition $X=x=\xi$, so that the point I is also on the curve d; for this curve it is an ordinary point of ordinate smaller than all the others; outside of the points N_1, I, N_2 the curve d is entirely below the curve* D.

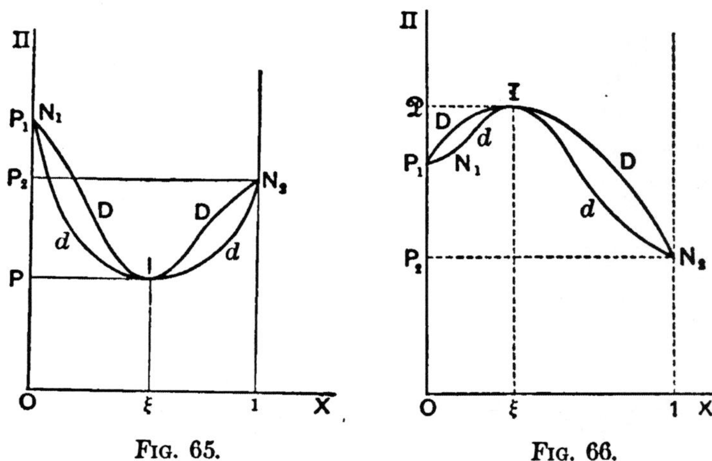

FIG. 65. FIG. 66.

THIRD CASE: BETWEEN THE POINTS N_1 AND N_2 (FIG. 66) THE CURVE D HAS A POINT *I*, OF ORDINATE 𝔓 GREATER THAN ALL THE OTHERS.—*The curve d also passes through this point I, which is an indifferent point ($X=x=\xi$), and has there an ordinate greater than all the others. Outside of the points* N_1 *I*, N_2 *the curve d is constantly below the curve* D.

200. Distillation of a mixture of two liquids at constant temperature.—This conclusion, furnished by thermodynamics, will permit us to discuss the phenomena of distillation which are produced when, without changing the temperature, the vapor emitted by the mixed liquid is constantly aspirated, provided that we make use of the following proposition, furnished likewise by thermodynamics:

When the liquid and the vapor which it emits have not the same composition, distillation may not be produced at a constant temperature, unless the pressure of the vapor constantly diminishes.

By the same sort of reasoning we used concerning distillation under constant pressure, we may without difficulty establish the following results:

FIRST CASE.—*During distillation the vapor tension constantly* *decreases and approaches the tension of saturated vapor P_2 of the substance 2 taken in a state of purity; the proportion of the substance 1 constantly diminishes, both in the liquid and in the vapor form; both tend to be no longer formed otherwise than of substance 2.*

SECOND CASE.—*Whether the value of X which represents the initial composition of the liquid be less than ξ or higher than ξ, the vapor tension constantly decreases and approaches P; the composition X of the liquid and the composition x of the vapor both vary in the same way and approach ξ. When the vapor tension reaches the value P there is established a permanent régime of distillation; the vapor tension changes no more; the vapor distilled and the liquid not distilled conserve a constant composition ξ. This permanent régime is stable.*

THIRD CASE.—*The system may possess a permanent régime of distillation, under the invariable pressure $\mathfrak{P}$, the liquid and the vapor having the same constant composition ξ; but this régime is unstable.*

If the value of X which represents the initial composition of the liquid is less than ξ, while the pressure diminishes and approaches the tension of saturated vapor P_1 of the fluid 1, the composition of the liquid and the composition of the vapor constantly vary in the same direction; these two fluids tend to be formed of the substance 1 alone.

If the value of X which represents the initial composition of the liquid is higher than ξ, while the pressure diminishes and approaches the tension of saturated vapor P_2 of the substance 2, the composition of the liquid and the composition of the vapor constantly change in the same direction; these two fluids tend to be formed of the same substance 2 alone.

In our second case, the mixture of concentration ξ which has, at the temperature considered, a definite vapor tension and passes over unaltered by distillation, may be confused with a definite compound; this confusion may be readily obviated if it is noticed that the composition of the mixture having these properties depends on the temperature.

The evaporation, at ordinary temperature, of an aqueous solution of hydrochloric acid always furnishes, after a certain time, a mixture of constant vapor tension and of constant composition, which Bineau had considered as a definite hydrate, represented by the formula $HCl \cdot 6H_2O$; Roscoe and Dittmar[1] have shown that the composition of this mixture varies with the temperature at which evaporation takes place.

201. Relation between distillation at constant pressure and distillation at constant temperature.—Between the permanent régime which may be established when a mixture is distilled at constant pressure and the permanent régime which may be established when evaporation takes place at constant temperature, there exists a relation.

Suppose that, under the pressure P, we may observe a condition of indifferent equilibrium where the liquid mixture and the saturated vapor have the same composition ξ; the boiling-point θ of the mixture of composition ξ is maximum or minimum among the boiling-points which the liquid mixture may have under constant pressure P.

At the constant temperature θ the liquid mixture of composition ξ will give off a saturated vapor of same composition with which it will be in indifferent equilibrium; the tension of the saturated vapor of this mixture of composition ξ will have a value P; this value must be a maximum or minimum among the tensions of saturated vapors which the liquid mixture may have at the constant temperature θ.

The following propositions may be demonstrated:

If the temperature θ is a *maximum* among the boiling-points which the liquid mixture may have when its composition is varied by keeping constant the pressure P, the pressure P will be a *minimum* among the tensions of saturated vapor from the liquid mixture when its composition is varied, leaving constant the temperature θ.

If the temperature θ is a *minimum* among the boiling-points possessed by the liquid mixture when its composition is varied keeping the pressure P constant, the pressure P will be a maximum among the tensions of saturated vapor from the liquid mixture when its composition is varied, leaving the temperature θ constant.

---

[1] Roscoe and Dittmar, *Liebig's Annalen*, v. 112, p. 327, 1859; *Annales de Chimie et de Physique*, 3d S., v. 58, p. 492, 1860.

The first of these two propositions is evidently the equivalent of the following:

Suppose that a liquid mixture is being distilled under constant pressure P; there comes a moment when the distillation lets pass over a vapor of constant composition ξ, *the boiling-point becoming fixed at the value* θ, *constant from this moment; conversely, if this mixture is evaporated at the temperature* θ, *the tension of the saturated vapor will finally become fixed at the value P, and the evaporation will furnish a vapor of constant composition, still equal to* ξ.

This law has been experimentally established by Roscoe and Dittmar[1] by studying mixtures of water and hydrochloric acid.

[1] ROSCOE and DITTMAR, *loc. cit.*

CHAPTER XII.

BIVARIANT SYSTEMS (*Continued*). TRANSITION AND EUTEXIA.

202. Common point to the solubility curves of two hydrates. Three cases to distinguish.—Suppose that a solution of a salt in water may give two different solid precipitates, both of definite composition: for example, a salt, anhydrous or hydrated, and ice, or an anhydrous salt and a hydrated salt, or again two different hydrated salts; let us operate under a pressure given once for all, and let us ask ourselves if, under this pressure, we may observe a system in equilibrium containing at once the solution and the two precipitates.

When the two precipitates coexist in contact with the solution, the system, always formed of two independent components, is divided into three phases; it is no longer bivariant, but monovariant; in general it cannot be in equilibrium under the pressure considered, except at a particular temperature which we shall denote by θ.

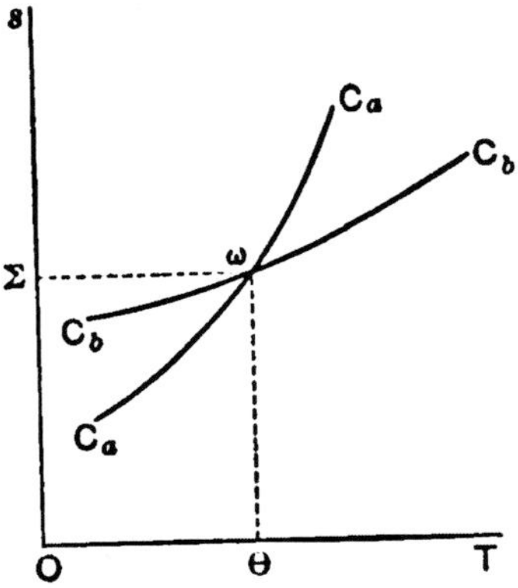

It is not difficult to define in an exact manner the temperature θ.

Fig. 67.

Let a and b be our two precipitates. Under the pressure considered, the precipitate a has a solubility curve, the curve C_a (Fig. 67); in order that the solution be in equilibrium in contact with the substance a, it is necessary and sufficient that the representative point, which has the temperature for abscissa and the concentration of the solution for ordinate, be on the line C_a. The precipitate b likewise has a solubility-curve C_b; in order that the solution remain in equilibrium in contact with the precipitate b, it is necessary and sufficient that the representative point be on the curve C_b.

It is therefore clear *that for a solution to remain in equilibrium, under the pressure considered, in contact with the two precipitates a and b, it is necessary and sufficient that the temperature has the value* θ, *and the concentration of the solution the value* Σ, θ *and* Σ *being the coordinates of the point* ω *common to the two solubility curves* C_a *and* C_b.

When the representative point is elsewhere than at ω, it is impossible for our bivariant system to remain in equilibrium; it must be transformed until the complete disappearance of one of its phases takes place. What laws govern these transformations? To determine these laws, we must distinguish three cases, as follows:

First Case.—*The solution of concentration* Σ *contains more water than either of the two precipitates a and b.*

This case may also be thus defined:

The two curves C_a *and* C_b *which intersect at the point* ω *are the lower branches of the solubility curves of the two precipitates a and b.*

Second Case.—*The solution of concentration* Σ *contains less water than either of the two precipitates a and b.*

We may state this case in the following way:

The two curves C_a, C_b, *intersecting at the point* ω, *are the upper branches of the solubility curves of the two precipitates a and b.*

Third Case.—*The solution of concentration* Σ *contains less water than the precipitate a and more than the precipitate b.*

Otherwise stated this case becomes:

The part of the curve C_a *which passes through the point* ω *is the upper branch of the solubility curve of the precipitate a; the part of the curve* C_b *passing through the point* ω *is the lower branch of the solubility curve of the precipitate b.*

This third case offers peculiarities which sharply distinguish it from the first two; on the contrary, the properties of the first two are so analogous that it will be sufficient for us to study one of them, the first for example.

203. Transition-point.—Of the two curves C_a, C_b, which intersect at the point ω, there is one of them, followed from left to right, which rises more sharply or descends less sharply than the other; suppose this to be the curve C_a. Then the branch C_a of this curve (Fig. 68) which corresponds to temperatures less than θ is below the corresponding branch C_b' of the curve C_b; on the contrary, the branch C_a' of the curve C_a which relates to temperatures above θ is above the corresponding branch C_b of the curve C_b.

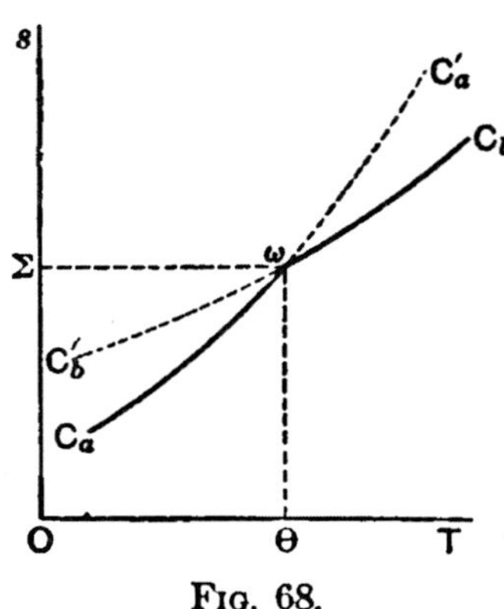

FIG. 68.

At a temperature less than θ can the solution remain in equilibrium in contact with the precipitate b? In order that the solution remain in equilibrium in contact with the precipitate b, it is necessary in the first place that it may neither dissolve nor abandon a certain mass of this precipitate, which requires that the representative point lie on the line C_b; but this is not sufficient; it is further necessary that the solution may not give birth to a certain mass of the precipitate a, which requires that the representative point lie below the line C_a, since this line is the solubility curve of a substance *less rich in water* than the solution. It is therefore evident that *at temperatures below θ the precipitate b cannot remain in equilibrium in contact with the solution.*

It may be shown similarly that *at temperatures above θ the precipitate a cannot remain in equilibrium in contact with the solution.*

As to the mixture of the two solid precipitates, exempt from liquid solution, it remains necessarily in equilibrium at temperatures near to θ, whether they are above or below θ. If, therefore, this mixture of two solid substances, each of which is less rich in water than the solution of concentration Σ, underwent aqueous fusion, the solution produced would have a concentration greater than Σ; the temperature being close to θ, the representative point of this solution would lie above the two curves C_a and C_b, so that the solution, supersaturated by each of the substances a and b, could not remain in equilibrium.

204. Various examples: sodium sulphate.—There are numerous opportunities in practice to apply these principles.

The case known for the longest time is that furnished by sodium sulphate, carefully studied by Lœwel.[1] Hydrated sodium sulphate, $Na_2SO_4 \cdot 10H_2O$, dissolves with absorption of heat, so that the branch of the solubility curve along which the solution is less concentrated than the hydrate (the only one known) rises from left to right following $C_a\omega$ (Fig. 69). Anhydrous sodium sulphate, Na_2SO_4, dissolves with liberation of heat, so that the solubility curve for this substance descends from left to right following ωC_b. These two curves intersect in a point ω, whose abscissa corresponds sensibly to the temperature +33° C. From what precedes, at temperatures less than +33° C. the only true equilibrium which can be observed is the equilibrium between the solution and the anhydrous sodium sulphate; this is, in fact, what experiment shows. By means of a phenomenon of false equilibrium one may observe systems in which a solution, supersaturated with respect to hydrated sodium sulphate, is in equilibrium in the presence of anhydrous sodium sulphate; the states of equilibrium thus obtained are represented by various points of the line $C_b'\omega$.

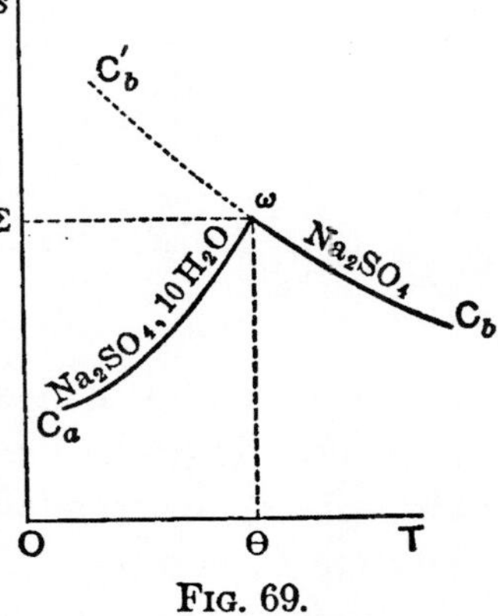

FIG. 69.

205. Thorium sulphate.—Roozboom has called attention to a case[2] analogous to that of sodium sulphate, but where the phenomena of supersaturation are produced with an exceptional facility; this case is that of thorium sulphate.

Thorium sulphate with 9 molecules of water, $Th(SO_4)_2 \cdot 9H_2O$, dissolves with absorption of heat and corresponds to a solubility curve $C_a\omega C_a'$ (Fig. 70) rising from left to right; on the contrary, thorium sulphate with 4 molecules of water, $Th(SO_4)_2 \cdot 4H_2O$,

[1] LŒWEL, *Annales de Chimie et de Physique*, 3d S., v. 29, p. 62, 1850.

[2] ROOZBOOM, *Archives néerlandaises des Sciences exactes et naturelles*, v. 24; *Zeitschrift für physikalische Chemie*, v. 5, p. 198, 1890.

dissolves with liberation of heat and corresponds to a solubility curve $C_b'\omega C_b$ which descends from left to right; these two curves intersect at a transition-point ω whose abscissa corresponds to the temperature $+43°$ C.

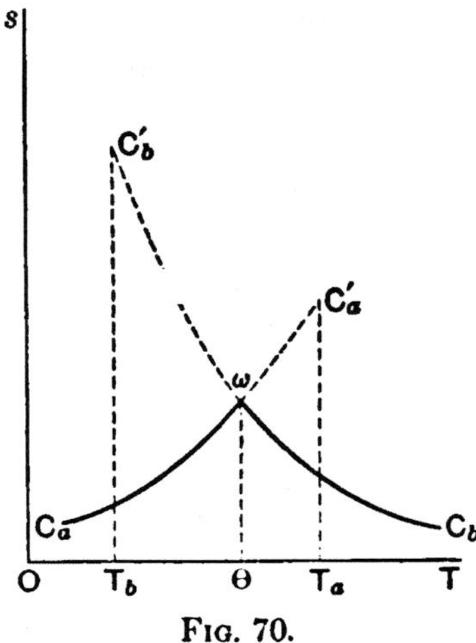

Fig. 70.

The two branches $C_a\omega$, $C_b\omega$ alone correspond to states of true equilibrium; nevertheless the line $C_a\omega$ might be prolonged beyond the point ω, to the point C_a', whose abscissa T_a corresponds to the temperature $+55°$ $C.$, but the segment $\omega C_a'$ represents solutions supersaturated with respect to the hydrate with 4 molecules of water; and the line $C_b\omega$ might have been extended, beyond the point ω, to the point C_b', whose abscissa T_b corresponds to the temperature $+17°$ C., while the segment $C_b'\omega$ represents solutions supersaturated with respect to the hydrate of 9 molecules of water.

When we meet thus a point common to two solubility curves of two hydrates of the same salt, and if at this point the solution is richer in water than either of the two hydrates or less rich in water than either of the two hydrates, we shall say that this point is a *transition-point.*

206. Eutectic point. — Things take place quite differently for the last of the three cases which we have enumerated; in this case the saturation curve C_aC_a' of the precipitate a and the saturation curve C_bC_b' of the precipitate b intersect at the point ω (Fig. 71) of abscissa θ and ordinate Σ; the solution of concentration Σ is less rich in water than the precipitate a and richer in water than the precipitate b. Otherwise expressed, the branch of the curve C_aC_a'

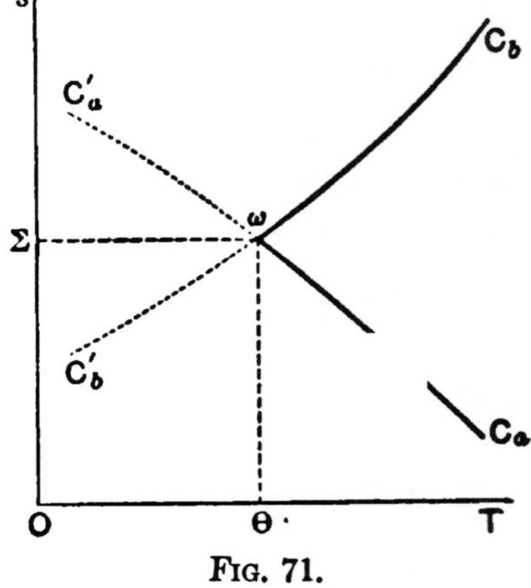

Fig. 71.

which passes through the point ω belongs to the upper branch of the solubility curve of the hydrate a; on the contrary, the branch of the curve C_bC_b' which passes through the point ω belongs to the lower branch of the solubility curve of the hydrate b.

To fix our ideas, we shall suppose that the two precipitates dissolve with *absorption* of heat; it follows from what we have seen in the preceding chapter (Art. 188) that the curve $C_a'C_a$ descends from left to right, while the curve $C_b'C_b$ rises from left to right.

Outside of the point ω, we cannot observe in equilibrium the system divided into three phases; but it may be that we may observe, in equilibrium, a system divided into two phases, which may occur in three ways:

1°. The system may be formed by the substance a in contact with a solution;

2°. The system may consist of the substance b with a solution;

3°. The system may be composed of the substances a and b in the absence of any solution.

In order that a system formed of the precipitate a in contact with a solution remain in equilibrium, it is necessary, in the first place, that the solution be saturated with the substance a, or, in other terms, that the representative point lie on the curve C_aC_a'; but this is not sufficient; it is further necessary that the solution may not give birth to the precipitate b, that it be not saturated with respect to this precipitate, consequently that the representative point does not lie above the curve C_bC_b'; whence the following conclusion:

In order that a system enclosing the precipitate a and a solution may remain in equilibrium, it is necessary and sufficient that the representative point lie on the branch ωC_a of the curve C_aC_a', starting from the point ω and extending to the right of this point.

In order that a system formed of the precipitate b in contact with a solution be in equilibrium, it is necessary in the first place that the solution be saturated with the substance b, that is to say, that the representative point lie on the line C_bC_b'; and furthermore it is also requisite that the precipitate a cannot be generated in the solution, and, as the saturated solutions of the substance a are represented by the various points of the plane located below the line C_aC_a', it is necessary that the representative point considered does not lie below this line C_aC_a'; whence the following conclusion:

For a system which contains the precipitate b and a solution to remain in equilibrium, it is necessary and sufficient that the representative point be on the branch ωC_b *of the line* C_bC_b', *starting from the point* ω *and extending to the right of this point.*

Finally, let us consider a system which contains the two solid precipitates a and b; is it going to remain in equilibrium or undergo aqueous fusion?

Imagine that a part of the two hydrates undergoes aqueous fusion and gives rise to a solution; this solution cannot be richer in water than the hydrate a, and consequently than the two hydrates; it cannot, either, be less rich in water than the hydrate b, and therefore than the two hydrates; it has necessarily a composition intermediate between that of the hydrate a and that of the hydrate b.

Suppose the temperature less than θ; the points situated below the line $C_b'\omega$ represent solutions supersaturated with respect to the precipitate a; the points situated between the lines $C_b'\omega$ and $C_a'\omega$ represent solutions supersaturated with respect to the two precipitates a and b; the points situated above the curve $C_a'\omega$ represent solutions supersaturated with respect to the precipitate b; whatever the composition of the solution formed, it is supersaturated with respect to at least one of the two precipitates a and b, so that it will give back a solid precipitate, and that until the solution has entirely disappeared; aqueous fusion is therefore impossible at temperatures less than θ.

At temperatures less than θ *a solid mixture of the two precipitates a and b cannot undergo aqueous fusion; a solution of any composition solidifies and forms a mixture of the two precipitates.*

Consider next a temperature higher than θ, can the two solid precipitates a and b remain in equilibrium at this temperature without undergoing aqueous fusion? Such an equilibrium, supposing it possible to realize, would be unstable. Thus imagine that a very small part of the mixture of the two solids undergoes aqueous fusion and gives a drop of solution.

If the point representing this solution is above ωC_a, the solution may dissolve the precipitate a; if it is below ωC_b, the solution may dissolve the precipitate b; therefore whatever be the position of the representative point, there is at least one of the two precipitates that the solution may dissolve. Hence, from the instant a drop of solution is produced in the system at a temperature higher than θ, equilibrium cannot be reestablished in the system unless at least one of the two precipitates has passed entirely into the solution.

At temperatures above θ a system containing the two solid substances a and b, in any proportion, passes into the liquid state until at least one of the two solid bodies has disappeared.

The temperature θ is the fusing-point of a system which contains the two solids a and b at once.

207. Formation of the eutectic mixture.—We shall show some new properties of the temperature θ and of the concentration Σ by examining the following question:

At a temperature above θ a solution is taken which is saturated neither with the substance a nor with the substance b, which nevertheless is in equilibrium; the temperature is gradually lowered; what are the precipitates from the solution?

There are two cases to distinguish according as the initial concentration of the solution is greater or less than Σ.

FIRST CASE.—*The initial concentration s is greater than Σ.*

The representative point for the initial state of the solution is a point M (Fig. 72) higher than the point ω.

When the temperature is lowered, starting from its initial value T, the representative point remains at first in the region included between the lines ωC_a and ωC_b; the solution, not being saturated either with the substance a or with the substance b, gives no precipitate and its concentration remains invariable; the representative point describes a parallel Mm to the straight line OT.

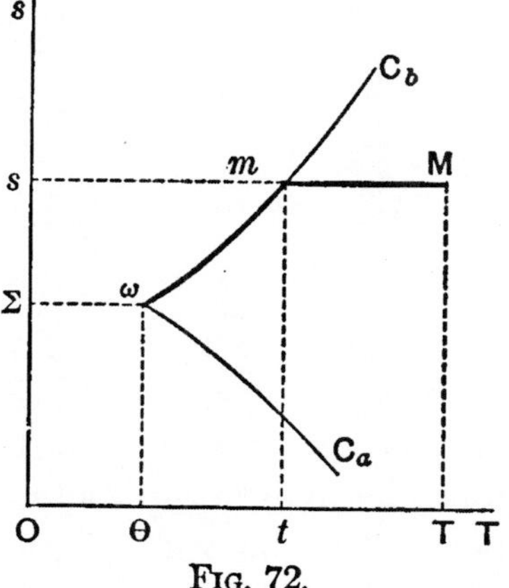

FIG. 72.

When the temperature is lowered to a certain value t, the representative point is situated on the curve ωC_b at m; the solution is then saturated with the substance b.

The temperature falling below t, the solution deposits a certain amount of the substance b, so as to remain saturated with this substance; the representative point describes the part $m\omega$ of the line ωC_b.

At the instant when the temperature, always falling, attains the value θ, the representative point is at ω and the concentration has the value Σ.

If we lower the temperature by a small fraction below θ, the solution solidifies; the solid deposit which it furnishes is not homogeneous; it is formed by a juxtaposition of particles of solids a and b; but its mean composition is well determined; it is the same as the composition of the solution of concentration Σ which has furnished it.

SECOND CASE.—*The initial composition of the solution is less than* Σ.

The representative point for the initial state of the solution is a point M (Fig. 73) lower than the point ω.

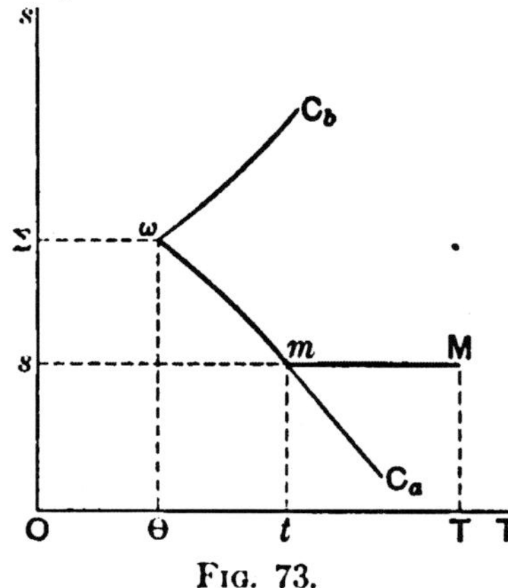

Fig. 73.

When the temperature, starting from the initial value T, commences to fall, the representative point remains, at first, included between the lines ωC_a and ωC_b; the solution, being saturated neither with the substance a nor with b, gives no precipitate; its concentration remains constant and the representative point describes a line Mm parallel to OT.

This parallel meets in a point m, of abscissa t, the line ωC_a; at the moment when the temperature reaches the value t the solution becomes saturated with the substance a. If the temperature is brought lower than t, the solution gives up some of the substance a in the solid state so as to remain saturated with this substance; the representative point describes the segment $m\omega$ of the line ωC_a.

The instant the temperature attains the value θ the representative point is at ω and the concentration of the solution has the value Σ.

If the temperature falls below θ, the solution solidifies; it gives a mixed precipitate, formed of particles of the substance a and of b, whose average composition is well determined; this composition is that of the solution of concentration Σ.

If a solution of any initial composition whatever is cooled, it lets first precipitate either the substance a or the substance b, in the pure state; but at the instant the temperature, in falling, passes through θ, the solution solidifies; the solid obtained is not homogeneous; it is formed by the juxtaposition of particles of the substances a and b; but its mean composition is perfectly determined, being identical with that of a solution of concentration Σ.

Guthrie has given the name *eutectic mixture* to the solid magma obtained in these conditions; the point ω is a *eutectic point.*

208. Particular case: ice and anhydrous salt.—The phenomena whose laws we have just sketched were first studied by taking for substance a ice and for substance b a salt, anhydrous or hydrated; the cooling mixtures that are obtained by mixing ice with a salt, such as sea-salt and saltpetre, had already been noticed by physicists near the end of the eighteenth century, and they had learned some of the properties of the point ω; it was known that the salt could no longer make the ice melt when the temperature is lower than θ, *temperature at which ice is formed in the midst of a solution saturated with the salt,* or, in other terms, *abscissa of the point of intersection of the curve ωC_a of the freezing-points and of the solubility curve ωC_b for the salt considered.* The temperature θ is thus the lower limit of the temperatures which may be produced by means of a cooling mixture formed of ice mixed with the salt considered.

Salt.	θ	Σ
Potassium sulphate.......	$-1°.9$	0.10
Potassium nitrate........	$-2\ .8$	0.13
Potassium chloride.......	$-10\ .9$	0.30
Ammonium nitrate.......	$-16\ .7$	0.45
Sodium chloride.........	$-21\ .3$	0.33

De Coppet,[1] first, and then Guthrie and other observers undertook to determine with accuracy the temperature θ for a number

[1] De Coppet, *Bulletin de la Société Vaudoise,* 2d S., v. II, p. 1, 1871.

of salts; some of these temperatures are given in the table; there is joined to them the value of the ordinate Σ of the point ω.

209. Non-existence of the cryohydrates.—Brought to a temperature below θ, the solution solidifies; the magma obtained is not a definite compound; particles of ice are mixed with salt crystals; but its composition is quite definite; it is identical with that of a solution of concentration Σ.

Guthrie,[1] who had observed with much care the formation of this solid of invariable composition, had at first regarded it as a definite hydrate for which θ would be the aqueous fusing-point; to this hydrate he gave the name *cryohydrate*.

One might decide between this opinion and the preceding theory by repeating Guthrie's experiments at a pressure quite different from atmospheric pressure; the composition of the solid furnished by the solution at the instant of solidification, in Guthrie's opinion, should be independent of the pressure exerted upon the system; in the opinion advanced here, on the contrary, the composition would in general depend upon the pressure.

In default of these experiments which would be conclusive, but which the extreme smallness of the variation to be recognized would render extremely difficult, other arguments may be invoked against the existence of the cryohydrates.

In the first place, no simple formula can be found for these substances; for example, here are some of the formulæ proposed by Guthrie:

$$Na_2SO_4 + 166H_2O;$$
$$KClO_3 + 222H_2O;$$
$$Ba(NO_3)_2 + 259H_2O;$$
$$AlNH_4(SO_4)_2 + 261H_2O.$$

In the second place, a microscopic examination of the pretended cryohydrates, either in natural light[2] if the salt is colored, or in polarized light if the salt is colorless but non-isotropic, shows that these substances are in no way homogeneous and that they are formed of crystals of salt mixed with ice crystals.

---

[1] GUTHRIE, *Philosophical Magazine*, 4th S., v. 49, pp. 206 and 266, 1875; 5th S., v. 1, pp. 49, 354, and 446, 1876; v. 2, p. 211, 1876.

[2] PONSOT, *Annales de Chimie et de Physique*, 7th S., v. 10, p. 79, 1897.

These arguments have led Guthrie [1] to renounce the hypothesis of definite cryohydrates and to propose, to denote these substances, the name of *eutectic mixtures,* which is generally employed to-day.

210. Eutectic point between ferric chloride hydrates. Investigations of Bakhuis Roozboom.—Of all the solid hydrates which a salt solution may form, ice is always the most hydrated and the anhydrous salt the least hydrated; always less rich in water than the first, the solution is always richer in water than the second; the solubility curve of the first is reduced to its upper branch, the solubility curve of the other to its lower branch; when these two branches meet their point of intersection is necessarily a eutectic point.

But these eutectic points are not the only ones which may be met with; when the upper branch of the solubility curve of a hydrate meets the lower branch of the solubility curve of another hydrate less rich in water, the point of meeting is a eutectic point.

The most remarkable example of such eutectic points has been furnished by Roozboom [2] from the study of the solubility of ferric chloride.

If we count the hydrate of zero concentration—ice—and the hydrate of infinite concentration—anhydrous ferric chloride—six different hydrates of ferric chloride may be obtained which are, in the order of increasing concentration:

H_2O (ice);
$Fe_2Cl_6 \cdot 12H_2O$;
$Fe_2Cl_6 \cdot 7H_2O$;
$Fe_2Cl_6 \cdot 5H_2O$;
$Fe_2Cl_6 \cdot 4H_2O$;
Fe_2Cl_6.

The first and last of these solids excepted, each of these substances corresponds to a solubility curve consisting of two branches meeting in an indifferent point; in the preceding chapter (p. 225) we have indicated the temperatures to which these four indifferent points correspond.

[1] GUTHRIE, *Philosophical Magazine,* 5th S., v. 17, p. 462, 1884.

[2] ROOZBOOM, *Archives néerlandaises des Sciences exactes et naturelles,* v. 27, 1892; *Zeitschrift für physikalische Chemie,* v. 10, p. 477, 1892.

Fig. 74 represents these various solubility curves; temperatures in centigrade have been taken as abscissæ; as ordinate the value N_2 of the number of molecules of anhydrous ferric chloride contained in 100 molecules of water in the solution has been taken.

It is seen from this figure that the upper branch of the solubility curve of each hydrate meets the lower branch of the solubility curve of the hydrate following immediately in the order of increasing concentrations; further, thanks to the phenomenon of

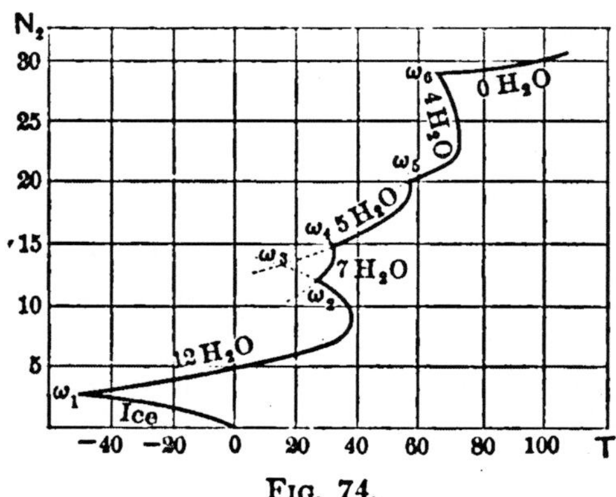

Fig. 74.

supersaturation, by avoiding the introduction of crystalline particles of the hydrate $Fe_2Cl_6 \cdot 7H_2O$, one may observe the intersection of the upper branch of the solubility curve for the hydrate with 12 molecules of water and the lower branch of the solubility curve for the hydrate with 5 molecules of water. One may, therefore, from the study of systems formed of ferric chloride and water, recognize the existence of six eutectic points. Among these points there are five whose properties are completely established by the researches of Roozboom; these properties are resumed in the following table:

Eutectic Point.	Hydrates between which Eutexia is Produced.		Temperature of Eutexia.	Value of N_2 for the Eutectic Mixture.
w_1	Ice	$Fe_2Cl_6 \cdot 12H_2O$	−55° C.	2.75
w_2	$Fe_2Cl_6 \cdot 12H_2O$	$Fe_2Cl_6 \cdot 7H_2O$	27.4	8.23
w_3	$Fe_2Cl_6 \cdot 12H_2O$	$Fe_2Cl_6 \cdot 5H_2O$	not studied	
w_4	$Fe_2Cl_6 \cdot 7H_2O$	$Fe_2Cl_6 \cdot 5H_2O$	30	6.66
w_5	$Fe_2Cl_6 \cdot 5H_2O$	$Fe_2Cl_6 \cdot 4H_2O$	55	20.32
w_6	$Fe_2Cl_6 \cdot 4H_2O$	Fe_2Cl_6	66	29.20

In the example we have just cited, eutectic points alone are met with; we shall next indicate some remarkable examples where both eutectic points and transition-points are encountered.

210a. Hydrates of perchloric acid. Van Wyk's investigations.—An example very similar to the preceding is furnished by the systems composed of perchloric acid and water.[1]

In the presence of liquid mixtures of perchloric acid and water six kinds of crystals may be observed, whose respective compositions are:

$$HClO_4\cdot H_2O;$$
$$HClO_4\cdot 2H_2O;$$
$$HClO_4\cdot 3H_2O;$$
$$HClO_4\cdot 4H_2O;$$
$$HClO_4\cdot 6H_2O;$$
$$H_2O.$$

If centigrade temperatures are taken for abscissæ, and for ordinates the ratio of the number N_2 of molecules of perchloric acid, $HClO_4$, to the sum (N_1+N_2) of the numbers of molecules of water and perchloric acid within the liquid mixture, the solubility curves have the following appearance (Fig. A).

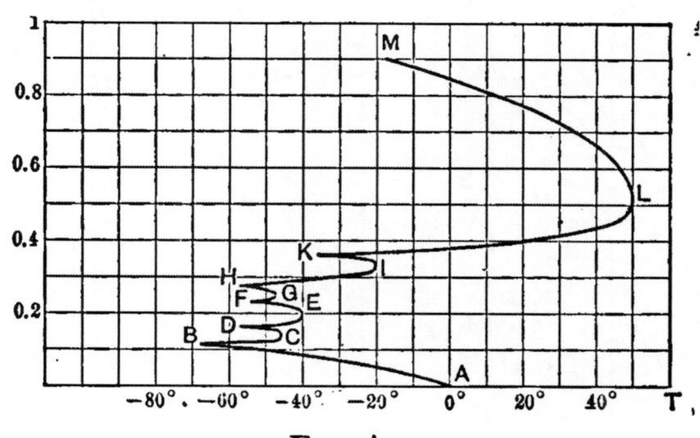

FIG. A.

Except the fusion curve of ice, which rises from right to left, each of the solubility curves is formed of two branches meeting in an indifferent point; one observes, therefore, in all, five indifferent points: *C*, *E*, *G*, *I*, L. The upper branch of the solubility curve of each hydrate meets the lower branch of the solubility curve of the hydrate which immediately precedes it in the above

[1] VAN WYK, *Zeitschrift für anorganische Chemie*, v. 32, p. 115, 1902.

table; there are thus obtained five eutectic points: *B, D, F, H, K.*

211. Researches of Van't Hoff and Meyerhoffer on magnesium chloride.—Van't Hoff and Meyerhoffer[1] have made a thorough study of the solubility of several hydrates of chlorine and of magnesium. Including ice, the hydrates furnished by magnesium chloride are six in number:

Ice;
$MgCl_2 \cdot 12H_2O$;
$MgCl_2 \cdot 8H_2O$;
$MgCl_2 \cdot 6H_2O$;
$MgCl_2 \cdot 4H_2O$;
$MgCl_2 \cdot 2H_2O$.

Further, the hydrate $MgCl_2 \cdot 8H_2O$ exists in two different forms which we shall denote by the symbols α and β.

The solubility curves of these various hydrates are arranged as shown in Fig. 75, the ordinates representing the number N_2 of molecules contained in 100 molecules of water.

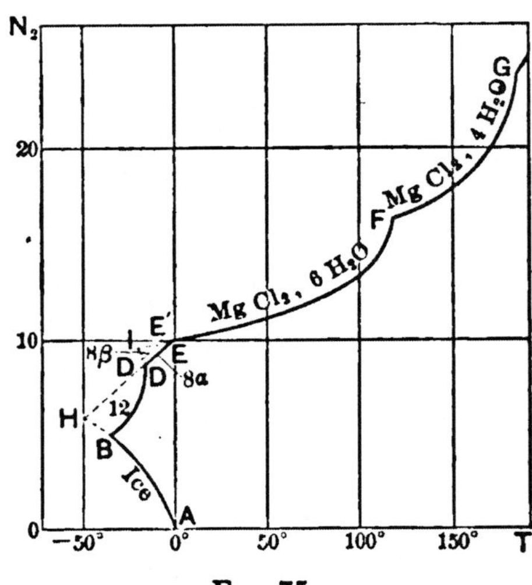

FIG. 75.

In their investigations, Van't Hoff and Meyerhoffer, instead of the concentration *s* of the solution, make use of the number *y*

---

[1] VAN'T HOFF and MEYERHOFFER, *Sitzungsberichte der Berliner Akad.*, Feb. 4 and Feb. 18, 1897; *Zeitschrift für physikalische Chemie*, v. 27, p. 75, 1898.

which put into the formula $MgCl_2 \cdot yH_2O$ would represent the constitution of the solution.

They have especially noted:

1°. The fusing-point of pure ice ($T=0°$, $y=\infty$);

2°. The point of intersection of the congelation curve and of the solubility curve of the hydrate $MgCl_2 \cdot 12H_2O$; this is a *eutectic point* for which

$$T=-33°.6, \qquad y=20.3;$$

3°. The point *C*, indifferent point for $MgCl_2 \cdot 12H_2O$:

$$T=-16°.3, \qquad y=12.0;$$

4°. The point D, intersection of the solubility curves for $MgCl_2 \cdot 12H_2O$ and $MgCl_2 \cdot 8H_2O\alpha$; it is a *eutectic point* for which

$$T=-16°.7, \qquad y=11.17;$$

5°. The point *E*, intersection of the solubility curves of $MgCl_2 \cdot 8H_2O\alpha$ and $MgCl_2 \cdot 6H_2O$; this is a *transition-point* for which

$$T=-3°.4, \qquad y=10.0;$$

6°. The point *F*, intersection of the solubility curves of $MgCl_2 \cdot 6H_2O$ and $MgCl_2 \cdot 4H_2O$; it is a *transition-point* for which

$$T=116°.67, \qquad y=6.18;$$

7°. The point *G*, intersection of the solubility curves of $MgCl_2 \cdot 4_2O$ and $Mg_2Cl \cdot 2H_2O$; this is a *transition-point* for which

$$T=181°.5, \qquad y=4.2.$$

They have, besides, studied a certain number of branches which the phenomena of supersaturation also render observable; among these branches, which are shown by dotted lines in Fig. 75, is the solubility curve *D′E′* of the hydrate $MgCl_2 \cdot 8H_2O\beta$. The eutectic point D′ between $MgCl_2 \cdot 12H_2O$ and $MgCl_2 \cdot 8H_2O\beta$ corresponds to

$$T=-17°.4, \qquad y=11.1,$$

while the transition-point *F′* between $MgCl_2 \cdot 8H_2O\beta$ and $MgCl_2 \cdot 6H_2O$ has the coordinates

$$T=-9°.6, \qquad y=10.3.$$

The phenomena of supersaturation also permitted the observation of the point H, eutectic point between ice and the hydrate $MgCl_2 \cdot 8H_2O\alpha$, for which

$$T = -50^\circ, \qquad y = 16.9,$$

as well as the point I, eutectic point between the hydrates $MgCl_2 \cdot 12H_2O$ and $MgCl_2 \cdot 6H_2O$, for which

$$T = -19^\circ.4, \qquad y = 10.6.$$

212. Roozboom's researches on calcium chloride.—Another remarkable example is given by the hydrates of calcium chloride, objects of an important memoir by Roozboom.[1]

Including ice, there are six hydrates of calcium chloride;

Ice;
$CaCl_2 \cdot 6H_2O$;
$CaCl_2 \cdot 4H_2O\alpha$;
$CaCl_2 \cdot 4H_2O\beta$;
$CaCl_2 \cdot 2H_2O$;
$CaCl_2 \cdot H_2O$.

The solubility curves of these hydrates are shown in Fig. 76;

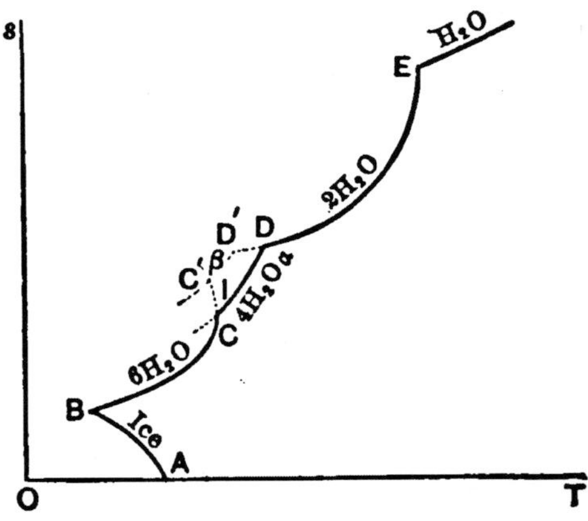

FIG. 76.

in this figure, where the scale is not given, the dotted lines represent the equilibrium states which may be observed for the cases of supersaturation.

---

[1] ROOZBOOM, *Recueil des Travaux chimiques des Pays Bas*, v. 8, p. 4; *Archives néerlandaises des Sciences exactes et naturelles*, v. 13, p. 199.

The following points on this figure are to be noted;

1°. The point A, fusing-point of pure ice:

$$T=0, \qquad s=0;$$

2°. The point B, in ersection of the congelation curve and the solubility curve of $CaCl_2 \cdot 6H_2O$; this is a *eutectic point* where

$$T=-55°, \qquad s=0.425;$$

3°. The point I, indifferent point of $CaCl_2 \cdot 6H_2O$:

$$T=+30°.2, \qquad s=1.027;$$

4°. The point C, intersection of the solubility curves of $CaCl_2 \cdot 6H_2O$ and $CaCl_2 \cdot 4H_2O\alpha$; this is a *transition-point* for which

$$T=+29°.8, \qquad s=1.006;$$

5°. The point C', interse tion of the curves of solubility for $CaCl_2 \cdot 6H_2O$ and $CaCl_2 \cdot 4H_2O\beta$; a *eutectic point:*

$$T=+29°.2, \qquad s=1.128;$$

6°. The point D, intersection of the solubility curves of $CaCl_2 \cdot 4H_2O\alpha$ and $CaCl_2 \cdot 2H_2O$; this is a *transition-point* for which

$$T=+45°.3, \qquad s=1.302;$$

7°. The point D′. intersection of the solbiulity curves of $CaCl_2 \cdot 4H_2O\beta$ and $CaCl_2 \cdot 2H_2O$; it is a *transition-point:*

$$T=+38°.4, \qquad s=1.275;$$

8°. The point E intersection of the solubility curves of $CaCl_2 \cdot 2H_2O$ and $CaCl_2 \cdot H_2O$; a *transition-point:*

$$T=+174°, \qquad s=2.757.$$

The indifferent point of $CaCl_2 \cdot 2H_2O$ would be very close to this point, for it would correspond to

$$T = +176^\circ \quad \text{and} \quad s = 3.08,$$

but this point would not be observed.

It is probable that in the neighborhood of 260° there would be found the intersection of the solubility curves of $CaCl_2 \cdot H_2O$ and of $CaCl_2$, which would be a new transition-point.

213. Stortenbeker's studies on iodine chlorides.—A mixture of iodine and of chlorine in the liquid state, studied with great care by Stortenbeker,[1] is comparable in all respects to the mixture formed by water and an anhydrous salt; the solids that this mixture will deposit in the circumstances in which Stortenbeker was working are the substances

$$\begin{gathered} I, \\ ICl\alpha, \\ ICl\beta, \\ ICl_3. \end{gathered}$$

Lay off temperatures as abscissæ, and as ordinates the number y which, put into the formula $I y Cl$, gives the composition of the solution; each of the four substances we have mentioned corresponds to a solubility curve; these four curves are as shown in Fig. 77; in this figure the dotted lines can be observed only on account of supersaturation phenomena.

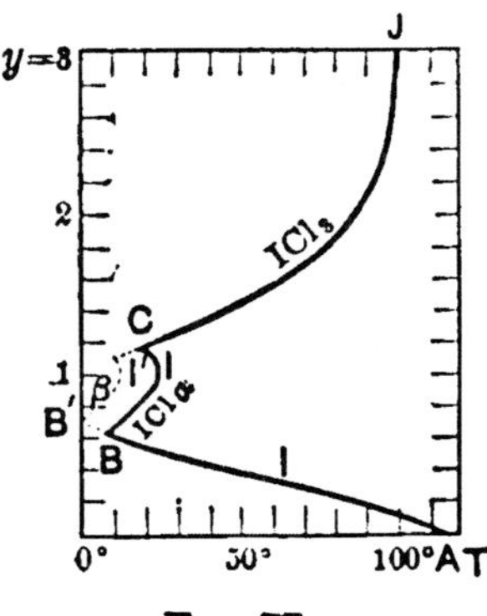

FIG. 77.

The following points are to be noted in this figure:

1°. The point A, fusing-point of iodine:

$$T = 114^\circ.1, \qquad y = 0;$$

2°. The point B, intersection of the curve of congelation for iodine and the

[1] W. STORTENBEKER, *Recueil des Travaux chimiques des Pays Bas*, v. 6, 1888; *Zeitschrift für physikalische Chemie*, v. 3, p. 11, 1888.

solubility curve of IClα; this is a *eutectic point* for which

$$T=+7°.9, \qquad y=0.66;$$

3°. The point B' of intersection of the congelation curve for iodine and the solubility curve of IClβ; also a *eutectic point* whose coordinates are not well known;

4°. The indifferent point I for IClα; the coordinates of this point are

$$T=27°.2, \qquad y=1;$$

5°. The indifferent point I' of the chloride IClβ, of coordinates

$$T=13°.9, \qquad y=1;$$

6°. The point of intersection C of the solubility curve of the monochloride IClα and of the solubility of the trichloride ICl_3; this is a *eutectic point* for which

$$T=22°.7, \qquad y=1.19;$$

7°. The indifferent point J of the trichloride ICl_3; the coordinates of this point are

$$T=101°, \qquad y=3.$$

214. Studies by Guthrie and by Le Chatelier on the mixtures of two salts.—Instead of observing the eutectic mixtures of ice and salt obtained by cooling an aqueous solution of this salt, one may cool the liquid obtained by dissolving a salt in another melted salt; in this way are obtained eutectic mixtures of the two salts.

In this case it is convenient to represent the composition of the liquid, not by the ratio $s=\frac{M_2}{M_1}$ of the mass M_2 of the salt 2 to the mass M_1 of the salt 1, but by the ratio $x=\frac{M_2}{M_1+M_2}$ of the mass of the salt 2 to the total mass (M_1+M_2) of the liquid; this ratio may vary from the value $x=0$, which corresponds to the salt 1 in the pure state, to the value $x=1$, which represents the salt 2 taken in the pure state.

Let us take the values of T for abscissæ and the values of x for ordinates. The congelation curve of the salt 1 in the mixture will be a curve C_1 (Fig. 78); this curve will start from the point F_1 of abscissa T_1, which is the fusing-point of the salt 1 in the pure state and whose ordinate is $x=0$; it will rise from right to left. The curve of congelation of the salt 2 will be a curve C_2 which, starting from F_2 of abscissa T_2, the fusing-point of the pure salt 2 whose ordinate is $x=1$, will descend from right to left.

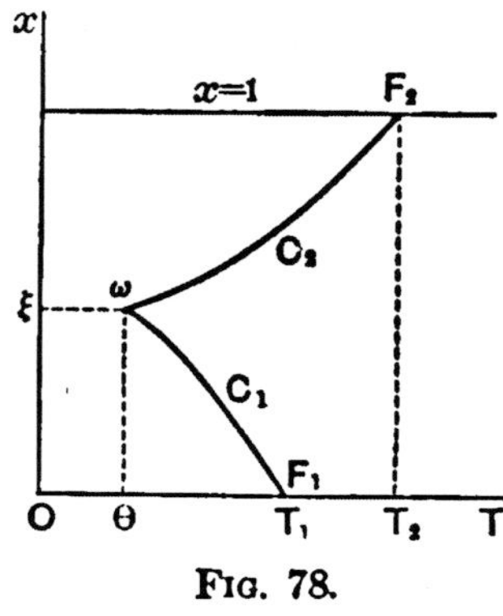

Fig. 78.

These two curves will intersect at the eutectic point ω, of abscissa θ and ordinate ξ.

Guthrie[1] has studied a number of eutectic mixtures formed by melted salts; below are the coordinates θ, ξ of the eutectic points of some of these mixtures:

Mixed Salts.	θ.	ξ.
1. Potassium nitrate...... 2. Lead nitrate..........	207°	53.14
1. Potassium nitrate...... 2. Calcium nitrate........	251°	74.64
1. Potassium nitrate...... 2. Strontium nitrate......	258°	74.19
1. Potassium nitrate...... 2. Barium nitrate........	278°	70.47
1. Potassium nitrate...... 2. Potassium chromate....	295°	96.24
1. Potassium nitrate...... 2. Potassium sulphate....	300°	97.64
1. Potassium nitrate...... 2. Sodium nitrate.	215°	67.10
1. Sodium nitrate........ 2. Lead nitrate..........	268°	57.16

Guthrie has also traced the curves C_1, C_2 for the mixture of potassium nitrate and lead nitrate.

Le Chatelier[2] has likewise studied the phenomena of eutexia

[1] Guthrie, *Philosophical Magazine*, 5th S., v. 17, p. 462, 1884.
[2] Le Chatelier, *Comptes Rendus*, v. 118, p. 709, 1894.

which are observed by cooling the mixture of two melted salts. With melted sodium chloride he has mixed sodium carbonate, the neutral pyrophosphate of sodium and barium chloride; with melted lithium sulphate he has mixed calcium sulphate, sodium sulphate and lithium carbonate; for all these cases he has found the phenomena of eutexia.

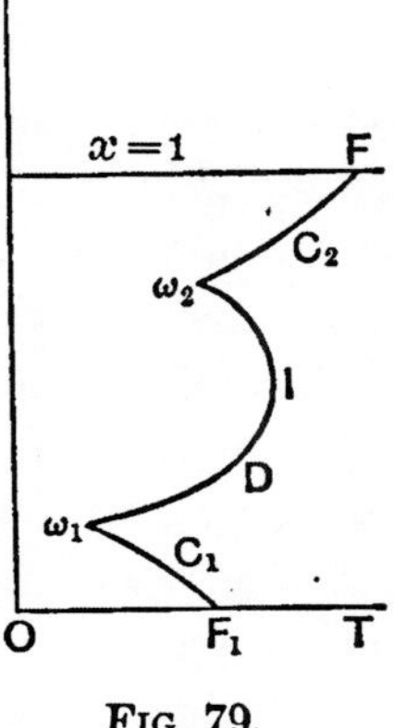

FIG. 79.

The phenomenon becomes complicated when the two melted mixed salts may form a double salt;[1] to the two curves C_1 and C_2 it is necessary in this case to join the solubility curve D (Fig. 79) of the double salt in the liquid mixture; this curve possesses in general an indifferent point I; its points of intersection ω_1, ω_2 with the curve C_1, C_2 are eutectic points.

[1] LE CHATELIER, *Comptes Rendus*, v. 113, p. 801, 1894.

CHAPTER XIII.

MIXED CRYSTALS. ISOMORPHOUS MIXTURES.

215. Isomorphous salts; Rüdorff's observations.—Let us take two salts incapable of reacting chemically, for example, ammonium chloride and ammonium nitrate; put them in contact with a quantity of water too small to entirely dissolve either one of the two salts; three independent components, water, ammonium chloride, ammonium nitrate make up the system; this system is besides divided into three phases, the two crystallized salts and the aqueous solution of these two salts; we have therefore to deal with a bivariant system which may be put in equilibrium under every pressure and at all temperatures, when equilibrium is established at a given pressure and temperature; the solution, saturated with each of the two salts, should have a perfectly definite composition, independent of the masses of ammonium chloride, of ammonium nitrate and water that coexist with this solution.

This is what experiments, made a good while ago, revealed to Rüdorff,[1] not only for what concerns the two salts o which we have spoken, but also for a certain number of pairs of salts incapable of any chemical reaction, either because they are from the same base or from the same acid.

But Rüdorff also found a certain number of saline couples which do not obey the rule above stated.

For instance, in place of ammonium chloride and ammonium nitrate. take potassium sulphate and ammonium sulphate; put them in the presence of a quantity of water incapable of dissolving them totally; at a given temperature and pressure the system

[1] Rüdorff, *Poggendorf's Annalen*, v. 148, p. 456, 1873.

still comes to an equilibrium condition; but the composition of the solution, for the system in equilibrium, is no longer determined by the knowledge of the pressure which the system supports and of the temperature to which it is brought; it still depends upon the relative values of the masses of the two alts and water which have been put together; if, without changing pressure or temperature, there is added a certain quantity of one or the other of the two salts, the solution changes in composition, enriches itself relatively to the salt added, and becomes poorer in the other salt.

216. Interpretation of the preceding facts; isomorphous mixtures are solid solutions.—These properties do not belong to a bivariant system; a multivariant system alone can possess them; it follows, therefore, that the calcu ation which has caused us to consider the system bivariant—ammonium chloride, ammonium nitrate, water—is false in some way when we try to extend it to the system potassium sulphate ammonium sulphate, water. Now the error evidently cannot be in the number of independent components, a number certainly equal to 3; it can therefore only be in the number of phases; the number of phases into which the system is divided when equilibrium is reached cannot be equal to 3; it cannot exceed 2. Whence comes this redu tion in the number of phases?

Potassium sulphate and ammonium sulphate are two isomorphous salts; when masses of these two salts are left a long time in contact with an aqueous so'ution the crystals of both cease to be distinct, and at last there rest only mixed crystals, containing both sulphate of potassium and ammonium sulphate.

Rüdorff's experiments, compared with the theorems of J. Willard Gibbs, show us that the mixed crystals should be considered not as two phases, but as a single phase these crystals are not therefore, as many writers have supposed simp'y mechanical mixtures, a juxtaposition or a mixing of crystalline particles of potassium sulphate and crystalline pa ticles of ammonium sulphate; in them the two component salts are physically mixed in a manner as intimate as for an aqueous solution; every volume, however small which may be cut from one of these crystals, contains a certain quantity of each one of these salts; these mixed crystals, formed by two isomorphous bodies, constitute, according to the expression created by Van't Hoff in considering other facts, a *solid solution.*

217. Theory of the solubility of two isomorphous salts.—This assimilation into a solid solution of mixed crystals formed by two isomorphous salts leads to a complete theory of the phenomena which are produced when two isomorphous salts are brought into the presence of water.

We have in fact here a system formed of three independent components water 0 and the two salts 1 and 2; this system is divided into two phases, the liquid solution for which we shall continue to indicate by s_1, s_2, the two concentrations, and the mixed crystals C; this system is therefore *trivariant;* when the temperature and pressure only are given, the composition of each of the two phases capable of remaining in equilibrium in contact with each other is not completely determined; it becomes entirely determined if to the temperature and pressure there is added another given quantity, for example, one of the concentrations s_1 of the solution.

Suppose the pressure π given once for all and equal, for instance, to the atmospheric pressure. Whenever there is given the temperature T and the first concentration s_1 of the liquid solution, the econd concentration, s_2, should have a well-determined value if the liquid solution is to remain in equilibrium in contact with mixed crystals; if, therefore, as in Art. 102, we lay off on the three coo dinate rectangular axes the values of the temperature T and of the concentrations s_1 and s_2, we shall find that the temperature and the concentrations of every solution capable of remaining in equilibrium in contact with mixed crystals C are the coordinates of a point M situated on a surface S, a conclusion similar to that reached for the case in which the solid C was a chemical compound o definite composition.

But, and in this the problem we are treating is more complicated than that of Art. 102, the solution whose properties (temperature and concentrations) are represented by the coordinates of a point M of the surface S does not remain in equilibrium with any mixed crystals whatever; the mixed crystals which may remain in equilibrium in contact with this solution have a well-determined composition, which varies as the point M assumes successively different positions on the surface S.

These principles, necessary consequence of Gibbs' theories, have been brought out for the most part by Roozboom [1] and his pupils.

218. Isomorphism of the sulphates of the magnesium series. Studies of Stortenbeker.—The most complete experimental researches which have been made on this subject are due to Stortenbeker.[2] They deal with the phenomena of isomorphism possessed by the different hydrates of sulphates of the magnesium series:

$$MgSO_4, \quad ZnSO_4, \quad FeSO_4, \quad CuSO_4, \quad MnSO_4, \quad CdSO_4.$$

These cases of isomorphism had already been studied by Mitscherlich. When two of these sulphates are dissolved in water, the solution may precipitate mixed crystals; but in general, according to the temperature and composition of the solution, one may obtain various kinds of mixed crystals.

Let us take, for example, the case so well studied by Stortenbeker [2] where the solution contains zinc sulphate and copper sulphate; three kinds of mixed crystals may be obtained, namely:

Triclinic crystals (anorthic) corresponding to the formula $(Zn,Cu)SO_4 \cdot 5H_2O$;

Monoclinic crystals (clinorhombic) of formula $(Zn,Cu)SO_4 \cdot 7H_2O$;

Orthorhombic crystals having the same formula.

Concerning the solubility surfaces of these crystals one may repeat all that has been said in Arts. **103, 104,** and **105** about the surfaces of solubility of double salts.

To each of the mixed crystals corresponds a solubility surface referred to the axes OT, Os_1, Os_2; but certain parts of this surface will represent, in general, states of equilibrium observable only in solutions supersaturated with respect to another kind of crystal; if these parts are suppressed so as to keep only the representation of equilibrium states where all supersaturation is excluded, a polyhedron will be obtained with curved faces having as many faces as there are kinds of mixed crystals. The edges

---

[1] ROOZBOOM, *Archives néerlandaises des Sciences exactes et naturelles*, v. 26, p. 137, 1891; *Zeitschrift für physikalische Chemie*, v. 8, p. 504, 1891.

[2] STORTENBEKER, *Zeitschrift für physikalische Chemie*, v. 22, p. 60, 1897.

of this polyhedron will represent the solutions which may rest in equilibrium in contact with two distinct kinds of mixed crystals.

For the mixtures of copper sulphate and zinc sulphate Stortenbeker has not constructed the entire surface of which we have just spoken, but only the points on this surface which correspond to the temperature $T=18°$. In the system of axes Os_1, Os_2 (Fig. 80), where s_1 represents the concentration in copper sulphate and s_2 the concentration in zinc sulphate, he obtains three lines which correspond to the three kinds of mixed crystals. The dotted portions of these lines can only be observed thanks to the phenomena of supersaturation.

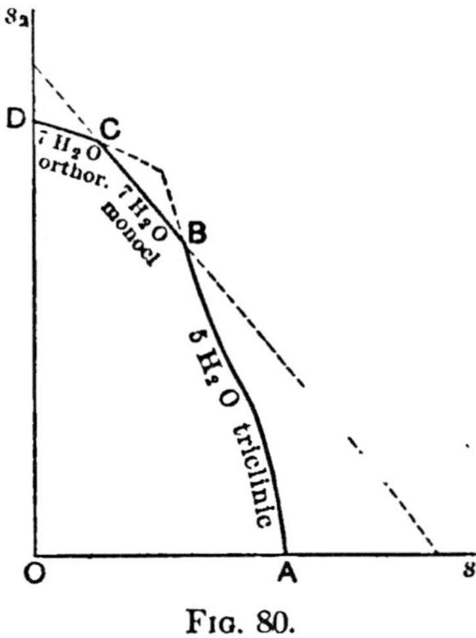

FIG. 80.

It is seen that at 18° the triclinic crystals with 5 molecules of water are obtained as long as the amount of zinc sulphate does not exceed a certain limit; there are next obtained clinorhombic crystals of 7 molecules of water; finally, the orthorhombic crystals of 7 molecules of water are deposited on y from solutions very poor in copper sulphate.

The information given by this figure does not suffice to make known all the properties possessed at 18° by isomorphous mixtures of copper sulphate and zinc sulphate; it is further necessary to know the proportion of zinc and of copper in the mixed crystals deposited from the solution represented by each of the points of the various lines shown in Fig. 80. Stortenbeker has made known this proportion; he has constructed curves which determine, for the temperature 18°, the composition of the crystals when one knows the composition of the solution which rests in equilibrium in contact with these crystals.

Take a number of grammes of each crystal equal to its molecular weight and determine the number n of copper atoms which are therein contained. When we follow the line AB from A to B within the triclinic crystals with 5 water molecules, n varies from

1 to 0.828; when we follow the line *BC* from *B* to *C* in the midst of the clinorhombic crystals with 7 water molecules, *n* changes from 0.319 to 0.149; finally, when we follow the line *CD* from *C* to *D* within the orthorhombic crystals with 7 molecules, *n* varies from 0.0197 to 0.

Therefore, at a given temperature, if all phenomena of supersaturation are excluded, the crystals of each kind which can be obtained have a composition which remains included between two given limits; between the limiting compositions of the crystals of two different kinds there exist gaps; certain compositions correspond to no kind of crystal susceptible of remaining in equilibrium, at the temperature considered, with a solution freed from all supersaturation.

The phenomena which we have just described and the curves which represent them change with the temperature. Stortenbeker has not fol owed, for the above ca·e, this influence of temperature; but he has examined it in studying the isomorphous mixtures of magnesium sulphate and of copper sulphate.[1]

There are here two kinds of mixed crystals: tri-clinic (anorthic) corresponding to the formula $(Cu,MnSO_4 \cdot 5H_2O$ and clinorhombic crystals of formula $(Cu,Mn)SO_4 \cdot 7H_2O$.

Take s_1 for the concentration of copper sulphate and s_2 for that of manganese sulphate and at each temperature draw a solubility curve for the two kinds of mixed crystals referred to the axes s_1Os_2.

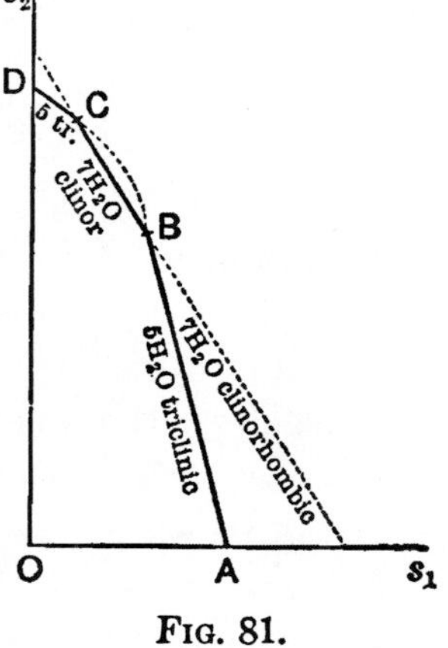

FIG. 81.

At 18° the solubility curves are arranged as indicated in Fig. 81; in this figure the dotted lines represent solutions saturated with respect to one kind of c ystal, but supersaturated with respect to the other.

If one supposes excluded all kinds of supersaturation, it is seen that the triclinic crystals with 5 water molecules are those obtained either in contact with solutions rich in copper,

---

[1] STORTENBEKER, *Zeitschrift für physikalische Chemie*, v. 34, p. 111, 1900.

or in contact with solutions rich in manganese; only the solutions of an intermediate composition can furnish clinorhombic crystals with 7 molecules of water.

If the number n is defined as in the preceding case, we see that for the triclinic crystals obtained in these conditions n is included between 1 and 0 229 (corresponding to the solubility curve AB), or between 0.105 and 0 (corresponding to the solubility curve CD); while for the clinorhombic crystals for which BC is the solubility curve, n is included between 0.235 and 0.16.

At 10° the arrangement of the solubility curves is that represented by Fig. 82; the solutions rich in copper continue to give triclinic crystals with 5 molecules of water, but the solutions rich in manganese give clinorhombic crystals with 7 molecules of water.

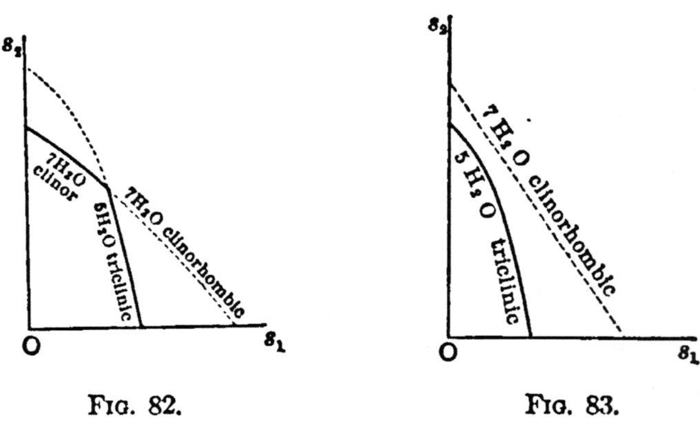

Fig. 82. Fig. 83.

Again, at 23° the solubility curves are arranged as indicated in Fig. 83. Avoiding all supersaturation, a solution can remain in equilibrium only in contact with triclinic crystals with 5 water molecules; the solutions saturated with respect to clinorhombic crystals with 7 water molecules are supersaturated with respect to the preceding crystals.

From these data it is not difficult to recognize the general arrangement of the solubility surface, limited by the plane s_1Os_2 (Fig. 84), which corresponds to $T=23°$; the line $ABCD$, intersection of the surface by the plane $T=18°$, is that represented by Fig. 81.

Stortenbeker[1] has also studied the mixed crystals formed by cadmium sulphate and ferrous sulphate; by zinc sulphate and magnesium sulphate; by magnesium sulphate and ferrous sulphate; by copper sulphate and manganese sulphate; by cobalt chloride and manganese chloride. We shall limit ourselves to referring the reader to these valuable memoirs.

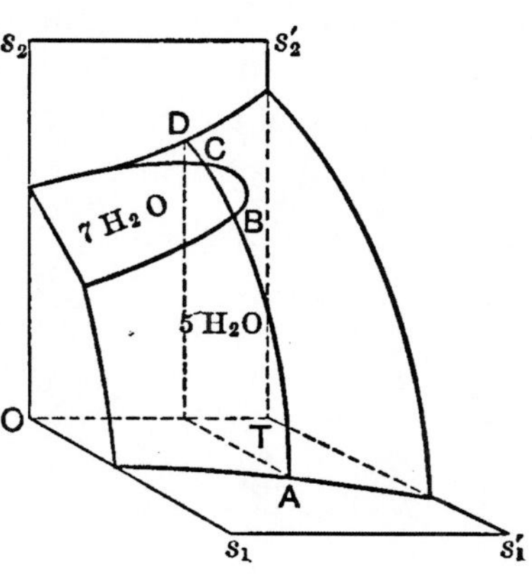

FIG. 84.

219. Solutions containing mixed crystals and definite compounds. Researches of Roozboom and of Retgers.—It may happen that a solution of two salts in water precipitates, according to circumstances, either mixed crystals or a definite compound such as a hydrate or a double salt. The solubility surface, referred as before to the axes OT, Os_1, Os_2, is composed of several regions; among these regions there are some which correspond to a definite compound, simple salt or double salt; there are others corresponding to mixed crystals; for each of the points of the latter there is a mixed crystal of given constitution; but this constitution varies according to the point chosen.

An important example has been studied[2] by Roozboom; it is furnished by the aqueous solutions of ferric chloride and of ammonium chloride. Although these two salts cannot be considered amorphous, their solutions, as had been previously shown by Lehmann, may furnish mixed crystals; they may also give as precipitates definite compounds, namely, the hydrate $Fe_2Cl_6 \cdot 12H_2O$ and the double salt $(NH_4)_2 \cdot FeCl_5 \cdot H_2O$.

On the solubility surface each of these kinds of precipitate has its region; if the solubility surface is cut by a plane perpendicular to OT, in such manner as to obtain an isothermal, this iso-

---

[1] STORTENBEKER, *loc. cit.* and *Zeitschrift für physikalische Chemie*, v. 16, p. 250, 1895, and v. 17, p. 643, 1895.

[2] ROOZBOOM, *Archives néerlandaises des Sciences exactes et naturelles*, v. 27, p. 1, 1892; *Zeitschrift für physikalische Chemie*, v. 10, p. 145, 1892; MOHR, *Zeit. f. phys. Chemie*, v. 27, p. 193, 1898.

thermal will be made up of three curves which will represent the solutions capable of remaining in equilibrium either in contact with the ferric hydrate or in contact with mixed crystals.

At the temperature of 15°, if the coordinate s_1 is taken as concentration in ammonia salt, and s_2 the concentration in ferric chloride, these three curves are arranged as shown in Fig. 85. According to this arrangement, the mixed crystals are precipitated from solutions rich in ammonia salt, the ferric hydrate from solutions very poor in ammonia salt, the double salt from solutions of intermediate composition.

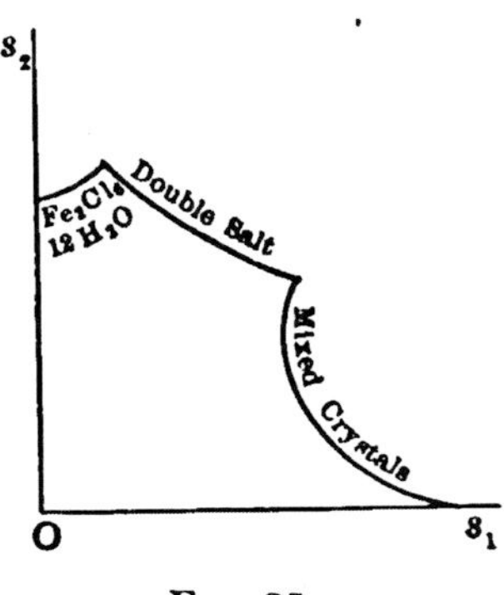

FIG. 85.

Retgers, whose researches[1] have contributed greatly to increase our knowledge concerning isomorphism, has shown that this property to precipitate, according to circumstances, either mixed crystals or a double salt, belongs very often to solutions of two isomorphous salts.

A disposition which seems to be realized frequently is the following:

The solutions for which the amount of salt 2 does not exceed a certain limit furnish mixed crystals, isomers of those given by the salt 1 in the pure state; the solutions for which the amount of salt 1 does not exceed a certain limit precipitate isomorphous crystals of salt 2; finally, solutions of intermediate composition furnish a double salt of definite composition.

In this way behave[2] the aqueous solutions of the two substances

$$K_2SO_4, \qquad Na_2SO_4.$$

The solutions rich in potassium sulphate give mixed isomorphous crystals of potassium sulphate; the solutions rich in sodium sulphate give mixed isomorphous crystals of sodium sulphate; finally, intermediate solutions give a double salt whose formula is

$$3K_2SO_4 \cdot Na_2SO_4.$$

[1] These researches are for the most part to be found in the *Zeitschrift für physikalische Chemie.*

[2] RETGERS, *Zeit. f. phys. Chemie,* v. 6, p. 226, 1890.

Solutions of calcium carbonate and magnesium carbonate behave in the same way;[1] one may obtain:

1°. Mixed crystals, isomers of *calcite*, containing from 0 to 0.025 of magnesium carbonate;

2°. Mixed crystals, isomers of *magnesite*, containing from 0 to 0.03 of calcium carbonate;

3°. A double salt, *dolomite*, having the formula

$$CaCO_3 \cdot MgCO_3.$$

Retgers was able, from considerations of this nature, to account for the peculiarities possessed by the mineralogical series of pyroxene, olivine and pyrite.[2]

220. Two melted isomorphous salts; case where there is produced a single kind of mixed crystals.—We have just studied the ormation of mixed crystals in an aqueous solution containing two isomorphous salts. By side of this generation of mixed crystals in the wet way, one may study their formation in the dry way; one may melt together two isomorphous substances and study the mixed crystals which the mixture furnishes on freezing; the problem is thus quite simplified, because we have to deal no longer with three, but with only two independent components.

Roozboom has given [3] a very complete theoretical study of the various peculiarities which may occur, and his pupils have added to this study remarkable experimental verifications; we shall limit ourselves to sketching some of the essential traits.

The most simple case which can oc ur is that where the liquid mixture formed of the sub tances 1 and 2 never furnishes, whatever its composition, other than a single kind of mixed crystals; from the point of view of composition, these latter may have all the intermediate states between the crystals of substance 1 in the pure state and the crystals of substance 2 in the pure state.

We have to do here with what we have called a *double mixture* (Art. 182); this double mixture is very comparable to that formed

---

[1] Retgers, *Zeit. f. phys. Chemie*, v. 6, p. 227, 1890.

[2] Retgers, *Annales de l'Ecole polytechnique de Delft*, v. 6, p. 186, 1891.

[3] Roozboom, *Archives néerlandaises des Sciences exactes et naturelles*, Series II, v. 3, pp. 414, 1900; *Zeitschrift für physikalische Chemie*, v. 30, p. 385, and 413, 1900.

by a mixture of two volatile liquids in the presence of a mixed vapor; in the discussion of this last double mixture it will suffice almost always to replace the words *liquid mixture* and *mixed vapor*, respectively, by the words *solid solution* and *liquid mixture* to obtain the theory of the first double mixture.

Let us represent the composition of each of our two mixtures as we have done in Art. 195; 1 gramme of solid solution or liquid contains X grammes of substance 2 and $(1-X)$ grammes of substance 1; X is variable from 0 to 1; $X=0$ represents the substance 1 in the pure state; $X=1$ represents the substance 2 in the pure state. Lay off this value of X as abscissa, and along the axis of ordinates lay off the value T of the temperature (Fig. 86). Suppose the pressure constant.

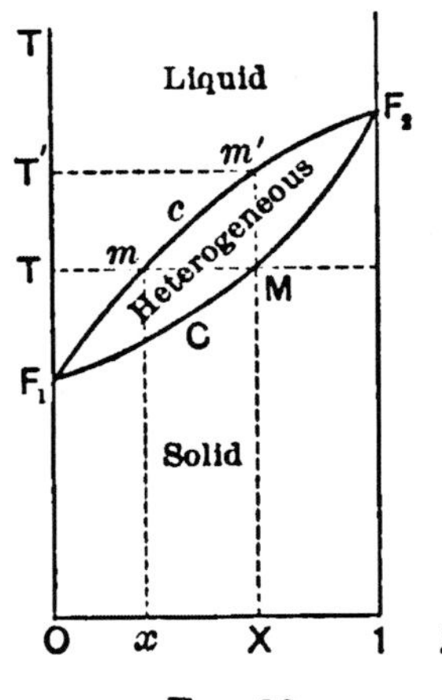

Fig. 86.

Take a solid solution of composition X and gradually raise its temperature; the point representing the state of the system will mount along XM parallel to OT.

As long as the temperature remains less than a certain value T the crystals will not undergo any trace of fusion.

The instant the temperature reaches the value T the representative point being then at M, there will appear the first liquid drop, which will not have the composition X.

The temperature increasing above T, the system which will conserve the *mean* composition X will be in part in the crystalline state, partly in the liquid state; neither the crystals nor the liquid will have the composition X.

When the temperature reaches a certain limit T', higher than T, the representative point being then at m', the system will have assumed entirely the liquid state; the liquid, whose composition will be assuredly X, will remain homogeneous at temperatures higher than T'.

If X is made to vary from 0 to 1, the point M describes a certain curve C; the point m' describes a certain other curve, c, situated entirely above the curve C. For $X=0$ the two curves C and c start from the same point F_1, whose ordinate OF_1 is the fusing-point of the substance 1 in the pure state; for $X=1$ the two curves C, c join in a point F_2, whose ordinate $1F_2$ represents the temperature of fusion of the pure substance 2.

The two curves C, c divide the plane into three regions. When the representative point is found in the region situated below the curve C the system is in the state of homogeneous solid; when the representative point is above the curve c the system is in a state of homogeneous liquid; when the representative point is between C and c the system of *mean* composition X is partly in the solid state and partly in the liquid.

If we draw a parallel TmM to OX, this line will meet the curve C in a point M, of abscissa X, and the line c in a point m of abscissa x; x represents the composition of the liquid, which at the temperature T can remain in equilibrium in contact with crystals of composition X.

According to the opinion pretty generally held among the chemists who have insufficiently meditated upon the laws of chemical statics, the two lines C, c would coincide for a great number of cases and would be reduced to a straight line joining the points F_1 and F_2. At a given temperature a fluid mixture of given composition would furnish crystals of the same composition.

G. Bruni [1] has very well shown this opinion to be inadmissible. We may, in fact, apply to the systems we are studying the theorems of Gibbs and of Konovalow (Art. 194), and particularly the first. It suffices to substitute for the words *mixed liquid, mixed vapor*, the words *mixed crystals, mixed liquid*.

If, at a certain temperature, the mixed crystals may remain in equilibrium in contact with a liquid mixture of the same composition, at this temperature the two curves C and c have a common point; they ought also, in virtue of the theroem indicated, to have a common tangent parallel to OX. If, therefore, the composition of the liquid which may remain in equilibrium in contact with mixed crystals is always identical with that of these crystals, not only the two curves C, c would coalesce, but their common tangent would be constantly parallel to OX; the two curves would therefore be reduced to a same straight line parallel to

[1] G. Bruni, *Rendiconti dell' Accademia dei Lincei*, v. 7, pp. 138 and 347, 1898.

OX. That this may be possible, it would be necessary that the two substances 1 and 2 had the same fusing-point, and that the same be true of all the mixed crystals which they may produce. We shall study an example of this last case in Art. 230.

This does not mean to say that the two curves C and c may not have, in certain cases, a common point I; at the temperature θ, which serves as abscissa to the point I, the mixed crystals may remain in equilibrium in contact with a liquid mixture of the same composition, so that this equilibrium state is indifferent. At the indifferent point I the two curves have a common tangent parallel to OX; this point is therefore, for the two curves, a point of maximum ordinate or of minimum ordinate.

A very good example of this last case is given us by the mixtures of mercury bromide and mercury iodide studied by Reinders.[1]

The liquid mixtures formed by melted mercury bromide and mercury iodide give on cooling a single sort of mixed crystals; these are the isomorphous orthorhombic crystals of yellow mercury iodide.

Let us denote by 1 the mercury bromide and by 2 the mercury iodide.

To the value $X=0$ corresponds the point F_1 (Fig. 87) whose ordinate OF_1 is the temperature of fusion of mercury bromide, that is to say, 236°; from this point start the two curves C and c which end at the point F_2, of abscissa $X=1$, of ordinate OF_2 equal to the temperature of fusion of yellow mercury iodide, that is, 255°; the two curves C, c meet at an indifferent point I of minimum ordinate; $\theta=216°.1$ is the ordinate of this point; at this temperature the mixed crystals have the same composition as the liquid in whose presence they exist; they contain 0.59 of a molecule of mercury bromide and 0.41 molecules of mercury iodide.

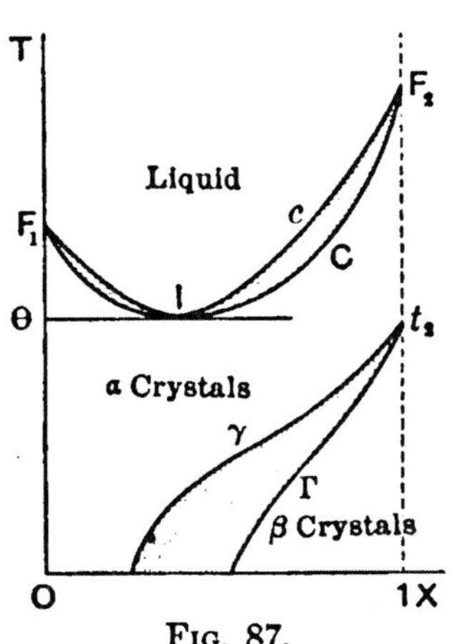

FIG. 87.

[1] REINDERS, *Zeitschrift für physikalische Chemie*, v. 32, p. 494, 1900.

221. Case in which there may be formed two kinds of mixed crystals.—In a great number of cases two isomorphous substances are *isodimorphous;* they may give birth to two different kinds of mixed crystals which we shall denote by the indices α and β.

Concerning the transformation of mixed β crystals into mixed α crystals, one may repeat almost textually what we have said regarding the transformation of mixed crystals into a liquid mixture.

Take a value of X corresponding to a given composition and suppose that, for this composition, the β crystals are in true equilibrium at low temperature.

If we raise gradually the temperatures of the mixed β crystals whose composition is X, these crystals remain unaltered so long as the temperature is below τ; the temperature exceeding τ, they commence to be transformed into α crystals; so long as the temperature lies between τ and τ' the system of mean composition X will be composed of mixed β crystals and mixed α crystals, having both a composition different from X; finally, when the temperature exceeds τ', the system will be entirely in the state of α crystals.

Let M be the point of abscissa X and of ordinate τ, and μ' the point of abscissa X and of ordinate τ'. When X varies, the point M describes a curve Γ, and the point μ' describes a curve γ. The points in the plane located below the curve Γ represent states where the system is homogeneous under the form of β crystals; the plane situated above the curve Γ represent states in which the system is homogeneous under the form of α crystals; finally, the points situated between the two curves Γ and γ represent heterogeneous states where the system is formed of α and β crystals.

The mixtures of mercury bromide and of mercury iodide, studied by Reinders, furnish us with another very simple example of these propositions.

It is known that when the temperature is lowered to about 126°, yellow mercury iodide changes over to the red iodide; similarly, by a lowering of temperature, the mixed crystals of mercury iodide and of mercury bromide, which are isomorphous with the yellow iodide and which play here the rôle of the α crystals, are transformed into mixed isomorphous crystals of red iodide, playing the rôle of the β crystals.

Reinders has drawn for these crystals the curves Γ and γ which are indicated in Fig. 87. These two curves unite for $X=1$ in a point t_2, whose ordinate $1t_2$ is equal to the temperature of transformation of yellow mercury iodide into the red iodide, that is, at 126°. These two curves do not extend to the line OT where $X=0$; indeed, beyond a certain content in mercury bromide, one observes only the mixed α crystals.

In Fig. 87 the cross-hatched regions correspond to the heterogeneous states of the system; in the region covered with cross-hatching parallel to OX the system is formed of liquid and mixed α crystals in the region cross-hatched parallel to OT the system is composed of α and β crystals.

222. The two kinds of mixed crystals may be furnished by the liquid mixture. Case of the transition-point.—In the case we have just examined the transformation of the α crystals into β crystals is produced at too low temperatures for the liquid mixture to be observable; the liquid cannot therefore deposit other than the α-salt crystals, which simplifies the study of these phenomena.

In a great number of cases it is quite otherwise; the liquid mixture may, according to circumstances, furnish either the α_1 crystals or the α_2 crystals; the α_1 crystals if it contains a large proportion of substance 1, case for which X has there a value near to 0; the α_2 crystals if it contains a large proportion of substance 2, case for which X has a value near to 1.

Let us consider, for example, a liquid mixture obtained by melting together silver nitrate and sodium nitrate, a mixture which has been studied by Hissink;[1] let us give the index 1 to silver nitrate, and the index 2 to sodium nitrate.

The liquid mixtures rich in silver nitrate (X near to 0) furnish mixed α_1 crystals, which are hexagonal crystals, isomorphous with those which are furnished by fused silver nitrate at its freezing-point.

The liquid mixtures rich in sodium nitrate (X near to 1) crystallize in mixed α_2 crystals, which are also hexagonal, but of different parameters from the others; these crystals are isomorphous

[1] Hissink, *Zeitschrift für physikalische Chemie*, v. 32, p. 537, 1900.

with those furnished by sodium nitrate fused in the pure state at its freezing-point.

To each of these kinds of mixed crystals corresponds a curve; these two curves are respectively analogous to those we have called C and $\mathfrak{c}$ (Art. 220); we shall call C_1 and c_1 the two curves which refer to the α_1 crystals; C_2, c_2 the two curves corresponding to the α_2 crystals.

The curves c_1 and c_2 have the appearance as shown in Fig. 88. The line c_1 rises from left to right starting from the point F_1, whose ordinate $OF_1=208°.6$ is the freezing-point of pure silver nitrate. The line c_2 descends from right to left beginning at F_2, whose ordinate $1F_2=308°$ is the freezing-point of pure sodium nitrate.

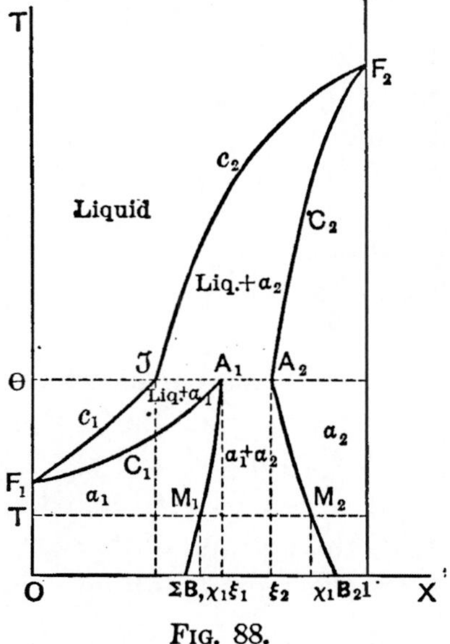

FIG. 88.

These two curves meet in a point $\mathfrak{J}$ of ordinate $O\theta=217°.5$.

When, therefore, the freezing-point increases from $OF_1'=208°.6$ to $O\theta=217°.5$, the liquid mixture deposits mixed crystals of the α_1 kind; when, exceeding $O\theta=217°.5$, the temperature of freezing increases to $1F_2=308°$, the liquid furnishes mixed crystals of the α_2 kind. One may say that the temperature θ is a *transition temperature* and that the point $\mathfrak{J}$ common to the two curves c_1, c_2 is a *transition-point*.

The curve C_1, starting from the point F_1, rises from left to right up to the point A_1, of ordinate $O\theta$, remaining below, and hence to the right of, the line c_1; the curve C_2, from the point F_2, descends from right to left as far as the point A_2, of ordinate $O\theta$, remaining below, and so to the right of, the line c_2; finally, the point A_2 is to the right of the point A_1.

If we designate by Ξ, ξ_1, ξ_2, the abscissæ of the points $\mathfrak{J}$, A_1, A_2, we have

$$\Xi<\xi_1<\xi_2.$$

At the temperature θ, ordinate of the point $\mathfrak{J}$, the same liquid, of composition Ξ, may be in equilibrium either in contact with the mixed α_1 crystals, of composition ξ_1, or in contact with the mixed α_2 crystals, of composition ξ_2; the principles of thermodynamics show then that the temperature θ, α_1 crystals, of composition ξ_1, and α_2 crystals, of composition ξ_2, put in presence of each other, remain in equilibrium; whence results an important property of the points A_1 and A_2.

Take a temperature T, less than θ. At this temperature one may observe mixed α_1 crystals which remain in equilibrium with mixed α_2 crystals; it is sufficient for this that the α_1 crystals have a composition $X = \chi_1$, and that the α_2 crystals have a composition $X = \chi_2$, χ_2 being greater than χ_1.

Let M_1 be the point of coordinates $(\chi_1 T)$, M_2 the point of coordinates $(\chi_2 T)$; when the temperature T is made to change keeping it less than θ, the point M_1 describes a line B_1M_1, and the point M_2 a line B_2M_2.

From the properties which we have recognized the points A_1 and A_2 to possess, the line B_1M_1 passes through the point A_1, and the line B_2M_2 through the point A_2.

It is now easy to find the properties possessed by the system when the position of its representative point (XT) is known.

If the representative point is above the lines c_1, c_2, the system is in the state of *homogeneous liquid.*

If the representative point is in the region $OF_1A_1B_1$ of the plane, the system is in the state of *mixed homogeneous crystals of the* α_1 *kind.*

If the representative point is in the region $F_2A_2B_2 1$, the system consists of *homogeneous crystals of the* α_2 *kind.*

If the representative point is located in none of these three regions, the system of *mean* composition X is *heterogeneous.*

It is formed of *liquid* and of α_1 *crystals* if the representative point is in the triangle $\mathfrak{J}F_1A$; of *liquid* and α_2 *crystals* if the representative point is in the region $B_1A_1B_2A_2$.

223. Case of a eutectic point.—The arrangement which we have just studied is not the only one that may be met with; the mixtures of sodium and potassium nitrates, also studied by Hissink, show another.

Let us give the index 1 to sodium nitrate, and the index 2 to potassium nitrate.

The mixtures rich in sodium nitrate furnish mixed α_1 crystals, which are hexagonal, isomorphous with the crystals denoted by α_2 of the preceding article. The mixtures rich in potassium nitrate furnish mixed α_2 crystals, which are orthorhombic.

Starting from the point F_1, whose ordinate $OF_1 = 308°$ is the freezing-point of pure sodium nitrate, the curve c_1 descends constantly from left to right (Fig. 89); from the point F_2, whose ordinate $1F_2 = 337°$ is the freezing-point of pure potassium nitrate, the line c_2 descends constantly from right to left.

These two curves meet in a point E, whose ordinate $O\theta = 218°$ is less than the fusing temperatures of pure sodium and potassium nitrates; Ξ is the abscissa of the point E.

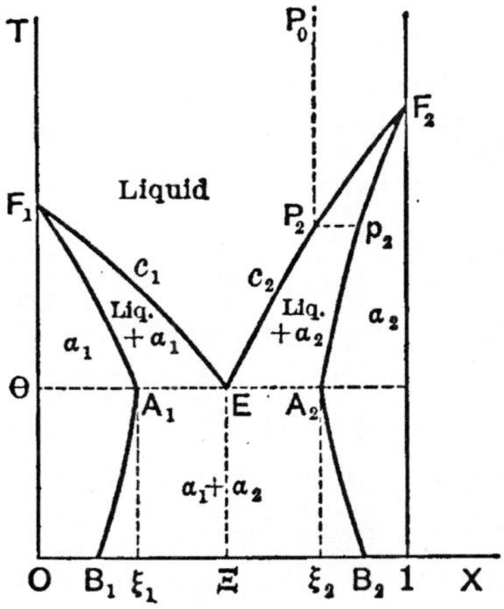

FIG. 89.

The two curves C_1, C_2, starting respectively from the points F_1, F_2, descend to the points A_1, A_2, which have the common ordinate $O\theta$; ξ_1 is the abscissa of the point A_1, ξ_2 is the abscissa of the point A_2, and it follows directly that

$$\xi_1 < \Xi < \xi_2.$$

Here are the remarkable properties which such an arrangement necessitate:

Let us take, at a sufficiently high temperature, a liquid mixture of the substances 1 and 2 and suppose, in order to be definite, that the composition of this mixture corresponds to a value of X greater than Ξ. The representative point is at P_0. Lower gradually the temperature of the system.

As long as this temperature remains above a certain limit the mixture will stay liquid and this liquid will have an invariable composition; the representative point of the state of the *liquid* will follow the line P_0P_2 parallel to TO.

It will thus attain the point P_2, located on the line c_2; at this instant mixed crystals of the α_2 form begin to deposit; in order to obtain the representative point p_2 of the state of these crystals, it will suffice to draw a parallel to OX through the point P_2, until it meets the curve C_2. These crystals being richer in potassium nitrate than the liquid from which they come, their precipitation causes the value of X to decrease for the liquid; the representative point for the liquid state is displaced towards the left; if the cooling is slow enough for the equilibrium to be at every instant established between the liquid and the mixed crystals, the representative point for the liquid descends the line c_1 and reaches the point E.

Consider, at the point E, the liquid of temperature and composition Ξ.

As soon as we decrease below θ the temperature of the system whose mean composition is Ξ, this system must form a heterogeneous mixture consisting of the mixed α_1 crystals of composition ξ_1, and of the mixed α_2 crystals of composition ξ_2; therefore, if we continue to cool our liquid, it will entirely freeze furnishing such a solid mixture; this mixture is produced in the same way as the *eutectic mixtures* studied in Art. 207; like them, it has a definite mean composition; like them, it is a heterogeneous mixture of two kinds of crystals; only these crystals, instead of being of definite chemical kinds, are mixed crystals; each of the two kinds of mixed crystals enclosed in the eutectic mixture has, furthermore, a fixed composition.

We should have reached analogous conclusions by taking to start with a liquid whose composition would have corresponded to a value of X less than Ξ.

We shall say for the case in hand that the point E is a *eutectic point*.

For the case studied by Hissink the eutectic mixture obtained at 218° had sensibly the chemical formula

$$0.507KNO_3+0.493NaNO_3.$$

It was formed of a conglomerate of α_1 crystals, having the formula

$$0.24KNO_3+0.76NaNO_3,$$

and of α_2 crystals, having the formula

$$0.85KNO_3+0.15NaNO_3.$$

224. Isotrimorphous and isotetramorphous substances; studies of Hissink and of van Eyk.—The silver and sodium nitrates studied by Hissink are *isotrimorphous* bodies; besides the α_1 and α_2 crystals of Art. **222** which may coexist with the liquid, one may observe other mixed crystals β_1 at temperatures at which the liquid cannot exist.

Pure silver nitrate, hexagonal at temperatures above 159°.5, is orthorhombic at temperatures below 159°.5; the mixed α_1 crystals are isomorphous with the hexagonal silver nitrate; the mixed β_1 crystals are isomorphous with the orthorhombic silver nitrate.

These mixed β_1 crystals are generated, by a sufficient lowering of temperature, at the expense of the mixed α_1 crystals very rich in silver nitrate.

One may, for the transformation of α_1 crystals into β_1 crystals, construct the curves Γ_1, γ (Fig. 90), analogues of the curves Γ, γ which were discussed in Art. **221**.

These curves start from the point τ_1, whose abscissa is $X=0$ and whose ordinate $O\tau_1=159°.5$ is the transformation temperature of crystals of pure silver nitrate. They both descend from left to right.

The line γ_1 meets the line A_1B_1 in a point B_1, of ordinate $O\theta_1=$ 138°; to this ordinate corresponds a point D_1 on the line T_1 The point D_1 is a *eutectic point;* by lowering the temperature, the α_1 crystals are transformed into a mixture of β_1 and α_2 crystals.

At the temperature 138° the α_1 crystals, whose composition is $X=\theta_1B_1$, remain in equilibrium in contact with the β_1 crystals, whose composition is $X=\theta_1B_1$; they remain also in equilibrium in contact with the α_2 crystals, whose composition is $X=\theta_1B_2$; thermodynamics shows without difficulty that the β_1 and α_2 crystals, whose composition we have just stated, remain in equilibrium at 138° in contact with each other.

At temperatures less than 138° one may observe equilibrium

states betwen the β_1 and α_2 crystals. The two points which represent the mixed β_1 and α_2 crystals capable of resting in contact at a given temperature have two curves for loci. From what we

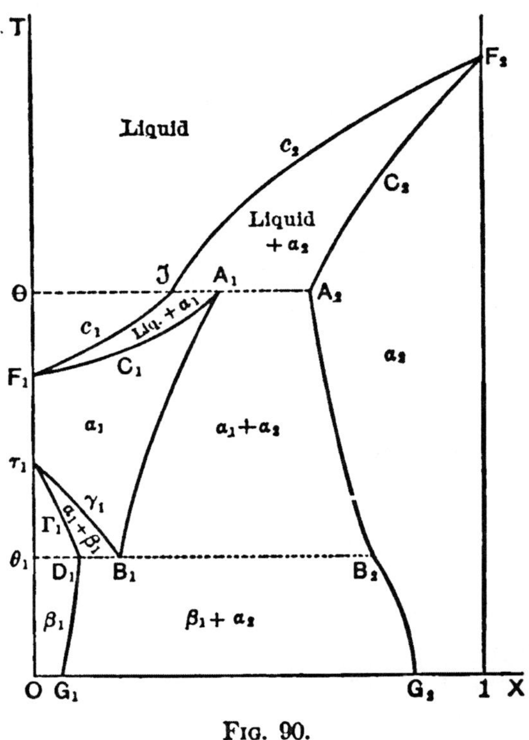

FIG. 90.

have just said, the first of these two curves, G_1D_1, passes through the point D_1, and the second, G_2B_2, ends at the point B_2.

There is thus obtained the arrangement of curves shown in Fig. 90; in this figure the scale is not preserved.

The admirable researches of van Eyk [1] on the mixtures of potassium nitrate and thallium nitrate have unravelled a still more complicated case, for the salts considered are *isotetramorphous;* Fig. 91, where thallium nitrate has been taken for substance 1

---

[1] VAN EYK, *Zeitschrift für physikalische Chemie*, v. 30, p. 430, 1899; *Archives néerlandaises des Sciences exactes et naturelles*, 2d S., v. 4, p. 118, 1901.

and potassium nitrate for substance 2, and where the scale is not kept, summarizes the results of these researches.

The liquid may coexist with the α_1 crystals if it is rich in thallium nitrate and with the α_2 crystals if it is rich in potassium nitrate; the α_1 and α_2 crystals both belong to the hexagonal system, but are not isomorphous with each other.

To the freezing into α_1 crystals correspond the curves c_1, C_1, which start from the point F_1, where $F_1=206°$, and descend from

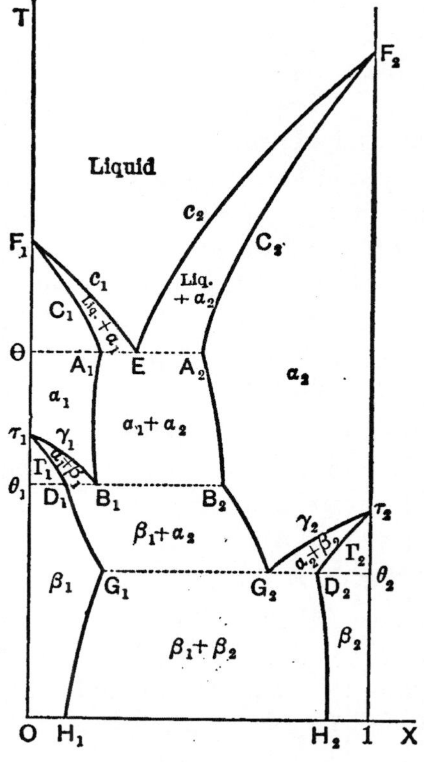

FIG. 91.

left to right; to the freezing into α_2 crystals correspond the curves c_2, C_2, which start from F_2, of ordinate 339°, and descend from right to left.

The lines meet at the point E, a eutectic point corresponding to the temperature $O\theta=182°$.

At a lower temperature the α_1 crystals are changed into β_1 crystals, which are orthorhombic; corresponding to this transformation are the curves γ_1, Γ_1, which start from the point τ_1, of ordinate 144°.3, and descend from left to right to the points B_1, D_1 of common ordinate $O\theta = 133°$.

The temperature being lowered still more, the α_2 crystals are changed into β_2 crystals, which are orthorhombic. To this transformation correspond the curves γ_2, Γ_2; starting from the point τ_2, whose ordinate is 129°.5, these curves descend from right to left as far as the points D_2, G_2, of common ordinate 108 .5.

These very complex cases, reduced to such clear and expressive representati ns, are quite fitting to give emphasis to the importance of thermodynamical princip es in the s udy of isomorphism.

224a. Sulphur and phosphorus. Researches of Boulouch.— The methods we have just exposed help us to decide if the crystals which grow in the presence of a liquid mixture are a definite compound or mixed crystals; they are also valuable for the discussion of certain questions in litigation; in particular they seem called to play a considerable rôle in the study of systems where two metalloids exist togeth r; more than one substance, obtained in such conditions and regarded as a definite ompound, is perhaps only a conglomeration of mixed crystals.

Boulouch [1] has applied this method to the study of bodies which are formed within a liquid mixture of sulphur and phosphorus.

Berzelius had described the sulphides P_4S, P_2S, P_2S_{12} as being formed in such conditions; according to the researches of Boulouch, none of these definite compounds really exists. By cooling liquid mixtures of sulphur and phosphorus, only two kinds of mixed crystals are obtained.

The first are fo·med in the mixtures rich in phosphorus; they are isomorphous with the crystals of white phosphorus; the freezing-point of the liquid giving rise to these crystals s the lower as the liquid is richer in sulphur; the fusing-point of these crystals is also the lower as the crystals contain more sulphur.

If the temperature is taken as ordinate, and for abscissa the

[1] R. Boulouch, *Comptes Rendus*, v. 135, p. 166, 1902.

ratio X of the mass of sulphur to the total mass of the mixture, we obtain (Fig. B) a congelation curve which descends from left to right along PE and a fusion curve which descends along PA; P is the freezing-point of pure phosphorus.

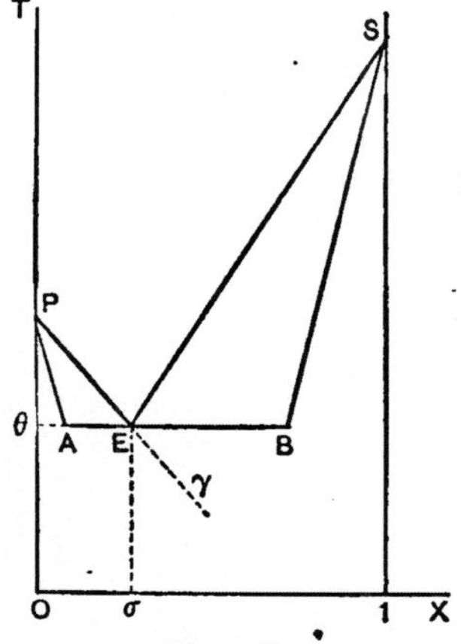

Fig B.

The second mixed crystals are formed in the mixtures rich in sulphur; they are isomorphous with the crystals of clinorhombic sulphur; the congelation curve descends from right to left along SE, and the fusion curve along SB; S is the freezing point of clinorhombic sulphur.

The point E furnishes us with an example of eutectic point as sha p as that studied by Hissink (Art. 223); the co-ordinates of this point are $\sigma=0.228$ and $\theta=9°.8$.

The temperature 9°.8 is the fusing-point of every solid system containing the two kinds of mixed crystals at once.

The liquid mixture remains readily in surfusion with respect to the mixed crystals of the second k nd; it may then furnish mixed crystals of the first kind; the corresponding freezing-points are located on the line $P\gamma$, extension of the line PE.

224b. Sulphur and selenium. W. E. Ringer's researches.—The conclusions reached by Boulouch studying the mixtures of sulphur and phosphorus are remarkably simple; much more complicated results follow the study o' the mixtures of sulphur and selenium, as has been found recently by W. E. Ringer.[1]

There are here formed four k nds of mixed crystals; the various peculiarities of their fusion, their crystallization, their transformation into each other are represented in Fig. C, where the scale has not been exactly kept Temperatures are taken as ordinates; as abscissæ are taken the ratio X of the mass of selenium to the total mass of the mixture; the full lines have been determined experimentally.

Two kinds of crystals rich in sulphur may be obtained; we shall denote them by the letters α and β; the α crystals are iso-

[1] W. E Ringer, *Zeitschrift für anorganische Chemie*, v. 32, p. 183, 1902.

morphous with orthorhombic sulphur, and the β crystals with clinorhombic sulphur.

The line AE is the congelation line of the liquid into the state of β crsytals; the line AB is the fusion line of these crystals; the point A is the fusing-point of clinorhombic sulphur.

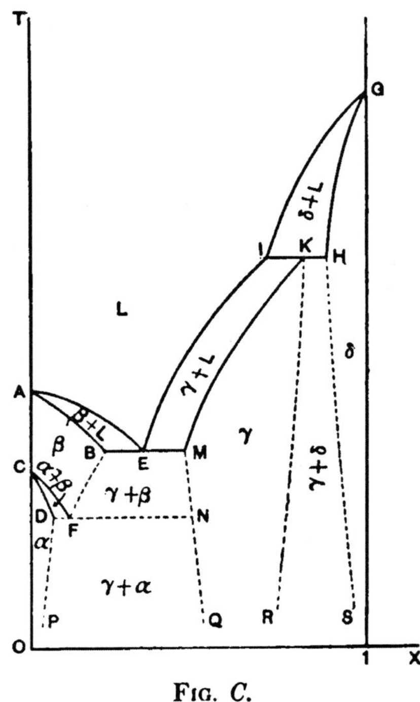

Fig. C.

The mixed β crystals may, when they are cooled, be transformed into mixed α crystals the transformation-points form the line CF; inversely, when they are heated, the mixed α crystals are transformed into β crystals, the transformation-points lie on the line CD; C is the transformation-point of cl norhombic into orthorhombic sulphur.

The liquid mixtures very rich in selenium furnish mixed δ crystals, isomorphous with metallic selenium; the corresponding freezing-points form the line GI; the fusing-points of the δ crystals form the line GH; G is the fusing-point of metallic selenium.

Finally, the liquid mixtures of mean composition deposit mixed crystals of a fourth form, γ, at the congelation-points, which are those of the line *EI*; the fusing points of these γ crystals are those of the line *MK*.

According to the position, in the plane, of the point representing the temperature and composition of the system, the system in equilibrium may contain a single one of the five phases α, β, γ, δ, and L (liquid), or be divided into two of these phases. The various circumstances which may occur have been marked on Fig. *C*.

The researches of Boulouch and of Ringer indicate clearly that the whole chemistry of the metalloids should be submitted to a revision guided by the methods of Thermodynamics.

CHAPTER XIV.

MIXED CRYSTALS (*Continued*). OPTICAL ANTIPODES. METALLIC ALLOYS.

I. Optical Antipodes.

225. Mixed crystals are not limited to mixtures of isomorphous bodies. Their frequency in organic chemistry.—Mixed crystals are constantly met with when there are crystallized together two substances of similar chemical formulæ, isomorphous in the sense Mitscherlich gave to this term. But frequently also substances which have not similar chemical formulæ show themselves capable of forming mixed crystals. Thus we have seen, in Art. **219**, ferric chloride forms mixed crystals with ammonium chloride. Facts of this sort indicate that prudence is needed when use is made of Mitscherlich's law in the appreciation of chemical analogies; the property of giving mixed crystals often accompanies the similarity of chemical formulæ, but it may be met with when this similarity is in default.

The compounds of organic chemistry, and especially the substances in the aromatic series, are, in a great number of cases, capable of forming mixed crystals two by two. This property is often correlative of a true crystallographic isomorphism; this is what takes place, for example, with azobenzol and stilbene, studied from this point of view by G. Bruni.[1] Furthermore, the symbols of these two substances

$$\begin{array}{c} N—C_6H_5 \\ \| \\ N—C_6H_5 \end{array} \qquad \qquad \begin{array}{c} HC—C_6H_5 \\ \| \\ HC—C_6H_5 \end{array}$$

Azobenzol Stilbene

[1] G. Bruni, *Rendiconti dell' Accademia dei Lincei*, v. 8, p. 570, 1899.

may be regarded as analogues, so that here is a case of complete isomorphism, in the sense given to this term by Mitscherlich.

In other cases it is more difficult to admit of an analogy between the chemical symbols of substances which mix in crystallizing; it is thus[1] that carbazol and anthracene both form mixed crystals with phenanthrene, while the chemical symbols of these three substances,

$$\begin{array}{c} H_4C_6—CH \\ \| \quad\quad \| \\ H_4C_6—CH \end{array} \qquad \begin{array}{c} CH \\ \diagup \mid \diagdown \\ H_4C_6 \mid H_4C_6 \\ \diagdown \mid \diagup \\ CH \end{array} \qquad \begin{array}{c} H_4C_6\diagdown \\ \mid \quad \rangle NH \\ H_4C_6\diagup \end{array}$$

Phenanthrene — Anthracene — Carbazol

can with difficulty be regarded as analogous.

The absence of analogy is still more striking between naphthalene and monochloracetic acid, whose mixtures have been studied by Cady.[2] Within these mixtures there are formed two kinds of mixed crystals; the first, rich in naphthalene, are isomorphous with pure naphthalene crystals; the others, rich in monochloracetic acid, are isomorphous with those furnished by this acid taken by itself.

The observed phenomena have the same characteristics as those described in Art. 223. There we found a eutectic conglomerate whose mean composition is fixed, and which is composed of two kinds of mixed crystals.

Organic chemistry furnishes innumerable examples of mixed crystals, among which several have been studied by Küster, Garelli, Bruni, and various other observers.[3]

226. Optical antipodes. Inactive substances to which they may give rise.—The idea of mixed crystals assumes great importance in the discussions relative to the properties of substances

---

[1] G. Bruni, *Rendiconti dell' Accademia dei Lincei*, v. 7, p. 138, 1898.

[2] Cady, *Journal of Physical Chemistry*, v. 3, p. 127, 1899.

[3] The reader will find interesting information on the whole of this question of solid solutions in the following: G. Bruni, *Ueber feste Lösungen* (Arhen's Sammlung, v. 6, part 12).

gifted with rotary power, discussions essential to the progress of stereochemical doctrines.

Everybody is acquainted with the researches of Pasteur on the tartaric acids and the tartrates.

There exist two tartaric acids which possess exactly the same physical and chemical properties save one: solutions of the first possess a certain rotary power to the right; solutions of the second possess exactly the same rotary power, but to the left. The first is the *right-handed* acid, the second is the *left-handed* acid.

The crystals furnished by the right-handed acid do not possess the rotary power, but they have a *non-superposable hemiedry;* the crystal is not superposable upon its image in a mirror.

The left-handed acid likewise furnishes crystals without action on polarized light and possessing a similar hemiedry. *A left-handed crystal is superposable upon the image of a right-handed crystal in a mirror and reciprocally.*

By representing each atom of quadrivalent carbon in the form of a regular tetrahedron, stereochemical notation attributes to these two acids two symbols as distinct as the crystals giving them. The symbol of the right-handed acid is not superposable upon itself in a mirror, but on reflecting it in a mirror it reproduces the formula of the left-handed acid.

These two acids have an *isomer*, the *inactive acid,* whose solutions are without action on polarized light; the crystals which it furnishes are *holoedrons;* each of them is superposable upon its image seen in a mirror; stereochemistry attributes to this inactive acid a formula which is reproduced, identical with itself by reflection in a plane mirror; no reaction separates this acid into right- and left-handed acids.

By combining molecule to molecule, right-handed tartaric acid and left-handed tartaric acid form a *polymer*, *racemic acid.* This acid, whose stereochemical formula is then superposable upon its image in a mirror, gives holoedric crystals gifted with the same property; by dissolving it, one obtains a liquid devoid of rotary power.

These properties are not peculiar to the tartaric acids and the tartrates; a great number of organic compounds likewise possess them.

Such a compound possesses two isomeric varieties which have exactly the same chemical and physical properties except one: the *right-handed* kind, in the state of fusion or of solution, possesses a right-handed power of rotation; the left-handed sort has exactly the same rotary power, but to the left. The two varieties of crystals are hemiedric; the crystals of the right-handed variety, on reflection in a mirror, reproduce crystals of the left-handed sort and conversely. These crystals are in general devoid of rotary power; when they are so gifted, the right and left crystals have equal rotary powers, in opposite directions. The stereochemical notation attributes different formulæ to these two isomers; one of the formulæ is the image of the other in a mirror. These two isomeric substances are said to be *enantiomorphous*, or to be *optical antipodes* of each other.

Often there is occasion to add a third inactive *isomer* to these two optical antipodes; devoid of rotary power in all its states, this inactive isomer furnishes holoedral crystals; stereochemistry assigns it a formula which is reproduced, identical with itself, by reflection in a mirror.

In a great number of cases a molecule of the right isomer may combine with a molecule of the left isomer to form a polymer which is without action on polarized light and which gives holoedral crystals; by analogy with racemic acid and the racemates, which are formed in this way, the name given to this polymer is *racemic compound.*

The racemic combination is not the only solid substance which, on melting or dissolving, furnishes a liquid *inactive by compensation.* The same property belongs to a mixture of right and left crystals where the two kinds of crystals appear in equal quantities. It belongs likewise to mixed crystals, which the two left and right varieties are often capable of furnishing, when the two sorts figure in the same proportion in these mixed crystals.

Not only may the two optical antipodes furnish mixed crystals, but it also happens that each of them may furnish mixed crystals with the inactive isomer. Thus Fock [1] has made the following curious observation:

---

[1] Fock, *Zeitschrift für Krystallographie*, v. 31, p. 479, 1899.

Inactive pinonic acid (pinonsäure), which is orthorhombic, forms, either with right pinonic acid or with left pinonic acid, mixed orthorhombic crystals. It forms also, with right pinonic acid, mixed quadratic crystals having hemiedry, and rigorously isomorphous with the crystals given by right pinonic acid when it is isolated. Finally, it gives, with left pionoic acid, mixed crystals symmetrical with the preceding

A racemic compound may perhaps be formed of the crystals mixed with each of the two optical antipodes, although as yet the fact has not been demonstrated with certainty.

227. Freezing of the mixture of two optical antipodes.—Suppose that two substances, optical antipodes of each other, are melted and mixed together. Let us study the freezing-point of this mixture and the nature of the precipitate obtained.

To express the composition of the liquid mixture or, if there is need, of the precipitate obtained, we shall lay off on the axis of abscissæ the mass X of the left antipode contained in a unit mass of the mixture; $(1-X)$ will be the mass of the right antipode which is associated with it. On the axis of ordinates lay off the temperature T.

The two right and left antipodes have exactly the same physical properties; if, therefore, the liquid mixture which contains X grammes of the right antipode and $(1-X)$ grammes of the left has a certain freezing-point, the liquid mixture containing X grammes of the left antipode and $(1-X)$ of the right should have identically the same freezing-point. The freezing-point curves will therefore be symmetrical with respect to the line $X=\frac{1}{2}$.

If the system can furnish mixed crystals, the freezing-point curve of these crystals will have the same axis of symmetry.

228. The congelation of the mixture furnishes neither racemic compound nor mixed crystals.—This is the simplest case.

The mixtures rich in he right-handed antipode should deposit crystals which enclose exclusively this right antipode; the phenomena may be compared in all respect· to the formation of ice within a salt solution. The freezing-point is the lower as the richness of the liquid mixture in the left antipode is the greater.

The freezing-point curve of right crystals ((Fig. 92) starts from the point F_1, fusing-point of the right crystals in the pure state, and descends from left to right. The congelation curve of the left crystals starts from F_2, fusing-point of the pure left crystals; the two points F_1. F_2 have the same ordinate equal to the common temperature of fusion of the right and left crystals.

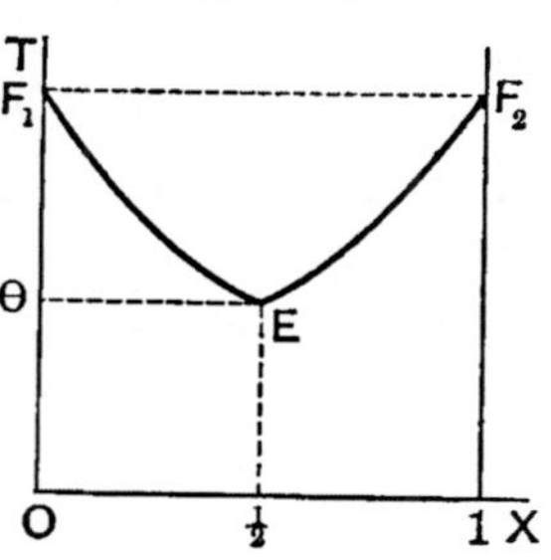

Fig. 92.

These two curves intersect at a point E, of abscissa ½ and ordinate θ; it is a eutectic point analogous to that observed (Art. 214) in studying the congelation of a mixture of two melted salts which do not form a double salt; the two cases diffe from one another only in he aspect of the two fusion curves which are anything for the case tre ted in Art. 214, and both symmetrical with respect to the line $X=\frac{1}{2}$ for the present case. The eutectic conglomerate has for mean composition $X=\frac{1}{2}$; it encloses in equal proportions right and left crystals; melted or dissolved, it will give an inactive mixture by compensation.

This case, theoretically possible, does not seem to have been met with as yet among those which have been carefully studied.

229. The congelation of the mixture may give a racemic compound.—We encounter here a particular case of the problem treated in Art. 214: freezing of a mixture of two melted salts capable of giving a double salt; the symmetry of the congelation curves with respect to the line $X=\frac{1}{2}$ alone distinguishes this from the general case.

The liquid mixtures which contain a large proportion of the right antipode deposit this substance in the pure state; on obtains a fusion curve $F E_1$ which descends rom left to right (Fig. 93); it is a portion of the line F_1E drawn in Fig 92.

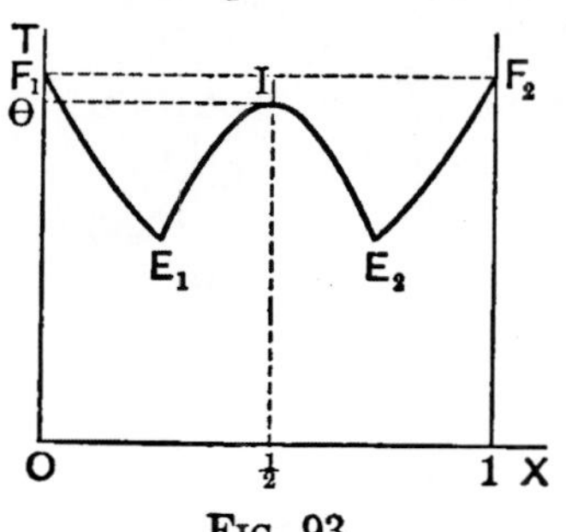

Fig. 93.

Similarly, the liquid mixtures rich in left antipode deposit this substance in the pure state; one obtains a fusion curve F_2E_2 descending from right to left; it is a portion of the line F_2E of Fig. 92.

The two points E_1, E_2 are joined to another by the congelation curve E_1IE_2 of the racemic compound; symmetrical with respect o the line $X=\frac{1}{2}$, this curve has, for the abscissa $X=\frac{1}{2}$, a point I where the tangent is parallel to OX; this point I is an indifferent point; the liquid mixture there has the same composition as the racemic compound; the ordinate θ of this point is the fusing-point of this compound.

This disposition is frequent; it is met with notably in the study of methylbenzoic ether. benzylaminosuccinic acid, aminosuccinic acid,[1] benzoyltetrahydroquinaldine.[2]

In certain cases the freezing-point curve E_1IE_2 of the racemic compound is extremely reduced and the arrangement of Fig. 94 is obtained. There is a tendency towards the case studied in the

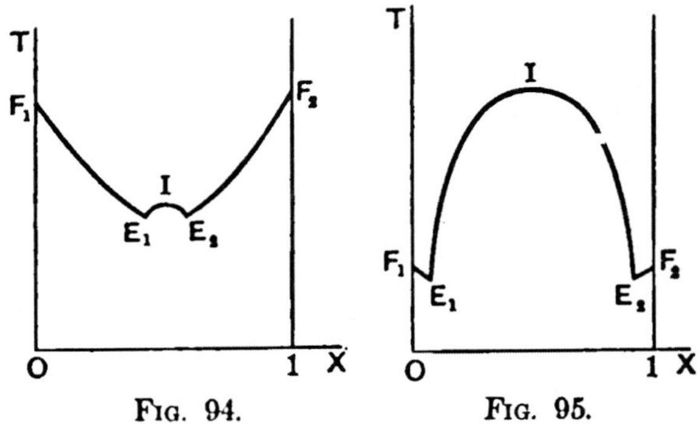

FIG. 94. FIG. 95.

preceding article. Phenylglycolic acid (Mandelsäure) and dimethylic ether of diacetyltartic acid, studied by Adriani, are two examples of this.

In other cases the two curves F_1E_1, F_2E_2 (Fig. 95) are greatly reduced and the freezing curve $E\,IE_2$ of the racemic compound occupies almost the whole field of congelation; this is also, according to Adriani, the case with dimethylic ether of tartric acid.

230. The congelation of the mixture gives mixed crystals.—The freezing of the mixture may give mixed crystals whose every element contains the right-handed substance and its left-handed isomer united in a certain proportion; let x be the mass of the

---

[1] CENTNERSZWER, *Zeitschrift für physikalische Chemie*, v. 29, p. 75, 1899.

[2] ADRIANI, *Zeit. f. phys. Chemie*, v. 33, p. 453, 1900.

right substance and $(1-x)$ the mass of the left substance in a unit of mass of these mixed crystals; for $x=\frac{1}{2}$ the crystals will be holoedral; by fusion or solution they will give a substance inactive by compensation; for two values of x equidistant from $\frac{1}{2}$ there will be two non-superposable crystalline forms both symmetrical with respect to a plane.

Let T be the freezing-point of the liquid of composition X, and M the point of coordinates X, T (Fig. 96); at this temperature T the liquid of composition X deposits mixed crystals of composition x; let m be the point of coordinates x, T.

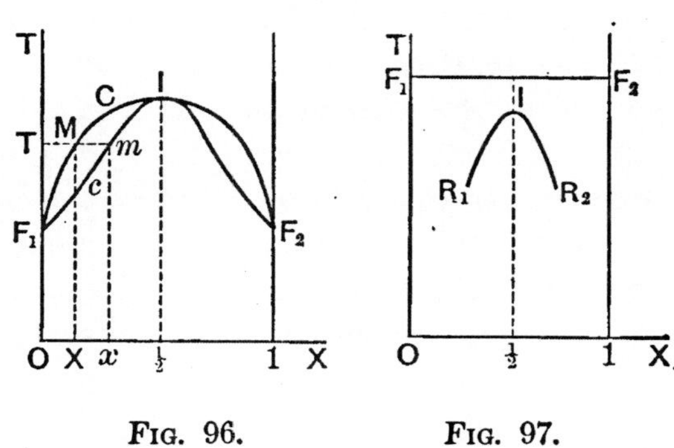

FIG. 96. FIG. 97.

While X varies from 0 to 1, the point M describes the curve C, and the point m the curve c, drawn below the curve C. These two curves pass through the fusing-point F_1 of the right crystals taken in the pure state and through the fusing-point F_2 of the left crystals in the pure state.

The curve C, formed necessarily by two branches both symmetrical with respect to the line $X=\frac{1}{2}$, has, for the abscissa $X=\frac{1}{2}$, a point of maximum ordinate. From the first theorem of Gibbs and Konovalow (Art. 194), which may be applied to the double mixture formed by the mixed crystals and the mixed liquid, this point belongs also o the line c, for which it is also a point of maximum or minimum ordinate. At this indifferent point I the mixed liquid, which is inactive by compensation, must give, on freezing, mixed holoedral crystals of composition $x=\frac{1}{2}$.

According to Adriani, this disposition is observed in the freezing of carvoxime, bihydrocarvoxime, and benzoic oxime.

Camphoric oxime offers a very curious particular case, represented in Fig. 97. Whatever is the composition X of the liquid mixture, Adriani found its freezing-point constant and equal to 118°.8; the line C is here reduced to a straight line F_1F_2 parallel to OX.

Each of the points on this line may be regarded, if so wished, as a point of maximum ordinate; the theorem of Gibbs and Konovalow may be applied to each of these points; whatever the composition of the mixed liquid, it deposits mixed crystals of the same composition.

We have here an example of the rule that various authors thought general for the congelation of mixed crystals (Art. **220**).

Another peculiarity renders this example, studied by Adriani, very interesting: when the temperature is lowered the mixed crystals are seen to transform themselves into crystals of a racemic compound; we may construct a portion R_1IR_2 of the curve, analogous to the congelation curve of a racemic compound within a mixed liquid, which corresponds to this transformation; the highest point I of his curve corresponds to the temperature 103°.

231. Formation in solution of a racemic compound.—The precipitation within a solution of one of the substances we have just studied leads to the study of the equilibrium of a system no longer bivariant, but trivariant; this study is, from the experimental point of view, much less advanced than the preceding; it has given rise nevertheless to several interesting researches; among this number is the analysis of the conditions of formation of the double racemate of sodium and ammonium, analysis for which we are indebted to Van't Hoff and van Deventer.[1]

The formation, within a solution, of a racemate at the expense of the right and left tartrates is comparable in all points with the formation of a double salt at the expense of two simple salts, formation which we have already studied (Arts. **102** *et seq.*). The study of the phenomenon will be somewhat simplified on account of the identity which exists between the physical properties of the two right and left isomers.

---

[1] Van't Hoff and van Deventer, *Zeitschrift für physikalische Chemie*, v. **16**, p. 173; Van't Hoff, Goldschmidt, and Jorissen, *ibid.*, v. **17**, p. 49.

Take, as was done in studying the double salts, three axes of rectangular coordinates OT, Os_1, Os_2 (Fig. 98); on the first lay off temperatures, on the second concentrations of the right-handed tartrate solution, on the third the concentrations of the left tartrate solution.

We shall be led to represent all the possible equilibrium states by a surface formed by the domain D of the right tartrate, the domain G of the left tartrate, the domain R of the racemate; this figure will be symmetrical with respect to the bisecting plane of the diedral TOs_1s_2.

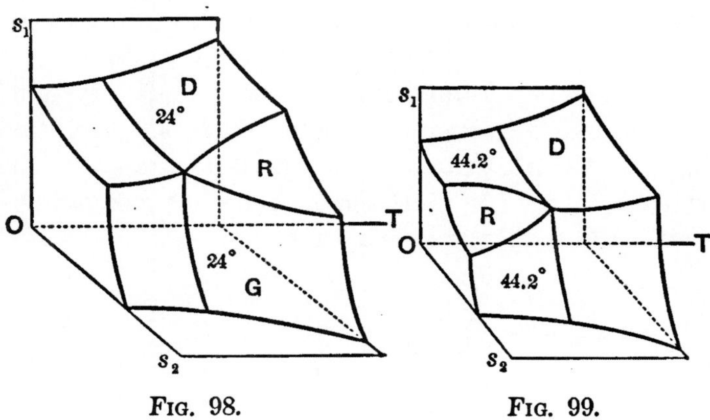

FIG. 98. FIG. 99.

The double racemate of sodium and ammonium is formed at the expense of the two double tartrates, according to the formula

$$NaNH_4C_4H_4O_6 \cdot 4H_2O\ (D) + NaNH_4C_4H_4O_6 \cdot 4H_2O\ (G) = Na_2(NH_4)_2(C_4H_4O_6) \cdot 2H_2O + 4H_2O.$$

This racemate forms in the solution only at temperatures higher than 24°; at temperatures below 24° the surface possesses only the domains of the right tartrate and of the left tartrate.

The two, right and left, tartrates of rubidium furnish a racemate according to the formula [1]

$$Rb_2C_4O_6H_4\ (D) + Rb_2C_4O_6H_4\ (G) + 4H_2O = Rb_4(C_4O_6H_4)_2 \cdot 4H_2O.$$

[1] VAN'T HOFF and MÜLLER, *Berichte der Deutschen Chemischen Gesellschaft*, v. 31, p. 2206.

Within the solution this racemate is formed only at temperatures below 40°.4; at higher temperatures the surface possesses only the domains of the right and left tartrates (Fig. 99).

II. The Metallic Alloys.

232. Liquid mixtures which deposit metals in the pure state or a definite compound.—The principles developed in the preceding chapters, and particularly the notion of mixed crystals, commence to throw some light on the constitution, obscure for so long a time, of the metallic alloys; the majority of alloys which were regarded as definite chemical compounds, having a fixed composition and a definite fusing-point, are considered to-day as eutectic conglomerates formed either of two solids, crystallized or not, or of two kinds of mixed crystals, or of two solid solutions.

G. Charpy [1] has studied with the greatest care (Art. **106**) the alloy formed by lead, tin, and bismuth; he has also studied, but in less detail, the following ternary alloys:

Sn,Cu,Sb;
Sn,Pb,Sb;
Pb,Cu,Sb;
Zn,Sn,Sb;
Cu,Sn,Pb.

Besides these cases, the only alloys which have been minutely studied are mixtures of two metals which we shall indicate by the indices 1 and 2.

The most simple case to be met with is that where a lowering of temperature imposed on the melted mixture of two metals always produces either the deposit of the metal 1 in the pure state, or the deposit of the metal 2 in the pure state.

This case is quite similar to the one treated in Art. **214**, where a liquid, formed by two fused salts, can furnish no other solid than one or the other salt in the pure state.

---

[1] G. Charpy, *Etudes sur les alliages blancs dits antifriction* (*Contribution à l'Etude des alliages*, published by the alloys committee of the Société d'Encouragement pour l'industrie nationale, p. 203, Paris, 1901). This is a very valuable contribution to our knowledge of alloys.

As abscissa (Fig. 100) take the value of X which represents the composition of the liquid mixture; as ordinate, take the temperature.

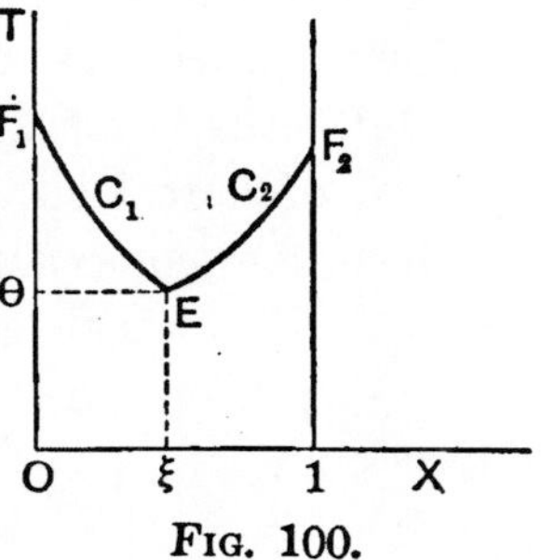

FIG. 100.

The congelation curve C_1 of the metal 1 starts from the point F_1, whose ordinate is the fusing-point of this metal, and descends from left to right; the congelation curve C_2 of the metal 2 starts from the point F_2, whose ordinate is the fusing-point of this metal, and descends from right to left. These two curves intersect in a eutectic point E, of abscissa ξ and ordinate θ; ξ and θ indicate the composition of the eutectic conglomerate and its fusing-point.

Guthrie[1] has studied several systems which enter into this category; here are the values he has found for the coordinates ξ and θ of the eutectic points:

Mixed Metals.	θ	ξ
1. Bismuth. 2. Zinc.	248°	0.0715
1. Bismuth. 2. Tin.	133°	0.539
1. Bismuth. 2. Lead.	122°.7	0.4442
1. Bismuth. 2. Cadmium.	144°	0.4081

Also belonging to this class is the system, studied by Sir Roberts-Austen,[2] formed of the two metals lead (1) and tin (2); for this case

$$\theta=183°, \qquad \xi=0.62.$$

But among the systems of this kind, none doubtless has been studied with as much care as the alloy formed of the two

[1] GUTHRIE, *Philosophical Magazine*, 5th S., v. 22, p. 46, 1884.

[2] ROBERTS-AUSTEN, *Proceedings of the Royal Society*, v. 63, p. 452, 1898.

metals lead (1) and antimony (2), object of the researches of Roland-Gosselin, H. Gautier,[1] and of Charpy.[2]

When the composition of the liquid mixture varies from $X=0$ to $\xi=0.13$, the freezing-point is lowered from $T_0=326°$, the freezing-point of pure lead, to $\theta=228°$.

When the composition of the liquid mixture changes from $\xi=0.13$ to $X=1$, the freezing-point rises from $\theta=228°$ to $T_1=632°$, freezing-point of pure cadmium.

The point

$$\theta=228°, \qquad \xi=0.13$$

is a eutectic point.

When a liquid mixture for which X is included between 0 and 0.13 is brought to the freezing-point, it furnishes crystals of pure lead; the proportions X in antimony increases, the freezing-point T is lowered; this goes on until X attains the value 0.13 and T the value 228°; at this moment the remaining liquid solidifies. Viewed with the microscope, the lingot obtained is seen to be formed of large lead crystals implanted in a finely grained eutectic mixture.

On the contrary, a liquid mixture for which X is included between 1 and 0.13, brought to the freezing-point, deposits antimony crystals; the proportions X in antimony decreases, until X reaches the value 0.13 and T the value 228°; then the rest of the liquid solidifies into a eutectic which cements the antimony crystals together, as may be shown with the microscope.

According to the same authors, the alloy formed of zinc (1) and aluminium (2) possesses properties in all points analogous to the preceding, X increasing from 0 to 0.05, the freezing-point decreases from $T_0=433°$, the zinc fusing-point, to $\theta=389°$; the solid produced is pure zinc. X continuing to increase from 0.05 to 1, the freezing-point increases from $\theta=389°$ to $T_1=650°$, fusing-point of aluminium. The point

$$\theta=389°, \qquad \xi=0.05$$

is a eutedic point.

---

[1] H. Gautier, *Bulletin de la Société d'Encouragement*, Oct. 1896, and *Contribution à l'Etude des alliages*, p. 93.

[2] G. Charpy, *Bull. de la Soc. d'Encouragement*, March 1897, and *Contrib. à l'Etude des alliages*, pp. 121 and 203.

A case more complicated than the preceding may occur: it is that where the liquids containing large proportions of the metal 1 deposit the metal 1 in the pure state, where the liquids containing large proportions of metal 2 deposit this metal in the pure state, finally where liquids of intermediate composition deposit a definite compound.

The congelation curves have then usually the shape we found (Fig. 79, p. 241) when studying a mixture of two melted salts where a double salt may be formed.

The type of these alloys seems to be the alloy formed of tin (1) and copper (2), studied [1] by H. Le Chatelier, by Sir Roberts-Austen and Stansfield, and by G. Charpy.

When X varies from 0 to 0.03, the solidifying-point is lowered from $T_0=232°$, fusing-point of pure tin, to $\theta=227°$; the solid deposited is pure tin.

When X varies from 0.72 to 1 the point of congelation rises from $W=770°$ to 1065°, fusing-point of pure copper; the solid deposited is pure copper.

When X increases from 0.03 to 0.72 the freezing-points increase constantly from $\theta=227°$ to $W=770°$; the solid deposited is a definite compound: $SnCu_3$.

From what we have just said, the freezing-point curve of this definite compound does not possess an indifferent point; the point

$$\theta=227°, \qquad \xi=0.03$$

is a eutectic point; the point

$$W=770°, \qquad X=0.72$$

is a transition-point.

In other cases, the freezing-point curve of the definite compound has an indifferent point; the three freezing-point curves have then, very exactly, the arrangement shown in Fig. 79, p. 241. Such would be the case realized, according to Le Chatelier,[2] by the alloys of copper and antimony within which the definite compound $SbCu_3$ may be formed.

233. Liquid metallic mixtures which give solid solutions.— The case we have just mentioned is the simplest, but it appears

[1] See *Contribution à l'étude des alliages*, pp. 99 and 139.

[2] H. Le Chatelier, *Bulletin de la société de l'Encouragement*, 1895, p. 573.

to be quite rare; the most often, when a mixture of two fused metals is cooled, a solid solution is obtained which contains the two metals in variable proportion.

The simplest case to be had is that where two metals, isomorphous with each other, form, whatever their proportions, a single kind of mixed crystals; all the freezing-points then range themselves along a single curve (Fig. 86, p. 261), joining the fusing-point of one of the metals to the fusing-point of the other.

This is the case with the alloys of gold and silver. The freezing-points all lie on a sensibly straight line extending from the fusing-point of gold to that of silver.

The alloys of bismuth and antimony, whose fusibility curve and microscopic structure have been studied by Roland-Gosselin and by Charpy,[1] are of this same type; the freezing-points lie along a single curve joining the fusing-point $T_0=268°$ of bismuth to the fusing-point $T_1=622°$ of antimony.

The particularly simple case realized by these alloys is quite rare; in general there may be formed two kinds of solid solutions, crystallized or not; the first, which include as a special case the metal 1 taken alone, are formed in liquid mixtures rich in this metal; the others, among which should be counted the pure metal 2, arise in liquid mixtures which enclose principally the metal 2.

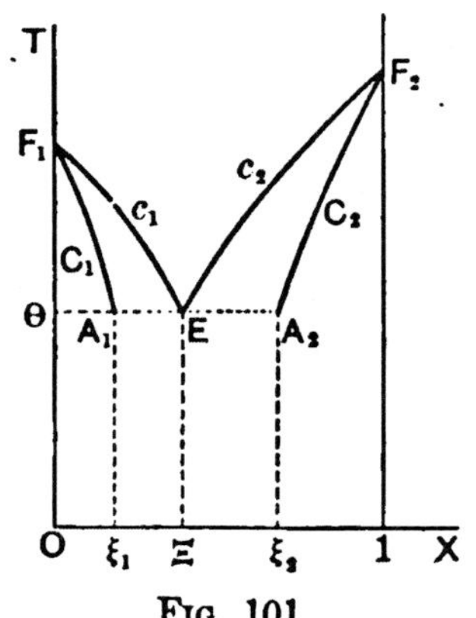

Fig. 101.

To these two kinds of solid solutions correspond two distinct congelation curves, c_1 and c_2, the first starting from the point F_1 (Fig. 101), whose ordinate is the temperature of fusion of the metal 1, the second starting from the point F_2, whose ordinate is the temperature of fusion of the metal 2.

In general the first of these curves descends from left to right and the second from right to left; the arrangement is similar to that found by Reinders in studying the mixtures in fusion of potassium nitrate and sodium nitrate (Art. 223).

The curves c, c_2 intersect in a certain point E of coordinates Ξ, θ.

[1] G. Charpy, *Contribution à l'étude des alliages*, pp. 114 and 138.

To these curves must be joined the fusion curves C_1, C_2 of the solid solutions. To the temperature θ corresponds on the first curve a point A_1 of abscissa ξ_1, and on the second curve a point A_2 of abscissa ξ_2.

A eutectic conglomerate of mean composition Ξ is produced at the temperature θ; this conglomerate is a juxtaposition of masses formed by the solid solution of the first kind whose composition is ξ_1, and of masses belonging to the solid solution of the second kind of composition ξ_2.

Their properties are similar to those of the alloys of silver and copper.

Solid alloys are known in which the copper is united to a proportion of silver varying from 0 to a certain limit, and alloys where silver is united to a proportion of copper variable from 0 to a certain limit. Besides, an alloy is known, Levol's alloy, in which copper and silver enter in a fixed ratio. This alloy has a definite fusing-point, which is $\theta = 777°$. It was for a long time considered as a definite compound, to which the formula Ag_3Cu_3 was attributed.

In studying the freezing of fused mixtures of silver and copper, Sir Roberts-Austen[1] and Heycock and Neville[2] have determined two congelation curves c_1, c_2; their point of intersection has for co-ordinates exactly the composition and fusing-point of Levol's alloy; the latter is a eutectic conglomerate; by a microscopic examination of Lavol's alloy Osmond[3] has corroborated this conclusion.

The alloys of copper and gold[4] give rise to considerations similar in all respects to the preceding.

233a. Alloys of zinc and antimony; Reinders' researches.— The two congelation curves c_1, c_2 of the two kinds of mixed crystals, instead of uniting in a eutectic point, as happens in the last two cases we have cited, may be joined in a transition-point. The arrangement they assume is then similar to that found by Hissink in studying the congelation of silver nitrate and of sodium nitrate (Fig. 88, p. 277).

A particularly remarkable example, and resembling this type,

---

[1] ROBERTS-AUSTEN, *Proceedings of the Royal Society of London*, 1875, p. 481; *Annual Mint Report*, 1900, p. 70.

[2] HEYCOCK and NEVILLE, *Philosophical Transactions*, v. 189, p. 25.

[3] OSMOND, *Comptes Rendus*, v. 124, p. 1094, 1897.

[4] ROBERTS-AUSTEN, *Annual Mint Report*, 1900, p. 70.

has recently been studied by Reinders;[1] it is furnished by the alloys of the two following metals: (1) zinc, (2) antimony.

Let ω_1 and ω_2 be the molecular weights of these two metals.

When $\frac{\omega_1}{\omega_2}X$ varies from 0 to 1, the mixture may deposit four kinds of distinct mixed crystals, which we shall indicate by the indices 1, 2, 3, 4; to these four kinds of crystals correspond four different congelation curves, c_1, c_2, c_3, c_4; each of these curves is joined to the following in a transition-point.

$\frac{\omega_1}{\omega_2}X$ increasing from 0 to 0.08, the freezing-point rises along the curve c_1 from the fusing-point $T_0=232°$ of pure tin to the temperature $W_{12}=243°$ of the first transition-point.

$\frac{\omega_1}{\omega_2}X$ increasing from 0.08 to 0.2, the freezing-point rises along c_2, from W_{12} to the temperature $W_{23}=310°$ of the second transition-point.

$\frac{\omega_1}{\omega_2}X$ increasing from 0.2 to 0.51, the freezing-point rises along c_3, from W_{23} to $W_{34}=430°$ of the third transition-point.

Finally, $\frac{\omega_1}{\omega_2}X$ increasing from 0.51 to 1, the freezing-point rises along the curve c_4, from W_{34} up to the point of fusion $T=622°$ of pure antimony.

In the same way as for the two congelation curves c_1, c_2 of Fig. 88 (p. 277) correspond two fusion curves C_1, C_2 joined to each other by a line A_1A_2 parallel to OX and having for constant ordinate the transition temperature W, so here we shall have four fusion curves C_1, C_2, C_3, C_4; each of these four curves will be joined to the next by a straight segment, parallel to OX, having for constant ordinate the ordinate of the corresponding transition-point.

According to Reinders, the first straight segment A_1A_2, which has the constant ordinate $W_{12}=243$, extends sensibly from $\frac{\omega_1}{\omega_2}X=$ 0.1 (point A_1) to $\frac{\omega_1}{\omega_2}X=0.065$ (point A_2).

The second straight segment $A_2'A_3$, of constant ordinate $W_{23}=310°$, runs sensibly from $\frac{\omega_1}{\omega_3}X=0.3$ (point A_2') to $\frac{\omega_1}{\omega_2}X=0.6$ (point A_3).

[1] W. Reinders, *Zeitschrift für anorganische Chemie*, v. 25, p. 113, 1901.

The third straight segment $A_3'A_4$, of constant ordinate $W_{34}=430°$, goes from the point A_3', whose abscissa is $\frac{\omega_1}{\omega_2}X=0.55$, to the point A_4 of abscissa $\frac{\omega_1}{\omega_2}X=0.9$.

The curves C_1, C_2, C_3, C_4, whose extremities are thus known, have not been determined.

233b. Amalgams of cadmium; Byl's researches.—Byl[1] has recently examined a system which comes within the type studied by Reinders; it is the system formed by mercury and cadmium.

The congelation of liquid amalgam may give rise, according to circumstances, to two kinds of mixed crystals.

The mixed crystals of the first kind, which we shall call α crystals, are isomorphous with crystals of pure mercury; they are deposited within liquid mixtures containing a proportion of cadmium less than a certain limit; if we attribute the index 1 to mercury and the index 2 to cadmium, and if we keep the notation of the preceding article, this limit corresponds to $\frac{\omega_1}{\omega_2}X \doteq 0.67$.

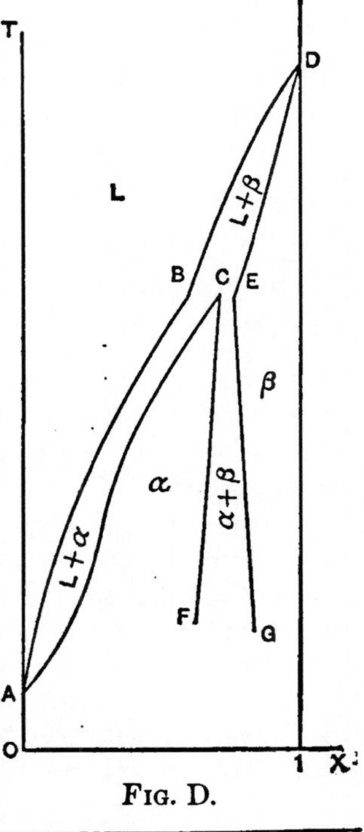

FIG. D.

The freezing-point is included between $-40°$, freezing-point of pure mercury, and 188°. The curve of congelation is the curve AB (Fig. D). The composition of the mixed crystals obtained is comprised between pure mercury and $\frac{\omega_1}{\omega_2}X=0.75$. The line of fusion for these crystals is the line AC.

The liquid mixtures whose proportion in cadmium exceeds $\frac{\omega_1}{\omega_2}X=0.67$ give other mixed crystals, isomorphous with pure cadmium, the β crystals. As the liquid becomes richer in cadmium, the freezing-point rises from 188° to 320°, freezing-point of pure cadimum, tracing the congelation line BD. The mixed crystals contain a proportion of cadmium which increases from the lower

[1] H. C. Byl, *Zeitschrift für physikalische Chemie*, v. 41, p. 641, 1902.

limit $\frac{\omega_1}{\omega_2}X=0.67$ up to totality; the fusion line of these crystals is the line ED.

The α and β crystals may be transformed into each other. When the point representing the state of the system is included between the lines CF and EG, the system in equilibrium incloses the two kinds of crystals; if the representative point is to the left of CF, the system is homogeneous and of form α; it is homogeneous and of form β if the representative point is to the right of the line EG.

234. Carburized iron. Roozboom's Theory.—A considerable number of alloys have been studied according to the principle stated above; this difficult study [1] has given in many cases results which are still but hypothetical; we shall not stop to describe all the results attained.

There are some, nevertheless, which we cannot pass over in silence; although still incomplete, they already throw much light on a subject of the first importance; we mean the researches which concern the constitution of carburized iron.

These numerous investigations have rendered it possible for Roozboom [2] to give a very atisfactory epresentation of the phenomena which are produced within a mixture of iron and carbon cooled with extreme slowness from the liquid state.

When the temperature of a me ted mixture of iron and carbon is lowered there are two cases to consider according as the amount of carbon present in the fused mass is less or greater than 4.3%.

Let us give the index 1 to iron and the index 2 to carbon, and

---

[1] See on this subject ROBERTS-AUSTEN and A. STANSFIELD, *La Constitution des alliages métalliques* (Reports presented to the international Physics Congress, Paris, 1900, v. I, p. 363). See also *Contribution à l'étude des alliages métalliques, loc. cit.*

[2] BAKHUIS-ROOZBOOM, *Eisen und Stahl vom Standpunkte der Phasenlehre*, Zeitschrift für physikalische Chemie, v. 34, p. 437, 1900; *Iron and Steel from the point of view of the "Phase Doctrine"* (Journal of the Iron and Steel Institute, No. 2, 1900); STANSFIELD, *The Present Position of the Solution Theory of Carburized Iron* (*ibid.*); BAKHUIS-ROOZBOOM, *Le Fer et l'acier au point de vue de la doctrine de phases* (Contribution, p. 327); OSMOND, *Remarques sur le mémoire précédent* (same, p. 370); H. LE CHATELIER, *Observations sur le mémoire de M. Bakhuis-Roozboom* (same, p. 379).

suppose, in the first place, that the value of x which indicates the composition of the casting is greater than 0.043.

In these conditions the cooled casting deposits pure carbon in the state of graphite; the temperature at which this deposit is made, at which, consequently, the melted casting may be regarded as a saturated solution of graphite in iron, depends on the proportion of carbon in the liquid mixture; it is the lower as x is smaller; when x is reduced to the value 0.043 this temperature descends to 1130°. The locus of points, which have for abscissæ values of x and for ordinates the temperatures, is the curve c_2 (Fig. 102), solubility curve of graphite in melted iron.

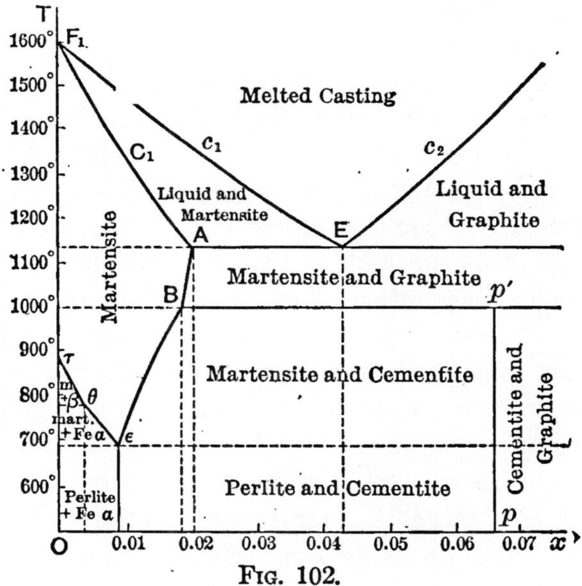

Fig. 102.

When, in the liquid mixture, the value of x is less than 0.043, things happen quite otherwise; by cooling the melted casting mixed crystals are obtained containing carbon and iron in variable proportion, and to which is given the name *martensite;* martensite is the principal constituent of the *white casting.*

The line c_1 is the congelation curve of martensite in the liquid mixture; it descends from left to right from the point F_1 ($x=0$, $T=1600°$) to the point E ($x=0.043$, $T=1130°$).

The study of this congelation is not complete so long as the composition of the martensite crystals which are formed at a given temperature is unknown; in order to know this composition, it is sufficient to trace the fusion curve C_1 of martensite; starting from the point F_1, this curve descends to the point A of ordinate $T=1130°$ and of abscissa $x=0.02$.

The point E is a eutectic point. When the temperature is lowered to this point the liquid part of the casting will certainly have for composition $x=0.043$.

By a new lowering of temperature, however slight, this liquid will solidify and form a eutectic conglomerate, containing on the average 4.3% carbon; this conglomerate will be formed by the juxtaposition of pure graphite crystals and of mixed martensite crystals of 2% carbon.

The solid conglomerates furnished by freezing may undergo, at temperatures below 1130°, various modifications.

In the first case, within the conglomerates of martensite and carbon, at a temperature below 1000°, there may be formed a definite compound, which separates from the mass; amorphous, it forms a cement between the crystals of martensite or of graphite; this compound, whose formula is Fe_3C, is *cementite.*

The system formed of the two independent components iron and carbon, divided into three phases, graphite, martensite, cementite, can, under atmospheric pressure, remain in equilibrium only at a single temperature, the which is in the neighborhood of 1000°; the composition of each of the three phases in equilibrium is likewise determined; this condition is fulfilled of itself for graphite and cementite; the martensite crystals which may be in equilibrium with these two substances contain about 1.8% carbon ($x=0.018$).

Outside of the conditions indicated, one of these phases will disappear from the system.

If the temperature is above 1000°, the cementite will decompose into graphite and martensite, which will remain alone in contact with each other.

When the temperature is below 1000°, the martensite and the graphite will combine to give cementite until one of the components has totally disappeared.

Cementite contains about 6.6% carbon ($x=0.066$); if, therefore the value of x which represents the average constitution of the conglomerate exceeds 0.066, case in which the representative point will be to the right of the line PP' ($x=0.066$), the conglomerate having attained the equilibrium condition will contain only cementite and graphite; if, on the contrary, the value of x representing the average composition of the conglomerate is less than 0.066, case for which the representative point will be to the left of the line PP', the conglomerate in equilibrium will be formed of mixed crystals of martensite embedded in amorphous cementite.

At temperatures included between 1130° and 1000° mixed crystals of martensite may be observed in equilibrium with graphite crystals; to each temperature corresponds an equilibrium state of this bivariant system at atmospheric pressure and the composition of each phase is given for this state of equilibrium; therefore at every temperature comprised between 1130° and 1000° the martensite crystals which may coexist with the graphite crystals have a given composition, and the law connecting this composition with the temperature may be represented by a certain curve.

From the meaning given to the point A this curve necessarily passes through this point; besides we have seen that at the temperature of 1000° there corresponds a point B whose abscissa is $x=0.018$.

If the point which represents the temperature and mean composition of the system lies to the left of the line AB, there may not be established, within the system, a state of equilibrium between the martensite and the graphite; the martensite crystals, too poor in carbon, dissolve the whole of the graphite.

At temperatures below 1000° no further equilibrium can be had between the martensite and the graphite, but there may be between martensite and cementite; the martensite crystals capable of figuring in such an equilibrium state have, at each temperature, a definite composition which corresponds to a point on the line $B\varepsilon$; this line starts necessarily from the point B ($x=0.018$, $T=1000°$). This line descends to the point ε ($x=0.0085, T=690°$), whose importance we shall see directly.

Between 1000° and 690°, within systems too poor in carbon to contain anything else than martensite, new transformations are produced; iron separates out in the pure state; this separation may take place in two different forms, which we shall denote with Osmond, as Fe_α and Fe_β.

The form Fe_α has a very great magnetic susceptibility; this quantity is very small for the Fe_β form. The Fe_α form changes into the Fe_β form when the temperature exceeds 770°; on the contrary, below this temperature iron passes from the Fe_β form to the Fe_α form.

A third form, Fe_γ, is that assumed by iron above 890°; martensite is isomorphous with this form; it may be said that Fe_γ iron is martensite with 0% carbon.

The solubility curve of Fe_β in martensite starts from the point τ ($x=0$, $T=890°$) which corresponds to the transformation-point of Fe_β into Fe_γ iron; it descends from left to right to the point θ ($x=0.0035$, $T=770°$).

From this point θ starts the solubility curve of Fe_α in martensite, curve which descends to the point ϵ of which we have spoken above.

A system whose representative point lies to the left of the line $\tau\theta$ is a conglomerate of Fe_β and of martensite crystals; a system whose representative point is to the left of $\theta\epsilon$ is a conglomerate of Fe_α and martensite.

The point ϵ is a eutectic point; the martensite capable of existing as far as this temperature ($T=690°$) has a well-defined composition ($x=0.0085$). The least lowering of temperature disintegrates it and it then forms a eutectic of the same average composition as formed by small particles of *ferrite* (Fe_α) embedded in cementite. Arnold and Sarby, who took this eutectic for a definite compound, gave it the name *perlite.*

It is clear that at a temperature below 690° a mixture of iron and carbon in equilibrium should possess a state determined solely by the knowledge of its average composition x. According to this value of x, this state is placed in one of the three categories which we are going to define:

1°. If x is included between 0 and 0.0085, the system is formed of *perlite* (eutectic of ferrite and cementite) with an excess of *ferrite* (Fe_α).

2°. If x is between 0.0085 and 0.0066, the system consists of *perlite* with an excess of *cementite* (Fe_3C).

3°. If x is greater than 0.066, the system is composed of *cementite* and *graphite.*

The theory of carburized iron, such as we have presented it as developed by Roozboom, cannot be regarded as definitely acceptable. Stansfield and also Le Chatelier have already pointed out various grave objections to the explanation by this doctrine. Other objections have been raised recently by Charpy and Grenet;[1] further, these chemists have put forth an opinion which merits putting to the test. According to them the only state of veritable equilibrium which a system composed of iron and carbon may have at low temperature would be formed of the pure ferric in contact with pure graphite; every other state would be observable only by a phenomenon of lag analogous to surfusion. The equilibrium states, whose laws Roozboom, Le Chatelier, Stansfield, and various other chemists have tried to give, may be compared to the state of equilibrium observed, at 44°, between liquid white phosphorus and solid white phosphorus; then, these two substances may be transformed into red phosphorus, the only form which is truly in equilibrium at low temperature. It is also thanks to this retardation of the transformation of white phosphorus into red that Boulouch was able to observe the states of equilibrium which are produced, below 100°, in the system sulphur-phosphorus (see Art. 224a).

For other reasons also, the theory of Roozboom cannot be regarded as sufficient to represent all the properties of carburized iron.

We know that a carburized iron possesses, at a given temperature, a state which is not determined by the knowledge of its composition alone; the permanent modifications known by the names of *tempering* and *annealing* may impress upon this state infinite variations; the preceding theory should be regarded, therefore, as a simplified and ideal theory, true for perfectly annealed systems; for the systems which do not enter into this ideal case it must give way before a theory which would be, doubtless, of an extreme complication.

---

[1] G. Charpy and L. Grenet, *Bull. de la Société d'Encouragement pour l'Industrie nationale,* v. 102, p. 399, 1902.

CHAPTER XV.

CRITICAL STATES.

235. The critical point in the vaporization of a single fluid.— Let us study the vaporization of a fluid, carbonic anhydride, for example.

There corresponds to every temperature T a pressure P, the *tension of saturated vapor at the temperature T,* which assures the equilibrium between the liquid and the vapor.

When, at the temperature T and under the tension of saturated vapor corresponding to this temperature, a unit mass of vapor condenses, there is a liberation of a quantity of heat L, which is the *heat of vaporization at the temperature T.*

At the temperature T and corresponding vapor pressure, unit mass of vapor occupies a volume v, and unit mass of the liquid occupies a volume v'; v and v' are *the specific volumes of the saturated vapor and of the liquid at the temperature T.*

Let us increase the temperature T and follow the variations of the four quantities P, L, v, v' whose definitions we have just recalled.

When the temperature T increases to a temperature θ, near to 31°.35 C., the tension of the saturated vapor increases to a value P near to 72.9 atmos. The heat of vaporization diminishes and approaches 0.

The saturated vapor becomes more and more dense, so that its specific volume v diminishes; the saturated liquid becomes less and less dense, so that its specific volume increases; the difference $(v-v')$ approaches 0; v and v' approach the same value which we shall denote by U.

Thus at a temperature below θ but differing very slightly from θ the saturated liquid and vapor are transformed one into the other without appreciable absorption or liberation of heat and without appreciable change of volume; the various physical properties, optica', capillary, etc., of one of the two phases cannot be distinguished from the analogous properties of the other phase.

Hence it may be said that when the temperature approaches the value θ, the liquid and vapor carbonic anhydride approach the same state, called the *critical state* of carbonic anhydride; θ, $\mathfrak{P}$, $\mathfrak{U}$ are called *critical temperature, critical p; essure, critical volume* of this fluid.

If, taking the temperatures for abscissæ and the pressures for ordinates (Fig. 103) we draw the curve C of tensions of saturated vapor, this curve will rise from left to right to the point γ, of abscissa θ and ordinate $\mathfrak{P}$, which bears the name *critical point.*

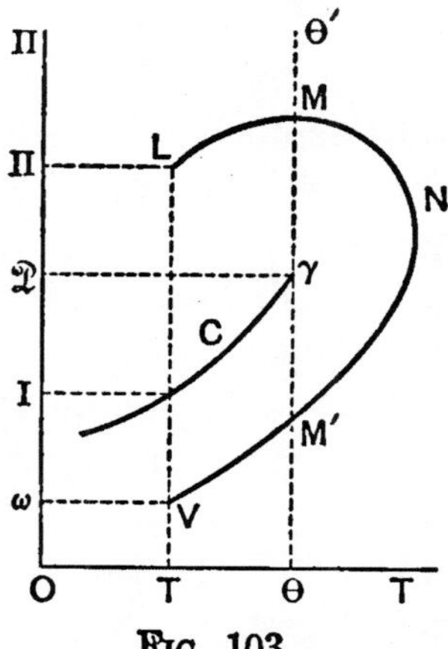

FIG. 103.

If the temperature T exceeds the critical temperature θ, it is impossible at any pressure to observe the carbonic anhydride divided between the two liquid and vapor phases; this substance is then constantly homogeneous in a state called the gaseous state.

At a temperature T, below θ, take the system under a pressure π which is above the tension $\mathfrak{P}$ of the saturated vapor at the same temperature; the representative point is then at L, and the system is a liquid.

We may cause the temperature and pressure to vary so that the representative point describes a path such as $LMNM'V$; this path cuts the line $\theta\theta'$ above the critical point γ, descends, remaining to the right of $\theta\theta'$, again cuts this line in a point M' located below the critical point γ, and arrives at a point V whose abscissa is the initial temperature T, but whose ordinate ω is less than the tension of saturated vapor for this temperature.

While the representative point goes from L to M the whole system is in the liquid state; it is in the gaseous state while the representative point follows the path MNM'; finally, the path $M'V$ corresponds to a state of homogeneous vapor.

Hence, while the representative point describes such a path, there is observed at no instant an abrupt change of any one of the properties of the system; the passage from the initial liquid state to the final state of vapor is made in a gradual and perfectly continuous manner.

One may therefore take a system at a temperature below the critical temperature, in the form of homogeneous liquid, and by a gradual transformation exempt from every abrupt change, bring it back to the same temperature in the vapor form; it is sufficient to cause the representative point of the system to describe a path which goes around the criti al point.

In 1869 Thomas Andrews made these very important observations on carbonic anhydride; since then many observers have repeated them on a great number of liquids; the idea of *critical state* dominates the study of transformations of a fluid of definite composition susceptible of existing in the two states of liquid and vapor.

We shall see that this principle is susceptible of very great extension.

236. **Double liquid mixtures. The temperature at which the two layers have the same composition does not correspond to an indifferent point.**—With W. Alexejew[1] and V. Rothmund,[2] let us take a mixture of water and phenol; the mixture of these two liquids is not always homogeneous; when the proportions of water and phenol are properly chosen, it separates into two layers of different composition and densities, thus forming a double liquid mixture which is a bivariant system.

At a given pressure and temperature each of the two phases in equilibrium has a definite composition, independent of the masses of water and phenol present. Thus at the temperature of $+34°.2$ *C.* a gramme of the denser layer contains 0.688 gr. phenol; a gramme of the rarer layer contains 0.93 g. of the same substance.

---

[1] W. Alexejew, *Wiedemann's Annalen*, p. 28, v. 305, 1886.

[2] V. Rothmund, *Zeitschrift für physikalische Chemie*, v. 26, p. 433, 1898.

Let us denote, in general, by x the number of grammes of phenol contained in 1 gramme of the rarer layer, and by X the number of grammes of phenol in 1 gramme of the denser layer; X is evidently greater than x. Keeping the pressure constant, say that of the atmosphere, let us study how x and X vary when the temperature T varies.

With this object in view take two rectangular axes (Fig. 104); lay off the temperatures T as abscissæ and the values of x as ordinates; at a given temperature T the composition of one of the two layers is represented by a point M, of coordinates T, X, the composition of the other layer by a point m, of coordinates T, x.

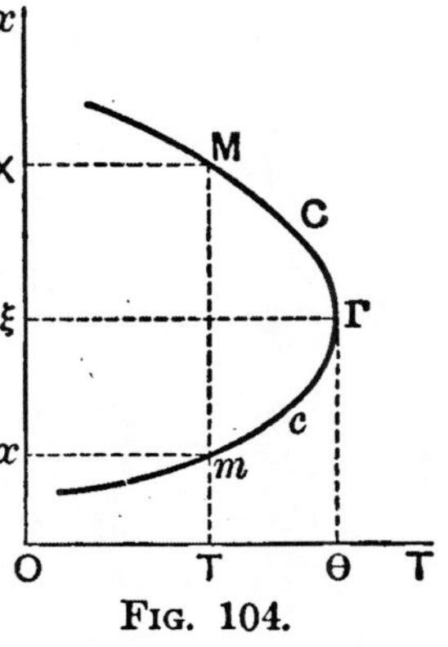

FIG. 104.

When the temperature T varies, the point M describes a certain curve C and the point m another curve c.

When the temperature T, in increasing tends towards a value $\theta = 69°$ C., the two curves C and c are seen to approach each other, the two points M and m approach a common point Γ, the two compositions X and x approach the same limit ξ; at the same time that the two curves C, c join in the point Γ, they so meet each other as to touch at this point a parallel to Ox.

One might think this phenomenon comparable to those studied in Chapter XI, and *that the point Γ is an indifferent point where two distinct liquid layers, but of the same composition, are in equilibrium with each other.*

It is easy to be convinced *that this supposition would be erroneous.*

If it were exact, we might say of the double liquid mixture formed by water and phenol all that has been said of the double mixture formed by a mixture of volatile liquids surmounted by the mixed vapor which it gives off; one of the two layers into which the double liquid mixture is separated would play the same rôle as the mixed liquid, the other layer would play the same rôle as the mixed vapor; if X and x were the compositions of the two layers capable of remaining in equilibrium in contact with each other at the temperature T, the point M (Fig. 105) of coordinates T, X, would describe *a curve with two branches* CIC'; the point m, of coordinates T, x, would describe *another curve of two branches*, cIc'; these two curves would pass through the point I, of coordinates θ, ξ, and touch there a parallel to Ox; at a given temperature T, below θ according to the masses of water and phenol employed, our double liquid mixture could possess two distinct states of equilibrium; in one, the two superposed layers would have the compositions X, x and the representative points M, m; in the other, these two layers would have the compositions X', x' and the representative points M', m'.

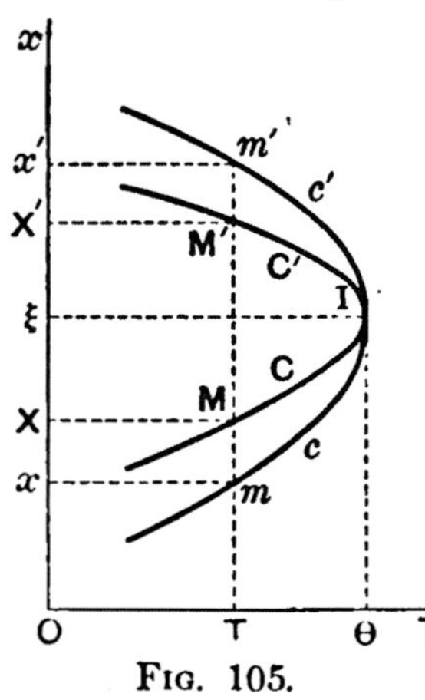

FIG. 105.

This disposition is in conformity with that shown in Fig. 61 for a mixture of volatile liquids whose boiling point passes through a maximum value; it has not, on the contrary, any analogy with that shown in Fig. 104. When, therefore, the temperature approaches 69°, we cannot suppose that the two layers into which the mixture of water and phenol is divided approach two distinct solutions having the same composition; and as there is no doubt that they tend to have the same composition, we are constrained to admit that they have as limiting state, not *two* distinct solutions, but *a single* solution.

237. This temperature is a critical temperature.—We may therefore formulate the following proposition:

When the temperature of a double liquid mix ure is gradually raised, it may happen that the composition and the various properties of the two layers into which it is divided differ less and less; when the temperature reaches a certain value θ *the two layers become identical in all respects.*

The analogy between this law and that discovered by Andrews in the study of the vaporization and liquefaction of a fluid of definite composition is evident; in virtue of this analogy, we shall say that θ *is the critical temperature and* ξ *the critical composition of the liquid mixture studied, under the pressure considered.*

At a temperature below the critical temperature the mixture is divided into two distinct layers, separated by a quite sharp surface of contact; when the temperature reaches the critical value the surface of separation becomes indistinct and disappears and, instead of two separate layers, there remains, at temperatures above θ, but one homogeneous mixture.

At a temperature T, lower than the critical temperature θ, the mixture is in a homogeneous state which we shall call *upper layer*, if it is rich enough in water for the representative point S (Fig. 106) to lie below the curve c; it is divided into two layers if the representative point is between the curves c and C; it is in the homogeneous state which we shall call *lower layer*, if it is rich enough in phenol for the representative point to be at I, above the curve C.

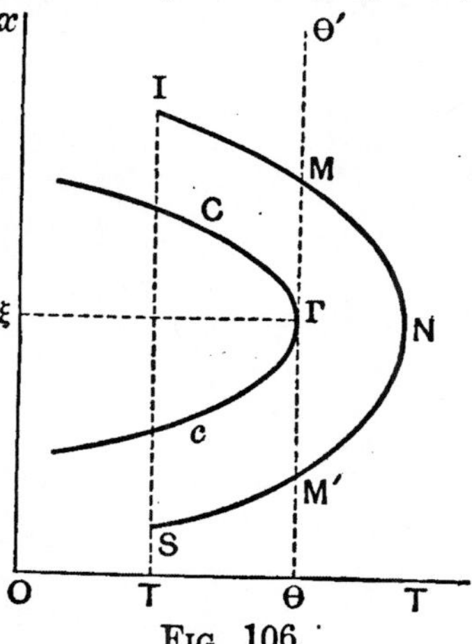

FIG. 106.

Take the system in the last state; raise the temperature and add water, so that the representative point describes the path IMN, which cuts the line $\theta\theta'$ in M, above the critical point Γ; then continue to add water, but lower the temperature so that the representative point describes the path $NM'S$ which cuts the line $\theta\theta'$ at M', below the critical point Γ; the system, taken at the temperature T and in the state of *upper layer*, returns to the same temperature, but in the state of *lower layer;* during the modification, it has remained homogeneous and has undergone no sudden change, but its various properties have altered gradually.

A liquid mixture may therefore pass from the state of upper layer to the state of lower layer by a continuous transformation; it suffices that the path of the representative point goes around the critical point.

238. Mixtures which separate into two layers at temperatures below the critical temperature.—The water-phenol mixture is not the only one which may separate into two layers below a certain critical temperature θ and which remains perforce homogeneous at a temperature above θ; the existence of such a critical temperature is a very general phenomenon, as may be seen from the following table.

Mixtures studied.	θ	Observers.
Phenol–water	+ 69°	W. Alexejew
Benzoic acid–water	116°	"
Phenolate of phenylammonium–water	140°	"
Aniline–water	166°	"
Secondary butyl aclohol–water	108°	"
Isobutyl-alcohol–water	132°	"
Proprionitrite–water	113°	V. Rothmund
Salicylic acid–water	95°	"
Furfurol–water	122°	"
Acetylacetone–water	88°	"
Isobutric acid–water	24°	"
Methylethylcetone–water	151°	"
Succinonitrite–water	55°	Schreinemaker [1]
Chlorobenzine–sulphur	116°	W. Alexejew
Essence of mustard–sulphur	124°	"
Aniline–sulphur	138°	"
Benzine–sulphur	164°	"
Toluene–sulphur	179°	"
Sulphide of carbon–methyl alcohol	40°	V. Rothmund
Hexane–methyl alcohol	42°	"
Resorcine–benzine	109°	"
Zinc–bismuth	Between 800° and 900°	W. Spring and Romanow [2]

239. **Mixtures which separate into two layers at temperatures higher than the critical point.**—In all the cases we have just cited, the mixture susceptible of being divided into two layers at temperatures below the critical point θ is of necessity homogeneous at temperatures higher than θ; we possess likewise examples (see accompanying table), less numerous it is

Mixtures studied.	θ	Observers.
Diethylamine–water	+122°	F. Guthrie [3]
β-Collidine–water	4°	V. Rothmund
Triethylamine–water	20°	"

[1] Schreinemakers, *Zeitschrift für physikalische Chemie*, v. 23, p. 417, 1897.

[2] Spring and Romanow, *Zeitschrift für anorganische Chemie*, v. 13, p. 29, 1897.

[3] F. Guthrie, *Philosophical Magazine*, 5th S., v. 18, pp. 29 and 499, 1884.

true, of mixtures which remain homogeneous at temperatures below the critical temperature θ, while at temperatures above θ it may separate into two layers; the two curves C and c are then arranged as shown in Fig. 107.

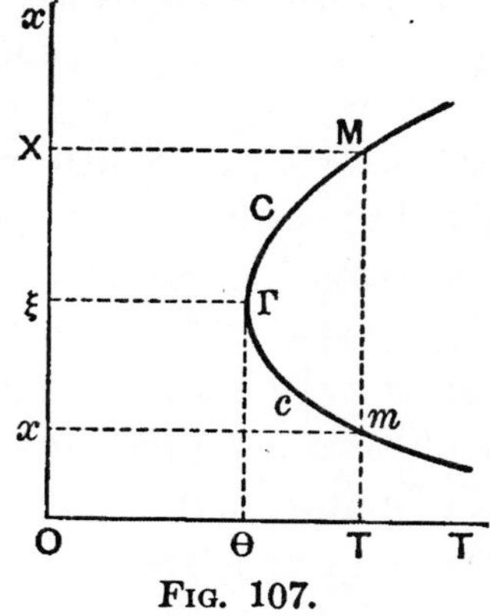

FIG. 107.

240. Influence of pressure on the critical temperature of a double liquid mixture.—The critical temperature θ and the critical pressure ξ may naturally depend upon the pressure at which the mixture is studied. Van der Lee [1] has found that the critical temperature of a water-phenol mixture rises at the same time as the pressure.

Further, it is necessary to impose upon the pressure a very considerable increase in order to obtain an appreciable variation of the critical point.

241. Vaporization of a mixture of two liquids; critical line; dew surface, ebullition surface.—It is the same when the two phases into which the liquid mixture is divided are a liquid phase (lower layer) and a vapor phase (upper layer); in this case every variation of the pressure imposes variations of the same order of magnitude on the critical temperature and on the critical composition of a mixture of given composition.

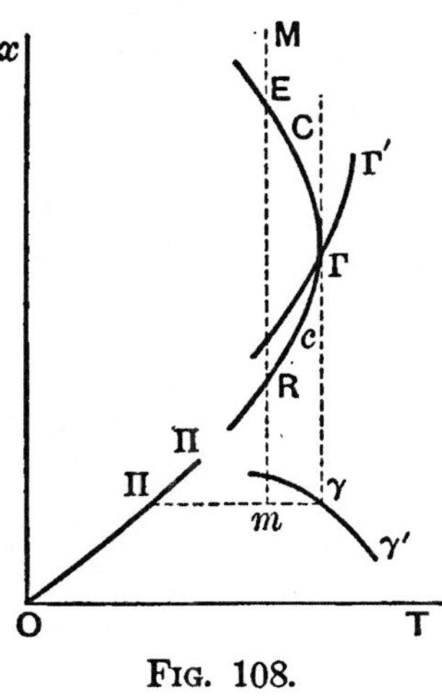

FIG. 108.

If, therefore, we lay off on three axes of rectangular coordinates the temperature T (Fig. 108), the pressure π, the composition x, to every pressure π will correspond a point Γ whose coordinates will give the critical t mperature and critical composition w th respect to this pressure. When the pressure π varies, thi point describes a line, the *critical line* $\Gamma\Gamma'$.

[1] VAN DER LEE, *Academy of Sciences of Amsterdam*, Oct. 29, 1898.

In varying the pressure, it is not merely the elements of the critical point which are varied; each of the two curves C, c is displaced at the same time, or, better, the continuous curve cC which they form together. This curve will generate a certain surface which we shall call the *limiting surface.* The limiting surface will be composed of two sheets, the one s generated by the curve c, and the other S generated by the curve C; we shall call the sheet s the *dew surface* and the sheet S the *ebullition surface;* these definitions will be justified shortly.

These two sheets meet each other along the critical line. If the limiting surface is projected on the plane $TO\pi$, it is clear that the projection $\gamma\gamma'$ of the critical line will make a part of the contour of the projection.

Through the point M, taken in the plane $TO\pi$, draw a parallel to Ox; it will meet the dew surface s at R and the ebullition surface S at E.

Choose a starting-point M located above E and, without changing pressure π or temperature T, add to the system a gradually increasing mass of the fluid 1; x will diminish and the point M will describe the line Mm.

As long as the representative point lies above E, the system will remain in the state of homogeneous liquid; the instant the representative point reaches the position E, the second phase will appear in the system under the form of a bubble of vapor; there will be *ebullition.*

The representative point lying between E and R, the system will be divided into two phases; the mixed liquid will be surmounted by a layer of vapor. When the representative point reaches R, the last drop of liquid will disappear; if the representative point continued to descend, the system would be in the state of homogeneous vapor.

If, at this moment, we introduced some of fluid 2 to increase x, the representative point would remount; the system would be at first in the state of homogeneous vapor, but at the moment the representative point attained the position R, a liquid drop would appear; there would be a *deposit of dew.*

To sum up, in order that the system may remain in equilibrium in a state where it is divided into two phases, liquid and vapor, it is necessary that the representative point lie between the two sheets s, S of the limiting surface; if it leaves this region by traversing the surface S, the system passes into the state of homogeneous liquid; if it leaves through the surface s, the system vaporizes entirely.

242. Dew line and ebullition line of a compound of given composition.—When it is desired to study the vaporization or condensation of a fluid mixture, it is enclosed in a Cailletet tube and the temperature and pressure are varied; during these operations the value of x which characterizes the average composition of the mixture rests invariable. It is therefore interesting to discuss the properties of a system for which is given the value x of the average composition.

The plane perpendicular to Ox which corresponds to this value of x cuts (Fig. 109) the ebullition surface S in the curve $\mathfrak{E}$ and the

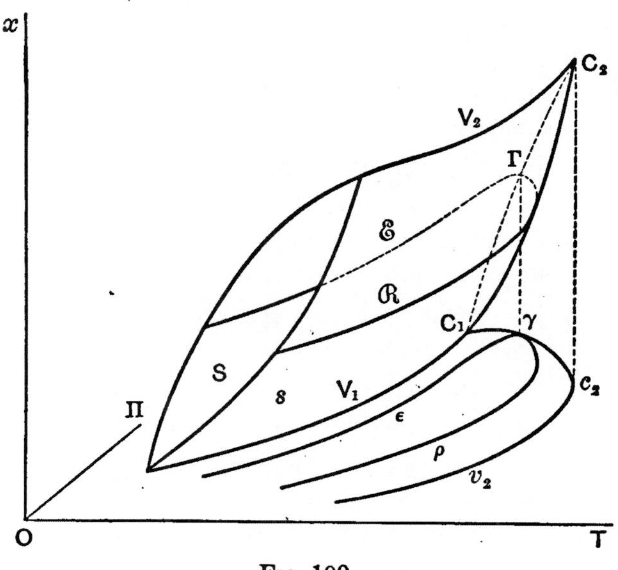

Fig. 109.

dew surface in the curve $\mathfrak{R}$; these two curves meet in the critical point Γ of the mixture of concentration x, so as to form a single curve, section of the limiting surface made by the plane considered.

For the particular case when $x=0$ the two lines $\mathfrak{E}$ and $\mathfrak{R}$ coalesce into a single line V_1, which is the curve of tensions of saturated vapor of fluid 1 in the pure state; this line ends at the point C_1, critical point of the fluid 1.

Similarly, for the case when $x=1$ the two lines $\mathfrak{E}$ and $\mathfrak{R}$ become a single line V_2, which is the curve of tensions of saturated vapor of fluid 2 in the pure state; this line ends at the critical point C_2 of the fluid 2.

The critical line $C_1\Gamma C_2$ unites the point C_1 to the point C_2.

Project the figure on the plane TO , where $x=0$.

The curve V_2 of tensions of saturated vapor of the fluid 2 is projected in its true magnitude along the line v_2, which ends at the point c_2, projection of the point C_2.

The critical line is projected along the line $C_1\gamma c_2$; this latter is a part of the contour of the projection of the limiting surface.

The line $\mathfrak{E}\Gamma\mathfrak{R}$ is projected in its true magnitude along the line $\varepsilon\gamma\rho$, which is the limiting line of the mixture of composition x; ε is its *ebullition line*, and ρ the *dew line;* they meet in the point γ, projection of the critical point Γ, and at this point they are tangent to the projection $C_1\gamma c_2$ of the critical line.

Take a mixture of composition x, at a temperature T and under a pressure π which serve as coordinates to a point in the plane $TO\pi$. When this representative point is in the interior of the limiting curve $\varepsilon\gamma\rho$, the mixture of mean composition x is divided into two phases, a mixed liquid and a mixed vapor. One of these two phases disappears and the system becomes homogeneous when the representative point passes beyond the limiting line. It is the vapor phase which disappears if the representative point passes beyond the limiting line at a point which belongs to the ebullition line; it is, on the contrary, the liquid phase which disappears if the representative point passes beyond the limiting line at a point belonging to the dew line.

243. Normal condensation. Retrograde condensation.—The consideration of limiting lines plays an important rôle in all the researches relative to the liquefaction and vaporization of fluid mixtures. The detailed analysis of these researches would exceed the plan of this work; so we shall not give it. We shall be content to notice a remarkable consequence of the preceding theories.

Suppose that the disposition of the limiting line be that represented in Fig. 110; *the point M, whose abscissa τ is a maximum, belongs to the dew-line;* experiment shows this to be so in a great number of cases.

Take first a temperature T less than the critical temperature θ of the mixture of composition x; at this temperature cause the pressure to increase gradually from a very low value to a very great value; the representative point will rise constantly along the straight line TT', which meets the dew-line in a point R and the ebullition-line in a higher point E.

So long as the pressure has not reached the value TR the system will be in a state of homogeneous vapor; the instant the

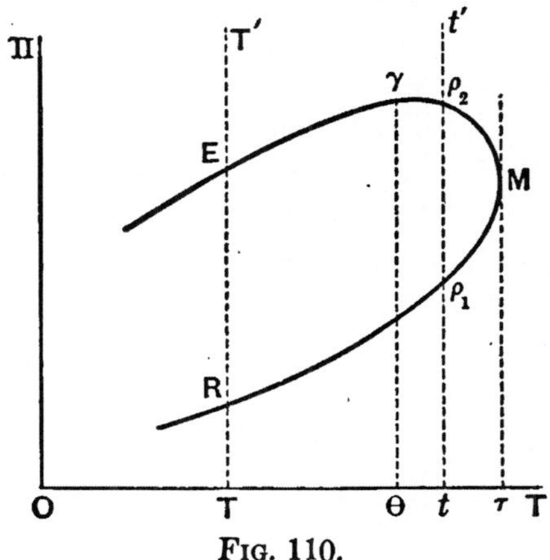

FIG. 110.

pressure attains this value which corresponds to the *dew-point*, a first drop of liquid will appear; the pressure continuing to increase, the mass of liquid will increase at the expense of the vapor; when the pressure reaches the value TE, which corresponds to the *ebullition-point*, the last bubble of vapor will disappear, and for higher pressures the system will be in a state of homogeneous liquid. If the pressure be decreased, the same phenomena will be produced in the inverse ord r.

When one observes the succession of phenomena which we have just enumerated, the system is said to undergo *normal condensation.*

The succession of observed facts is quite otherwise when the system is compressed keepi g co stant a temperature t, higher than the critical temperature θ, and also less than the temperature τ.

The representative point rises along the line tt', which cuts the dew-line in a first point, ρ_1, then in a second point, ρ_2, of ordinate greater than the first.

While the pressure is less than $t\rho_1$ the system remains in a state of homogeneous vapor; at the instant the pressure reaches the value $t\rho_1$, which corresponds to the *first dew-point*, a liquid drop appears; the pressure continuing to increase, the mass of liquid increases at first, but afterwards passes through a maximum, then diminishes, and at the instant the pressure attains the value $t\rho_2$, which corresponds to the *second dew-point*, the last drop of liquid disappears; if the system is compressed further, it remains in the state of homogeneous vapor.

This series of phenomena constitutes *retrograde condensation*.

Retrograde condensation was discovered in 1880 by L. Cailletet, who was studying the liquefaction of a mixture of air and carbonic acid gas; the following year Van der Waals independently made this observation which he thought new; confirmed by the researches of Cailletet and Hautefeuille, and of Andrews, the phenomenon of retrograde condensation plays a great part in the theoretical and experimental studies relative to the liquefaction of gaseous mixtures; these studies, developed by numerous physicists, among whom we may mention Van der Waals, Kuenen, and Caubet, cannot be analyzed here; the interested reader of this important question can refer to Caubet's treatise.[1]

244. Critical states in the mixtures of three substances.—A great number of liquid mixtures formed by the reunion of three substances are susceptible of separating themselves into two layers, of different composition, which remain in equilibrium in contact with each other; in certain conditions, these two layers take on the same composition, the same density, the same physical properties; in other terms, when these conditions are nearly realized the two distinct phases into which the system is divided tend towards a common limiting state which is a *critical state*.

---

[1] F. CAUBET, *Liquéfaction des mélanges gaseux* (*Mém. de la Soc. physiques et naturelles de Bordeaux*, 5th Series, v. I, 1901), and A. Hermann, Paris, 1901.

The existence of a critical state in the mixtures formed of three liquid substances was noticed in 1876 by Duclaux when studying the following mixtures:

Amyl alcohol–alcohol–water;
Alcohol–ether–water;
Acetic acid–ether–water.

For some years past, the study of ternary liquid mixtures and their critical states has been the object of numerous important investigations, both theoretical and experimental, due to Schreinemakers,[1] Snell,[2] and G. Bruni;[3] we must limit ourselves here to the mentioning of these investigations.

245. **Limiting crystalline forms.**—The existence of critical states appears, therefore, to be very general; it is possible to verify it in the majority of cases where a fluid system is divided into two phases, whether the system be formed of one, two, or three components; whether the two phases be liquid, or one a liquid and the other a vapor.

It is probable also that the notion of critical state should not be restricted to fluid systems.

Pasteur noticed that the two crystalline forms of a dimorphous substance are, in general, little different from each other; when the temperature and pressure are increased, these forms are modified in such a way that the characteristics which distinguish them are attenuated; we may imagine that, the temperature and pressure[4] tending towards certain well-determined values, which would be the *critical temperature* and the *critical pressure*, the two crystalline forms would approach a common limiting form, which would be the *critical form;* the dimorphism of a substance would then be comparable to the coexistence in a fluid of the two forms liquid and vapor, the limiting crystalline form replacing

---

[1] SCHREINEMAKERS, numerous memoirs published since 1897, in the *Archives néerlandaises des Sociétés exactes et naturelles* and in the *Zeitschrift für physikalische Chemie.*

[2] SNELL, *Journal of Physical Chemistry*, v. 2, p. 457, 1898.

[3] G. BRUNI, *Rendiconti dell' Accademia dei Lincei*, 5th Series, v. 8, p. 141, 1899.

[4] To be exact, it would be necessary to speak here not of the pressure, but of the six components of the elastic actions.

the gaseous state, common limit of the liquid state and of the vapor state.

Two isodimorphous substances may furnish two kinds of mixed crystals; the crystalline forms of these two kinds of crystals are, in general, but slightly different. Thus, as was shown by Retgers (Art. 219), one may obtain mixed crystals of calcic carbonate and of magnesium carbonate which are isomorphous with calcite; some may be obtained which are isomorphous with magnesite. Now, the primitive form of calcite and the primitive form of magnesite are two rhombohedrons having slightly different angles; these angles, further, vary with the temperature and the elastic actions. One may suppose, therefore, that *mixed critical crystals* may exist possessing a common limit for the two forms of mixed crystals which are observed under ordinary conditions. These limiting mixed crystals would be comparable to the critical states in which the two phases of a mixture of two liquids blend.

CHAPTER XVI.

CHEMICAL MECHANICS OF PERFECT GASES.

246. Necessity of new hypotheses in order to penetrate further into the study of chemical systems.—All we have said as yet on the subject of the various problems of chemical mechanics is of very great generality; a single hypothesis, which neglects the capillary actions, was used to specify the form of the internal potential of the systems studied (see Chapter VI, Art. 89).

This great generality, which gives value to the considerations developed in the preceding chapters, is not without some inconvenience; from the fact that they are of wide extent, the principles set forth are less able to aid in penetrating into the detail of phenomena. We know, for instance, that, under a given pressure, it suffices for the temperature to be determined in order that the saturated solution of a certain salt may have a definite concentration; we know that this concentration varies in the same sense as the temperature or in the opposite sense, according as the salt is dissolved, in saturated solution, with absorption or liberation of heat; but there our information ceases; now, it is clear this is far from being complete and we may legitimately desire more; we may, for example, ask to know, in an exact or approximate way, the form of the law which connects the concentration of the saturated solution with the temperature.

But to obtain, by means of thermodynamical principles, more detailed propositions than those we have as yet secured, it is necessary to join to the hypothesis we have already introduced new hypotheses of a more particular nature; the difficulty consists in discovering such hypotheses as will furnish in cases of a sufficient generality exact consequences or, at least, those of a sufficient approximation.

247. Properties of the systems to be studied.—Gibbs and Horstmann have succeeded in defining such cases and in formulating such hypotheses.

The chemical systems to which the theory developed by these physicists is applied may contain solids, liquids, and gases.

The solids and the liquids do not mix with each other and do not dissolve the gases, so that *each of the solid or liquid phases which the system contains is a definite and pure chemical compound.*

The specific volume of each of the solid or liquid phases is negligible with respect to the specific volumes of the gases considered.

The specific heat of each of these solid or liquid substances is sensibly independent of the temperature.

The gases contained in the system are in the perfect state.

248. Hypotheses which characterize a mixture of perfect gases.—These various suppositions would suffice to put into equations in a complete manner the many problems of chemical mechanics if the gaseous phase which the system is supposed to contain was formed by a single gas; but in a great number of important cases this phase is a mixture of several gases which must, according to the principles just stated, be regarded as perfect gases; we are therefore led to ask ourselves the following question: From the view-point of thermodynamics, how will the mixture of two or more perfect gases be characterized?

There is a certain number of propositions which all physicists are agreed to regard as characterizing the mixtures of two or more perfect gases; let us recall these propositions.

The first is this: *A mixture of two or more perfect gases, taken in known proportions, acts in all circumstances like a single perfect gas.* Thus, for example, air, which for the chemists is a mixture of several gases, is constantly cited and studied by physicists as type of a gas near to the perfect state.

The second is the one discovered by Berthollet and which is known under the name of *law of the mixture of gases: To keep in equilibrium at a given volume and pressure a mixture of perfect gases, it is necessary to submit it to a pressure equal to the sum of the pressures which would be maintained separately, at the same volume and temperature, by each of the mixed gases.*

The third may be stated as follows:

If two vessels, which contain at the same pressure and temperature two different perfect gases susceptible of mixing, are put into communication with each other, the two gases diffuse into one another without absorbing or giving out heat.

Finally, the fourth proposition is the classic *law of the mixture of gases and vapors*, well known in the following form:

When a liquid is in equilibrium, at a given temperature, with its vapor mixed with a gas, the tension of the gaseous mixture is the sum of the tensions which is attained, at the same temperature, by the saturated vapor of the same liquid in a space previously empty and of the pressure which would be maintained by the gas, at the same temperature, of volume equal to the volume of the mixture.

These several laws completely characterize, from the thermodynamic point of view, the properties of a mixture of perfect gases; they lead in fact to the following proposition, which allows calculating all these properties when those of mixed gases are known:

The internal potential of a mixture of perfect gases is always equal to the sum of the internal potentials which it would be proper to attribute to each of the mixed gases if it occupied alone, at the same temperature, the entire volume of the mixture.

249. **Notations.**—We know now the source of Gibbs' and of Horstmann's theory; let us leave the steps in the development of this theory and state the succession of results to which it leads.

These consequences are condensed into three essential formulæ.

Imagine a system in which either a dèfinite chemical reaction or an inverse reaction may be produced; this will be, for instance. a system containing hydrogen, silver, hydrogen sulphide, silver sulphide; the reaction

$$Ag_2S + H_2 = H_2S + Ag_2$$

may be produced or the inverse reaction.

Let us write, as we have just done for this example, the chemical equation which represents the former reaction. In the first member figure certain gaseous substances a_1, a_2, . . . and also certain solids or liquids A_1, A_2, . . .; in the second member appear certain gaseous substances a_1', a_2', . . . and certain solid or liquid bodies A_1', A_2', . . .

Let us denote in the following manner the molecular weights and the numbers of molecules of these various substances which figure in the chemical equation:

Substance.	Molecular Weight.	Number of Reacting Molecules.
a_1 a_2	ω_1 ω_2	n_1 n_2
A_1 A_2	π_1 π_2	n_1 n_2
a_1' a_2'	ω_1' ω_2'	n_1' n_2'
A_1' A_2'	π_1' π_2'	N_1' N_2'

The first member of our chemical equation represents a mass:

$$n_1\omega_1+n_2\omega_2+\ .\ .\ .\ +N_1\pi_1+N_2\pi_2+\ .\ .\ .$$

The second member represents a mass:

$$n_1'\omega_1'+n_2'\omega_2'+\ .\ .\ .\ +N_1'\pi_1'+N_2'\pi_2'+\ .\ .\ .$$

These two masses are equal to each other. Let P be their common value:

$$(1)\qquad \begin{aligned} &n_1\omega_1+n_2\omega_2+\ .\ .\ .\ +N_1\pi_1+N_2\pi_2+\ .\ .\ . \\ &=n_1'\omega_1'+n_2'\omega_2'+\ .\ .\ .\ +N_1'\pi_1'+N_2'\pi_2'+\ .\ .\ .=P. \end{aligned}$$

Let σ_1, σ_2, . . . σ_1', σ_2', . . . be the volumes occupied, in the normal conditions of temperature and pressure, by 1 gramme of each of the gases a_1, a_2, . . . a_1', a_2', . . .; the masses of these gases which figure in the chemical equation occupy respectively, in the normal conditions of temperature and pressure, the volumes

$$n_1\omega_1\sigma_1=U_1,\qquad n_2\omega_2\sigma_2=U_2,\ .\ .\ .\ ;$$
$$n_1'\omega_1'\sigma_1'=U_1',\qquad n_2'\omega_2'\sigma_2'=U_2'.\ .\ .$$

Finally, in the same conditions, 1 gramme of hydrogen occupies a volume Σ. The ratios

$$\frac{U_1}{\Sigma}=V_1, \qquad \frac{U_2}{\Sigma}=V_2, \ldots$$

$$\frac{U_1'}{\Sigma}=V_1', \qquad \frac{U_2'}{\Sigma}=V_2' \ldots$$

are integers which simplify greatly the chemical equations; if, for example, the gas a_1 obeys the laws of Avogadro and of Ampère, the mass ω_1 of this gas occupies, in the normal conditions of temperature and pressure, the same volume as 2 grammes of hydrogen; hence we have

$$\omega_1\sigma_1=2\Sigma, \quad \text{or} \quad V_1=2n_1$$

More generally, V_1, V_2, . . . V_1', V_2', . . . are the numbers written by chemists when they wish to express *by volume* the reaction which the chemical equation considered expresses *by weight.*

250. Law of equilibrium for the systems studied.—Suppose the system in equilibrium at the *absolute* temperature T in a vessel where it is either alone or in the presence of perfect gases which do not take part in the reaction; let $p_1, p_2, \ldots p_1', p_2' \ldots$, be the partial pressures of the gases $a_1, a_2, \ldots a_1'a_2', \ldots$ in this mixture; these pressures satisfy the EQUILIBRIUM CONDITION

$$(3) \quad V_1 \log p_1 + V_2 \log p_2 + \ldots \\ - V_1' \log p_1' - V_2' \log p_2' - \ldots = \frac{M}{T} + \text{N} \log T + Z,$$

M, N, and *Z being three constants.* The symbols log represent the common logarithms.

The pressures $p_1, p_2, \ldots p_1', p_2', \ldots$ are expressed in terms of any unit desired; the choice of this unit influences only the value of the constant Z.

251. Heats of reaction at constant pressure and at constant volume.—Imagine now that in a system sensibly in equilibrium a small mass μ passes, without change of temperature, from the state represented by the first member of the chemical equation to the state represented by the second member of the same equation; if the modification takes place at *constant pressure*, it liberates a quantity of heat $L\mu$; if it takes place at *constant volume*, it sets free a quantity of heat $\lambda\mu$; the quantities L and λ, or, better, the quantities PL and $P\lambda$, which are the quantities usually made use of in treatises on thermochemistry, are determined by the two following expressions:

$$(4)\ PL=\frac{\pi_0}{T}\frac{\Sigma}{E}(NT-0.4301M);$$

$$(5)\ P\lambda=\frac{\pi_0}{T_0}\frac{\Sigma}{E}[(N+V_1'+V_2'+\ldots-V_1-V_2-\ldots)T-0.4301M].$$

π_0 is the *normal pressure* (as atmospheric pressure) and T_0 the *normal temperature* (for instance, that of melting ice) which give to the specific volume of hydrogen the value Σ.

The constant quotient $\frac{\pi_0\Sigma}{T_0E}$ is in the C.G.S. system very nearly equal to 1. If, therefore, the quantities PL and $P\lambda$ are expressed in *small calories*, we may replace the formulæ (4) and (5) by the formulæ

$$(4')\ PL=NT-0.4301M;$$

$$(5')\ P\lambda=N+V_1'+V_2'+\ldots-V_1V_2-\ldots-0.4301M.$$

252. Tensions of saturated vapor. A. Dupré's formula.— Let us indicate by some examples the applications which may be made of the preceding formulæ.

The simplest case is that where a single gaseous substance figures in the system; the type of a transformation of this kind is furnished by the condensation of water vapor, represented by the chemical equation

$$H_2O\ (\text{vapor})=H_2O\ (\text{liquid}).$$

Water vapor obeying Avogadro's law, and a single molecule of water vapor figuring in the first member of the chemical equation, we shall have $V=2$ and the condition of equilibrium (3) will become

$$2\log p=\frac{M}{T}+N\log T+Z;$$

or in putting

$$m=\frac{M}{2}, \qquad n=\frac{N}{2}, \qquad z=\frac{Z}{2},$$

(6) $$\log p=\frac{m}{T}+n\log T+z.$$

Athanase Dupré[1] was the first to propose representing by such a formula the relation which exists between the tension of saturated vapor of water or of any other liquid and the temperature.

Can this formula (6) represent with sufficient exactness the tensions of saturated vapors measured by observers? This is a question treated by many authors and which has been recently examined in a very thorough manner by J. Bertrand.[2]

Three observations of vapor tensions, at different but known temperatures, allow determining the three constants m, n, z; it is then easy to calculate the value of p that formula (6) gives corresponding to each value of the temperature T, and to compare the numbers thus obtained with the results of observation.

Let us take as example water vapor.

If the pressure p is calculated in millimetres of normal mercury, the constants m, n, and z have the following values:

$$m=-2795;$$
$$n=-\quad 3.8682;$$
$$z=+\quad 17.44324.$$

Bertrand compared in five-degree steps, for temperatures included between $T=243°$ ($-30°$ C.) and $T=273°$ (0° C.), then in ten-degree steps from $T=273°$ (0° C.) to $T=503°$ (230° C.), the numbers given by the formula

$$\log p=17.44324-\frac{2.795}{T}-3.8682\log T$$

with the results of Regnault's experiments; in the following table are some of the numbers from this comparison:

[1] ATHANASE DUPRÉ, *Théorie mécanique de la chaleur*, p. 97.
[2] J. BERTRAND, *Thermodynamique*, p. 101.

T	$T-273$.	Observed.	Calculated.
243	− 30	0.39	0.39
273	0	4.60	4.59
323	+ 50	91.98	91.96
373	+100	760.00	763.04
423	+150	3 581.2	3 608.48
473	+200	11 689.0	11 701.72
483	+210	14 324.8	14 297.12
493	+220	17 390.4	17 306.72
503	+230	20 926.4	20 957.88

The maximum error, adds Bertrand, is 169 millimetres for $T=503°$ (230° *C.*) less than 0.01 of the calculated quantity, and corresponding to an error of 0°.47 in the temperature.

Bertrand obtained analogous results for the following liquids; p is still reckoned in millimetres of normal mercury:

Name of Liquid.	m	n	z
Water	−2 795	−3.86 82	17.44 324
Ether	−1 729.97	−1.97 87	13.43 3115
Alcohol	−2 743.84 2	−4.22 482	21.44 6868
Chlorohydric ether	−1 747.13	−3.87 21	17.04 235
Chloroform	−2 179.14 2	−3.91 584	19.29 7930
Carbon bisulphide	−1 684	−1.76 89	12.58 852
Carbon chloride	−2 226.8	−3.94 567	19.28 670
Sulphurous acid	−1 604.8	−3.21 98	16.99 036
Ammonia	−1 449.83	−1.87 26	13.37 156
Nitrogen protoxide	+ 328.05	+8.71 19	−17.98 7082
Carbonic acid	− 819.77	+0.41 861	6.41 443
Turpentine	−2 674.9	−3.72 83	18.88 373
Sulphuretted hydrogen	− 992.6	−0.51 415	8.80 739
Methyl alcohol	−2 661.25	−4.63 36	22.43 1907
Mercury	−2 010.25	+3.88 06	− 4.79 892
Sulphur	−4 684.49 2	−3.40 483	19.10 740

E. Riecke [1] has shown that the tensions of vapor of liquid white phosphorus measured by Troost and Hautefeuille could be represented by Dupré's formula.

253. Dissociation tensions.—Dupré's formula must apply likewise, as is evident, to dissociation phenomena, when a single gas is concerned in the reaction. Bertrand showed, in fact, that formulæ of this type could represent in a satisfactory manner the

[1] E. Riecke, *Zeitschrift für physikalische Chemie*, v. 7, p. 115, 1891.

dissociation tensions of certain ammoniacal chlorides studied by Isambert. Nevertheless, Isambert's determinations being not accurate, this verification had only a doubtful value. Joannis and Croizier[1] have taken up the study of the dissociation of ammoniacal silver salts; they find that the laws of this dissociation are very exactly expressed by formula (6), with the following values of the constants m, n, and z; the pressures p are reckoned in centimetres of normal mercury:

Salt studied.	m	n	z	Temperature Limits (C).
$AgBr,3NH_3$	$-$ 1787.1294	+ 1.075	+ 5.7148	0° to +21°
$AgBr,\frac{3}{2}NH_3$	$-$ 6650.6086	$-$35.239	+111.1904	+30 to +55
$AgBr,NH_3$	$-$ 4033.0512	$-$13.2489	+ 47.5847	+18 to +64
AgI,NH_3	Same formula as for $AgBr,NH_3$			
$AgI,\frac{1}{2}NH_3$	$-$ 3438.3604	$-$ 8.8803	+ 34.0799	+26 to +100
$AgCN,NH_3$	$-$12497.1255	$-$58.7176	+186.3546	+81 to +117
$AgNO_3,3NH_3$	$-$ 5864.6826	$-$26.1384	+ 85.3665	+15 to +83

Joannis[2] has also studied with much care the dissociation of sodammonium and potassammonium.

He found that the dissociation tensions of sodammonium, expressed in centimetres of mercury, were very exactly represented between the temperatures of $-78°$ C. and $+26°.21$ C. by the formula

$$\log p = -\frac{619.9625}{T} + 5.055364 \log T - 7.814314.$$

Between the temperatures of $-20°$ C. and $+35°.15$ C. the dissociation tensions of potassammonium, measured in the same unit, are very exactly represented by

$$\log p = \frac{243.06}{T} + 11.775 \log T - 27.7003.$$

254. Guldberg and Waage's law.—Let us return to the general case where the system encloses any number whatever of gases taking part in the reaction.

---

[1] Joannis and Croizier, *Mémoires de la Société des Sciences physiques et naturelles de Bordeaux*, 4th S., v. 5, p. 41, 1895.

[2] Joannis, *ibid.*, 4th S., v. 5, p. 218, 1895.

If the temperature rests constant, the second member of equation (3) remains constant; the same is therefore true of the first member; now the first member of equation (3) may in virtue of the elementary properties of logarithms be regarded as the logarithm of the following number:

$$\frac{p_1^{\nu_1} p_2^{\nu_2} \cdots}{p_1'^{\nu_1'} p_2'^{\nu_2'} \cdots}$$

This number remaining constant at the same time as its logarithm, we have the following proposition:

If various systems, in which a reaction may be produced which, for all these systems, is represented by the same equation, are in equilibrium at the same temperature, the ratio

$$\frac{p_1^{\nu_1} p_2^{\nu_2} \cdots}{p_1'^{\nu_1'} p_2'^{\nu_2'} \cdots}$$

has the same value for all these systems.

This law is known under the name of NAUMANN'S LAW or GULDBERG AND WAAGE'S LAW; it is actually a special case of the law discovered by the two last named, and was not stated by the first.

255. Various examples: Ammonium carbonate.—Chemists have made several interesting verifications of this law; thus H. Pélabon has applied this relation to certain systems in which four gases enter; such is the system composed of realgar, sulphuretted hydrogen, hydrogen, and arsenic;[1] or again the system formed by mercury sulphide, hydrogen, sulphuretted hydrogen, and mercury.[2] We shall limit ourselves here to more simple cases.

The first phenomenon to which we shall apply this law is the formation of solid ammonium carbonate at the expense of ammonia gas (1) and carbonic acid gas (2)

The following is the reaction equation:

$$2NH_3 + CO_2 = NH_4OCONH_2.$$

---

[1] H. PÉLABON, *Comptes Rendus*, v. 132, p. 774, 1901.

[2] *Ibid.*, v. 132, p. 1411, 1901.

Two molecules of ammonia ($n_1=2$) and one molecule of carbonic anhydride ($n_2=1$) take part in it; the ammonia and carbonic anhydride obeying Avogadro's law, we have

$$V_1=4, \qquad V_2=2.$$

If, therefore, several systems enclosing solid ammonia carbonate, ammonia gas whose partial pressure is p_1, and carbonic acid gas of partial pressure p_2 are in equilibrium at the same temperature, the product $p_1^4p_2^2=(p_1^2p_2)^2$ will have the same value for all these systems; it amounts to the same thing as saying that the product

$$p_1^2p_2 \tag{8}$$

has the same value for all these systems.

Suppose to start with that the carbonate dissociates in an enclosure empty at first; the system is divided into two phases, the solid carbonate and the gaseous mixture; it contains, furthermore, but a single independent component, for each molecule of carbonic anhydride which it contains is accompanied by two molecules of ammonia; if the mass of the first substance it contains is known, in the free state or in combination, the mass of second substance contained in it is also known; the system is therefore monovariant; at each temperature T equilibrium is maintained by a well-determined pressure π.

This pressure π is the sum of the two partial pressures p_1, p_2; also, since the mixed gas encloses, in this case, two molecules of ammonia gas for one molecule of carbonic acid gas, the pressure p_1 is twice the pressure p_2, so that

$$p_1=\tfrac{2}{3}\pi, \qquad p_2=\tfrac{1}{3}\pi,$$

and the value of the product $p_1^2p_2$ is, in this case equal to $\dfrac{4\pi^3}{27}$; this will be also its value under all circumstances at the same temperature.

Suppose the system studied encloses not only two molecules of ammonia per molecule of carbonic anhydride, but also an *excess of ammonia gas;* at the temperature of the experiment this extra gas, distributed throughout the volume given over to the gaseous mixture, would exercise a pressure there of value ω_1; if, in the system in equilibrium at the temperature considered, the carbonic acid gas has a partial pressure p_2, the ammonia gas exercises a partial pressure $p_1=2p_2+\omega_1$; as the total pressure ϕ of the gaseous mixture is equal to the sum (p_1+p_2), it is evident that we have

$$p_1=\tfrac{1}{3}(2\phi+\omega_1);$$
$$p_2=\tfrac{1}{3}(\phi-\omega_1).$$

The product $p_1^2p_2$ has the value $\frac{(2\phi+\omega_1)^2(\phi-\omega_1)}{27}$; we also know that it must have the value $\frac{4\pi^3}{27}$; hence the following equality is obtained:

$$(2\phi+\omega_1)^2(\phi-\omega_1)=4\pi^3. \tag{9}$$

Suppose now that for two molecules of ammonia the system encloses not only one molecule of carbonic anhydride, but *an excess of carbonic acid gas* besides; the excess of carbonic anhydride distributed, at the temperature of the experiment, in the volume occupied by the gaseous mixture, will exert a pressure ω_2 there; if, with the system in equilibrium, the ammonia gas exerts a partial pressure p_1, the carbonic acid gas exerts a partial pressure $p_2=\frac{p_1}{2}+\omega_2$; and as the sum (p_1+p_2) must always be equal to the total pressure ϕ of the gaseous mixture, we have

$$p_1=\tfrac{2}{3}(\phi-\ \omega_2);$$
$$p_2=\tfrac{1}{3}(\phi+2\omega_2).$$

The product $p_1^2p_2$ has the value $\frac{4(\phi-\omega_2)^2(\phi+2\omega_2)}{27}$, and is also equal to $\frac{4\pi^3}{27}$; therefore it follows that

$$(\phi-\omega_2)^2(\phi+2\omega_2)=\pi^3. \tag{10}$$

The values of the quantities ω_1, ω_2, π, being readily accessible to obervation, equations (9) and (10) may be verified.

Naumann and Horstmann were the first to attempt this verification; more exact experiments have been performed by Isambert.[1]

Five barometer tubes, graduated in tenths of a cubic centimetre, and containing ammonia carbonate, were placed side by side in a heating-chamber; the first, containing no excess of carbonic acid gas nor of ammonia, gave *directly* the dissociation tension π of ammonia carbonate in vacuo at the temperature of the oven; these results are found in column I of the following table.

The four other tubes contain either an excess of carbonic anhydride or of ammonia.

Tube II had received an excess of carbonic acid gas occupying, in the normal conditions of temperature and pressure, 16.9 cc.; tube III had received similarly 6.1 cc. of carbonic anhydride; tube IV, 6 cc. of ammonia gas; finally, tube V, 11.4 cc. of the same gas; the readings of tubes II and III, joined to equation (10), give, at each temperature, two *indirect determinations* of π, put in columns II and III of the table; the readings of tubes IV and V, taken with equation (9), give for each temperature two other *indirect determinations* of π, inscribed in columns IV and V of the table.

The values of π determined indirectly by the last four tubes differ in general but slightly from the value of π observed directly by means of the first. We may therefore regard Isambert's observations as confirming very exactly the law as stated.

Temperatures Centigrade.	I. mm.	II. mm.	III. mm.	IV. mm.	V. mm.
34°.0	169.8	170.4	164.5	166.8	181.3
37 .2	211.0	210.8	204.6	205.9	215.5
39 .1	234.1	234.4	228.5	229.4	236.9
41 .8	269.4	271.7	267.7	265.6	274.5
42 .5	288.3	289.2	284.2	286.2	291.9
43 .9	313.8	314.5	311.8	313.5	318.4
46 .9	375.7	375.3	372.0	375.6	378.3
50 .1	453.8	452.9	452.2	454.1	455.0
52 .6	526.2	523.7	522.3	523.8	526.2

256. Ammonium cyanide.—Isambert has likewise studied the dissociation of certain solid non-volatile substances, formed by

[1] Isambert, *Comptes Rendus*, v. 93, p. 731, 1881; v. 97, p. 1212, 1883.

the union *molecule by molecule* of two component gases; the following equations represent such reactions:

$$NH_3+H_2S=HNH_4S;$$
$$PhH_3+HBr=PhH_4Br;$$
$$NH_3+HCN=NH_4CN.$$

In this case $n_1=1$, $n_2=1$, and as the gases studied obey Avogadro's Law, $V_1=2$, $V_2=2$; at a given temperature the product $p_1^2p_2^2=(p_1p_2)^2$ has a definite value, as has also the product p_1p_2.

If, for instance, the solid considered dissociates in vacuo, its dissociation tension attains a value π, well determined for each temperature; it is clear, besides, that in this case the two partial pressures p_1, p_2 are equal to each other and equal to $\frac{\pi}{2}$, so that the product p_1p_2 is equal to $\frac{\pi^2}{4}$. We may therefore state the following proposition:

Suppose that, at a given temperature, one of the solids we have cited exists in equilibrium with a gaseous atmosphere where the gaseous components have the partial pressures p_1, p_2; we shall have the relation

$$p_1p_2=\frac{\pi^2}{4}, \tag{11}$$

π being the dissociation tension of the solid, at the temperature considered, in an enclosure containing no gas at the start.

Isambert has verified this relation in studying the dissociation of ammonia bisulphydrate[1] and of the bromhydrate of phosphoretted hydrogen;[2] he has made an especially careful verification for the case of ammonium cyanide[3] in the presence of an excess of ammonia gas.

Let ω_1 be the pressure exerted by the ammonia gas in excess, at the temperature of the experiment, if it occupied alone the

[1] ISAMBERT, *Comptes Rendus*, v. 93, p. 731, 1881; v. 94, p. 958, 1882; v. 95, p. 1355, 1882.

[2] *Ibid.*, v. 96, p. 643, 1883.

[3] *Ibid.*, v. 94, p. 958, 1882; *Annales de Chimie et de Physique*, 5th S., v. 28, p. 382, 1883.

volume given up to the gaseous mixture; in this mixture the cyanhydric acid has a partial pressure p_2, and the ammonia gas a partial pressure p_1 which is evidently equal to $(p_2+\omega_1)$; the total pressure ϕ of the gaseous mixture being equal to the sum (p_1+p_2), we have

$$p_2=\tfrac{1}{2}(\phi-\omega_1). \tag{12}$$

Further, p_1 being equal to $(p_2+\omega_1)$, equation (11) gives

$$p_2(p_2+\omega_1)=\frac{\pi^2}{4}.$$

The measurement of ϕ, joined to the knowledge of ω_1, allows deriving from equation (12) a first value of p_2; again, the measurement of the dissociation tension π of ammonium cyanide in an enclosure empty at first allows finding from equation (13) another vale of p_2, which we shall denote by p_2'; if the law we are concerned with is true, the two pressures p_2, p_2' should be equal to each other.

Here are the values of p_2 and p_2' obtained by Isambert:

Temperatures Centigrade.	π mm.	ϕ mm.	ω_1 mm.	p_2 mm.	p_2' mm.
7°.3	175.0	358.0	314.2	21.2	22.7
7 .4	176.7	365.2	327.7	18.7	21.3
9 .2	196.0	369.8	317.0	26.4	27.8
9 .3	200.0	370.0	329.0	25.0	28.0
9 .4	202.0	373.4	323.2	25.1	26.9
10 .2	214.0	378.4	316.0	31.2	32.8
11 .0	227.4	393.3	323.0	35.1	35.8
11 .2	232.9	390.0	311.2	39.4	38.7
11 .2	234.0	395.6	320.6	37.5	38.2
12 .0	246.2	397.8	309.2	44.3	42.9
14 .3	265.5	413.2	308.8	52.2	49.1
15 .5	296.9	425.8	294.8	65.4	61.8
15 .7	300.9	426.1	295.1	65.5	63.2
15 .7	300.5	432.2	299.8	66.2	62.6
17 .0	322.4	441.1	287.3	76.9	72.2
17 .2	326.2	442.9	286.0	78.4	74.0

These figures place beyond doubt the exactitude of Guldberg and Waage's Law.

Isambert has also studied the dissociation of ammonium cyanide for the case where the cyanhydric acid is in excess, but there this substance condenses in part to the lipuid state; the liquid formed dissolves ammonium cyanide and ammonia, and the conditions proposed at the beginning of the present chapter are no longer verified.

257. Influence of temperature. Dissociation of mercuric oxide.—The various observations we have mentioned in the preceding article show us that formula (3) represents very exactly the law according to which the composition of the gaseous mixture within a system in equilibrium varies when, without changing the temperature, there is introduced into the system an excess of one or of the other of the gases which take part in the reaction; it remains to be seen if this formula represents as exactly the influence the temperature exerts on the degree of dissociation; this question has as yet received reply only in the case (Art. 253) where the system contains a single gaseous substance; it is important to examine it for more complicated cases.

Here is an elegant test to which this law has been submitted by Pélabon:[1]

The red oxide of mercury may dissociate into mercury vapor and oxygen according to the equation

$$HgO = Hg + O$$

Mercury vapor being monatomic and oxygen diatomic, if we give the index 1 to mercury and index 2 to oxygen we shall have

$$V_1 = 2, \qquad V_2 = 1,$$

and equation (3) may be written

$$2 \log p_1 + \log p_2 = \frac{M}{T} + N \log T + Z. \tag{14}$$

Suppose, in the first place, that an excess of mercury is maintained in the system; the partial pressure p_1 of the mercury vapor in the gaseous mixture will be equal, by the law of the mixture of gases and vapors whose truth is one of our fundamental hypoth-

---

[1] H. Pélabon, *Comptes Rendus*, v. 128, p. 825, 1899; *Mémoires de la Société des Sciences physiques et naturelles de Bordeaux*, 5th S., v. 5, p. 59, 1899.

eses, to the tension of the saturated vapor F of mercury at the temperature of the experiment; further, F is given by Dupré's formula,

$$\log F = \frac{m}{T} + n \log T + z. \tag{15}$$

If the pressures are measured in millimetres of normal mercury, the constants m, n, z have the values, as we have seen, Art. 252,

$$m = 2\,010.25, \qquad n = 3.88\,06, \qquad z = -4.79\,892. \tag{16}$$

If, in formula (14), we replace $\log p_1$ by the value of $\log F$ given by equation (15), and if we put

$$\mu = M - 2m, \qquad \nu = N - 2n, \qquad \zeta = Z - 2z, \tag{17}$$

we may write

$$\log p_2 = \frac{\mu}{T} + \nu \log T + \zeta. \tag{18}$$

Pélabon has shown that the values of p_2, expressed in millimetres of mercury, could be very exactly expressed by a formula of this type, on condition of taking

$$\mu = -27\,569, \qquad \nu = -57.58, \qquad \zeta = +203.94\,711. \tag{19}$$

Suppose now that mercury oxide dissociates in an enclosure which is at first a vacuum, and that the mercury resulting from this decomposition remains entirely in a state of vapor; it will follow necessarily, denoting by p_2' the partial pressure of oxygen in this case, that

$$p_1 = 2p_2',$$

and

$$\begin{aligned} 2 \log p_1 = 2 \log p_2' + 2 \log 2 &= 2 \log p_2' + \log 2^2 \\ &= 2 \log p_2' + \log 4. \end{aligned}$$

Substituting this value of $2 \log p_1$ in equation (14), we find

$$\log p_2' = \frac{\mu'}{T} + \nu' \log T + \zeta'. \tag{20}$$

if we put

$$\mu' = \frac{M}{3}, \quad \nu' = \frac{N}{3}, \quad \zeta' = \frac{Z - \log 4}{3}.$$

or, from equations (17), if

$$(21) \qquad \mu' = \frac{\mu + 2m}{3}, \quad \nu' = \frac{\nu + 2n}{3}, \quad \zeta' = \frac{\zeta + 2z - \log 4}{3}.$$

Equations (16), (19), and (21) permit us to calculate the values of μ', ν', ξ'; we thus find

$$(22) \qquad \mu' = -10\,529.8, \quad \nu' = -16.61, \quad \zeta' = +64.58\,240.$$

When, therefore, mercuric oxide dissociates in an enclosure at first a vacuum, the partial pressure of the oxygen must, if the preceding theory is correct, be represented by formula (20), the constants μ', ν', ζ' having the values given in (22).

Pélabon has determined experimentally a certain number of values of p_2' and has compared them with the values calculated as we have just indicated; the following table gives us an idea of the very satisfactory concordance for the results of the two sets of determinations:

Temperatures.	p_2' obs. mm.	p_2' calc. mm.
500° C.	995	972
520	1 392	1 403
580	3 610	3 589
610	5 162	5 308

258. Dissociation of selenhydric acid.—Another verification of formula (3), verification of not less importance than the preceding, was also obtained by Pélabon [1] in studying the formation of selenhydric acid from liquid selenium and hydrogen, according to the formula

$$H_2 + Se = H_2Se.$$

[1] H. Pélabon, *Comptes Rendus*, v. 121, p. 401, 1895; *Mémoires de la Société des Sciences physiques et naturelles de Bordeaux*, 5th S., v. 3, p. 207, 1898.

To simplify matters, let us neglect the volatility of selenium; we shall then have in the system but two gaseous bodies, hydrogen, whose partial pressure is p, and selenhydric acid, of partial pressure p'; then

$$V_1=2, \qquad V_2=2,$$

and equation (3) becomes

$$2 \log p - 2 \log p' = \frac{M}{T} + N \log T + Z,$$

or, in putting

(23) $$m=\frac{M}{2}, \quad n=\frac{N}{2}, \quad z=\frac{Z}{2},$$

we have

(24) $$\log \frac{p}{p_1} = \frac{m}{T} + n \log T + z.$$

Pélabon found that the va ues o $\frac{p}{p_1}$ determined experimentally between the temperatures 320° C. and 720° C were very well represented by a formula of the typ (24) on the condition of giving to the constant m, n, and z the following values;

(25) $$\begin{cases} m = 13\,170.3 \times 0\ 43\,01; \\ n = 15.53; \\ z = -119\ 88 \times 0.43\,01. \end{cases}$$

But Pélabon has gone farther not content with having verified equation (3), he has sought to verify equations (4') and (5'), which for the present case become identical with each other. By means of the equations (23) and (25) he was able to calculate:

1°. The tempe·ature for which L becomes equal to 0, temperature which he found equal to 575° C.;

2°. The heat PL *absorbed* by the formation at 15° C. of a molecule (81 grammes) of selenhydric acid at the expense of liquid selenium and hydrogen, quantity of heat which he found equal to 17300 calories.

The temperature at which L becomes equal to 0 must, according to the law of the displacement of equilibrium by change in temperature correspond to a minimum of dissociation or to a minimum of the ratio $\frac{p}{p'}$; Pélabon has found, from direct experiment, that such a minimum is produced at a temperature included between 550° *C*. and 600° *C*.

Further, a direct calorimetric determination gave Fabre[1] 18000 calories for the value of the product *PL* at 15° *C*.

Another verification, analogous to the preceding, has been obtained by Jouniaux by studying the action of hydrogen on silver chloride and the inverse action of hydrochloric acid on silver.[2] The study of states of equilibrium which are established at temperatures included between 525° and 700° allowed him to determine the coefficients *M* N, *Z* of the formula (3 He was then able by formula (44) to calculate the heat which is *absorbed* when hydrochloric acid transforms a molecule of silver into a molecule of silver chloride. He has found, for the value of this quantity of heat, 6790 calories, while the direct calorimetric determinations due to Berthelot gave 7000 calories.

A similar investigation concerning the action of hydrogen on silver bromide and the inverse action gave Jouniaux[3] 13700 calories for the heat of formation of silver bromide at the expense of hydrobromic acid and silver, while Berthelot's measurements give 14800 calories

259. Variations of the density of perchloride of phosphorus vapor.—We have not applied as yet formula (3) to equilibrium phenomena which may be produced in homogeneous gaseous systems.

Some work has been done on the dissociation of a gaseous body into its gaseous components by cooling briskly the strongly heated vessel enclosing the mixture and analyzing the cooled mixture, whose composition is supposed identical with that of the mixture not yet cooled; prolonged experiments by numerous investigators on the dissociation of hydriodic acid have been made

[1] FABRE, *Annales de Chimie et de Physique*, 6th S., v. 10, p. 482, 1887.

[2] A. JOUNIAUX, *Comptes Rendus*, v. 132, p. 1270, 1901; *Actions des hydracides hydrogènes sur l'argent et réactions inverse*, p. 59 (Thesis, Lille, 1901).

[3] A. JOUNIAUX, *Actions des hydracides . . .*, p. 96.

by this method; unfortunately, the decomposition of glass, at high temperatures, by the reacting bodies has rendered this experiment untrustworthy.

The method of sudden cooling seems to be the only one which may be applied to the study of the dissociation of a gas which is formed without condensation from its gaseous elements; but in the case where the gas which dissociates is formed from its gaseous elements with condensation, every decomposition of this gas has for effect the decrease of the density with respect to air of the gaseous mixture where it exists in presence of gases coming from its decomposition; the study of the variations undergone by the density referred to air of this mixture when the temperature and pressure are varied, when an excess of one or the other component is introduced into the system, furnishes, on the subject of the dissociation of the compound considered, indirect but exact information

Phosphorus perchloride is formed by the union of chlorine and protochloride, according to the formula

$$PhCl_3 + Cl_2 = PhCl_5,$$

and the reaction is accompanied by a condensation which at constant temperature and pressure reduces by half the volume of the system.

When one determines, as did Cahours,[1] the density of phosphorus perchloride vapor at higher and higher temperatures, this density is seen to diminish more and more; at the same time, the vapor assumes a darker and darker coloration, resembling that of chlorine; also, as soon as H. Sainte Claire Deville had made known the phenomena of dissociation, chemists agreed, with Cannizaro and H. Kopp, to admit that the temperature in increasing brings about a gradual dissociation of phosphorus perchloride into chlorine and trichloride; this opinion has been confirmed beyond dispute by Wanklyn and Robinson;[2] these observers have found that in diffusing through a porous substance, the vapors emitted

[1] CAHOURS, *Comptes Rendus*, v. 21, p. 625, 1845; *Annales de Chimie et de Physique*, 3d S., v. 20, p. 369, 1847.

[2] WANKLYN and ROBINSON, *Philosophical Magazine*, v. 26, p. 545, 1863.

by phosphrous perchloride furnish a mixture which contains chlorine in excess.

If, in the mixture where the perchloride of phosphorus exists in the presence of its elements, p_1 denotes the partial pressure of trichloride, p_2 the partial pressure of chlorine, and p_1' the partial pressure of perchloride, and since we have

$$V_1=2, \qquad V_2=2, \qquad V_1'=2,$$

equation (3) becomes

$$2\log p_1+2\log p_2-2\log p_1'=\frac{M}{T}+N\log T+Z;$$

or, in putting

$$\mu=\frac{M}{2}, \qquad \nu=\frac{N}{2}, \qquad \zeta=\frac{Z}{2},$$

$$\log p_1+\log p_2-\log p_1'=\frac{\mu}{T}+\nu\log T+\zeta. \tag{26}$$

Furthermore, an elementary computation shows that if we denote by δ_1, δ_2, δ_1' the densities of the three gases referred to air, the density Δ of the mixture, referred to air, has the value

$$\Delta=\frac{p_1\delta_1+p_2\delta_2+p_1'\delta_1'}{p_1+p_2+p_1'}. \tag{27}$$

Cabours has experimentally determined the value of Δ at pressures near to atmospheric pressure and for temperatures between 182° C. and 336° C.; Troost and Hautefeuille [1] on the one hand and Würtz [2] on the other have made analogous determinations at pressures below an atmosphere; finally Würtz determined the vapor density of a mixture which, instead of enclosing a molecule of phosphorus trichloride for one molecule of chlorine, contained an excess of trichloride by joining to all these experimental determinations an old observation of Mitscherlich, forty-three values of the density Δ are obtained under most diverse conditions; all these values, save one, are, as shown by Gibbs,[3] very exactly repre-

[1] TROOST and HAUTEFEUILLE, *Comptes Rendus*, v. 83, p. 977, 1876.

[2] WÜRTZ, *Comptes Rendus*, v. 76, p. 601, 1873.

[3] J. WILLARD GIBBS, *American Journal of Arts and Sciences*, v. 18, p. 381 1879.

sented by formula (26) and (27), on condition of giving the constants μ, ν, and ζ properly chosen values.

260. Dissociation of a gas into a vacuum space.—The study of vapor densities of phosphorus perchloride furnishes also a remarkable confirmation of the theory of dissociation within a system containing a mixture of perfect gases. Let us discuss the consequences to which this theory leads for the case where the compound dissociates in a vacuum enclosure, so that the gases coming from this decomposition exist in the system in just the proportion as in the compound itself.

Let us denote by x the mass of non-dissociated gas contained in 1 gramme of the gaseous mixture; x will be equal to 1 when the combination is entire and to 0 when the decomposition is complete; suppose that, leaving the pressure at a constant value π, the absolute temperature T increases from 0 to $+\infty$, and let us see how x varies according to formula (3).

The problem is particularly simple for the case in which the compound is formed without condensation from its elements; in this case the heat of formation under constant pressure and the heat of formation at constant volume of the compound substance are equal to each other, by definition; besides, we know (Chap. III, Art. 44) that they are both independent of the temperature, by virtue of the equality (4); we must have, therefore

$$N=0.$$

There are then two cases to distinguish: either M is negative and *the compound is unceasingly exothermic*, or else M is positive and *the compound is constantly endothermic.*

In the one case as in the other, the second member of equation (3) reduces to $\left(\frac{M}{T}+Z\right)$, which permits establishing the three following laws:

FIRST CASE: THE COMPOUND IS EXOTHERMIC.—Draw two rectangular axes OT, Ox (Fig. 111); lay off as abscissæ the absolute temperatures T, and as ordinates the values of x; for $T=0$, x has the value 1; the curve representing the variations of x starts from the point A, in very close contact with the straight line AA', which has the constant ordinate $x=1$; it is only after a considerable distance AB that it separates appreciably from this straight line; it then descends from left to right along BC, and when T increases beyond all limits, it approaches, without reaching, a straight line LL′ parallel to OT; this straight line has a constant ordinate greater than 0, so that, *the temperature increasing without limit, the system does not approach the state of complete dissociation.*

The curve representing the variations of x does not change if the value of the constant pressure changes under which the experiment is supposed to be made.

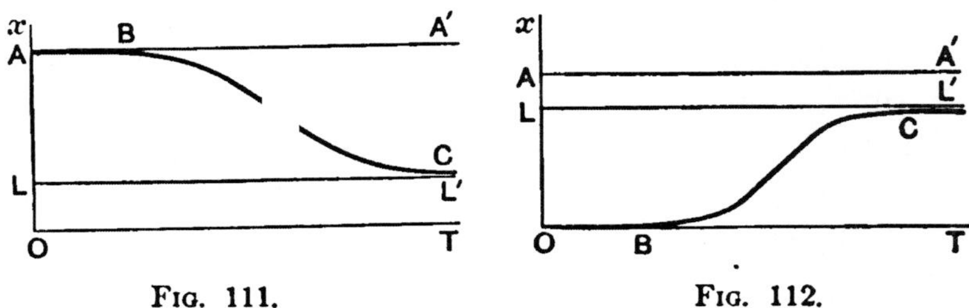

FIG. 111. FIG. 112.

SECOND CASE: THE COMPOUND IS ENDOTHERMIC.—The ratio x starts for $T=0$ from the value 0; the curve (Fig. 112) representing the variations of x starts from the point 0; it has close contact with the line OT so that it does not sensibly leave this curve until after the considerable distance OB; it then rises from left to right along BC, and when the temperature becomes indefinitely great, it approaches without reaching a straight line LL' parallel to OT; this line LL′ lies below the straight line AA', whose constant ordinate is $x=1$; consequently *when the temperature increases indefinitely the state of the system does not approach a complete combination.*

The curves representing the variations of x do not change if the value of the constant pressure changes under which the experiment is supposed to be made.

For the case in which the compound studied is formed with condensation the results become slightly more complicated; the path of the representative point is no longer independent of the pressure; it is, on the contrary, the higher as the pressure is greater.

Suppose in particular, following Gibbs, that the constant N be here also equal to 0; the heat of formation under constant pressure will no longer depend upon the temperature; it will be a simple constant; consider only the case where, M being negative, the COMPOUND IS CONSTANTLY EXOTHERMIC.

When the temperature T starts from zero and increases beyond limit, the pressure keeping an invariable value π, x starts from the value 1 and the curve representing the variations of x (Fig. 113) begins at the point A, whose ordinate OA is equal to unity.

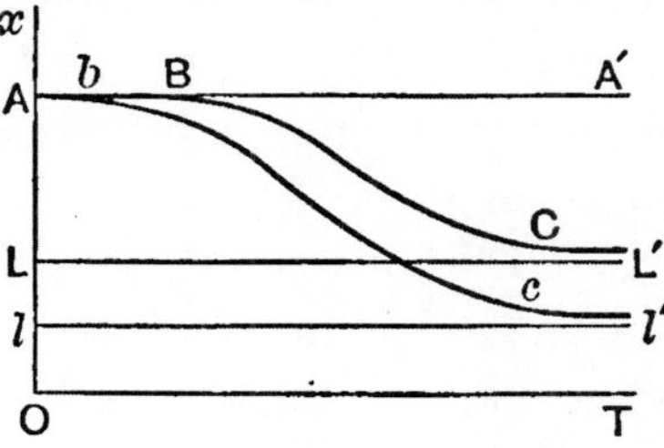

FIG. 113.

When the temperature rises, the curve remains for quite a long distance AB identical with the straight line AA' drawn through the point A parallel to OT; it then detaches itself to descend from left to right along BC; when the temperature increases without limit, it approaches more and more a line LL' parallel to OT, but situated above OT.

Under another constant pressure ω, less than π, things go on in a similar manner, but—

1°. The curve detaches itself from the straight line AA' at a point b, situated to the left of the point B;

2°. The curve bc is constantly below the curve BC

3°. When the temperature T increases without limit, the line bc approaches more and more the line ll', parallel to LL′, but situated between OT and LL′.

261. Variations of vapor density. Are they due to the dissociation of polymers?—It is in the study of the great variations of density of certain vapors that the preceding principles find their chief use.

If, in various conditions of temperature and pressure, the density of a sensibly pe fect gas referred to air is determined, the same number is always found; this density is a constant. This will not be true for a gas which is appreciably other than a perfect gas; for example, the density of carbonic acid gas referred to air, at atmospheric pressure, is somewhat less at 100° than at 0°.

The density referred to air of certain gases or of certain vapors undergoes very great variations when the temperature and pressure are changed; the first observation of such variations was made in 1844 by Cahours, who found the vapor density of acetic acid, taken at atmospheric pressure, to vary from 3.20 to 2.08, while the temperature rose from 125° *C.* to 338° *C.*

Since this time, analogous facts have multiplied; formic acid and nitrogen peroxide show similar variations to those manifested by acetic acid; Troost and Hautefeuille have proved that the density of sulphur vapor, at atmospheric pressure, passed sensibly from the value 6.6 to the value 2.2 when the temperature went from 500° *C.* to 1000° C.; the experiments of Crafts and Meier, performed by the method of displacement of air, showed that iodine vapor density nearly constant and equal to 8.8 while the temperature remained less than 700° C., rapidly decreased beyond, reaching a value slightly greater than 4.4 when the temperature exceeded 1600°.

We may be content with establishing these facts and saying that the gases or vapors whose density referred to air undergoes great changes, caused by variations in pressure and temperature, are far removed from the state of perfect gas.

Certain physicists have thought a more complete and fruitful interpretation of these variations could be sought: they have regarded the gases where they are manifested as capable of existing in two different states; when each of these two gaseous states is sensibly perfect, its density referred to air is sensibly independent of the temperature and pressure; but the densities referred to air of these two gases are different; they are in a simple ratio to each other; a gas whose density changes considerably with the temperature and pressure is in reality a mixture of two gases one of which is a *polymer* of the other, and the proportions of this mixture vary with the temperature and pressure.

Thus, according to this hypothesis, there exist two iodine vapors, one of which alone would have the density 8.8 and to which Avogadro's Law would assign the symbol I_2, while the other would have the density 4.4 and the symbol I by the same law; there would exist two acetic acid gases, two formic gases, two peroxides of nitrogen, the density of one of these two gases being double the density of the other.

There are a great number of cases where the existence of two forms of the same gas, polymers of each other, is incontestable; everybody knows, for instance, that oxygen exists in the state of ordinary oxygen and ozone; according to Soret's researches the density of ozone is $\frac{3}{2}$ that of oxygen; Avogadro's Law, which gives oxygen the symbol O_2, gives ozone the symbol O_3. Thus, in this case, we may observe, at the same temperature and pressure, samples of oxygen which have different densities, different chemical and physical properties, so that there is no doubt of the existence of an *allotropic* oxygen.

Organic chemistry furnishes us with a great number of analogous facts; thus every chemist knows that acetylene may be transformed into a polymer of triple density, benzine.

But if, in these various cases, we can put beyond doubt the existence of the same gas in two distinct polymeric forms, we are indebted to the phenomena of false equilibrium; in the conditions of temperature and pressure for which the states of false equilibrium would not be produced, oxygen, taken in definite conditions, would always enclose a determined amount of ozone; at a given pressure and temperature its properties would be perfectly determined; but its density taken with respect to a perfect gas would vary with the pressure and with the temperature; oxygen would behave, in terms of the variation of density produced by a rise in temperature, just as do sulphur vapor, iodine vapor, acetic acid vapor; one may not, therefore, argue from this fact that at a given pressure and temperature each of these gases exists in a perfectly determined state in order to deny, for each of them, the coexistence of two polymers; one may merely conclude there are not produced phenomena of false equilibrium in the conditions of temperature and pressure for which the experiments have been performed.

262. Comparison of experimental facts with the theory of dissociation.—To explain the variations undergone by the density referred to air of certain gases, when the temperature and pressure change, by regarding each of these gases as a mixture of two states one of which is a polymer of the other, is to make an hypothesis in no way unacceptable; this hypothesis will assume a high degree of probability if it be shown that in applying to such a mixture the thermodynamic properties of a mixture of gases near to the perfect state, one succeeds in taking account of these changes of density in a complete manner.

Suppose the polymer formed with liberation of heat and, to simplify, admit, as does Gibbs, that the constant N equals 0; if x denotes the mass of the polymer contained in 1 gramme of the gaseous mixture, the variations undergone by x when the temperature increases, keeping the pressure constant, will be represented by one of the curves ABC or Abc of Fig. 113. Observe now that the density Δ of our complex gas referred to air increases constantly with x, starts for $x=0$ from the density d of the non-polymerized gas, and attains for $x=1$ the density D of the polymer, and we may state the following results:

Take two rectangular axes (Fig. 114); on the axis of abscissæ lay off the absolute temperatures T, and on the axis of ordinates the densities Δ; take an invariable value π of the pressure, and let the temperature T increase from 0 to $+\infty$. The representative point starts from D, whose ordinate OD is measured by the density D of the polymer; the curve which it describes remains, for a conside able distance DB, almost identical with the straight line DD′ drawn through the point D parallel to OT; it then descends from left to right along BC; when T increases indefinitely, the representative point approaches a line RR', parallel to OT, but whose constant ordinate surpasses the density d of the non-polymerized gas.

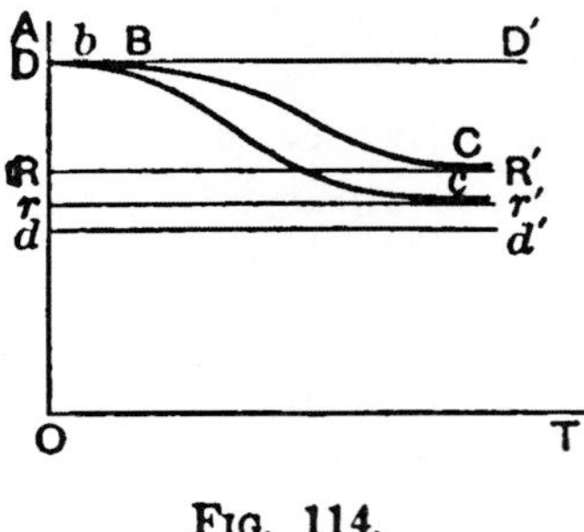

Fig. 114.

If the same observations are repeated under constant pressure ω, less than π, one finds repersenting them an analogous curve Dbc, but—

1°. The point b, where the curve Dbc quits appreciably the line DD′, is situated to the left of the point B;

2°. The line bc is constantly below the line BC

3°. When T increases indefinitely, the representative point approaches a line rr', parallel to OT, and situated between RR' and dd'.

263. Density of iodine vapor.—The arrangement we have just described is exactly that of the curves by which Crafts and

Meier [1] have represented the variations undergone by iodine vapor when the temperature varies from 500° to 1600°, at constant pressure, to which Crafts and Meier have given successively the values 0.4 atm., 3.0 atm., 0.1 atm.

264. Gibbs's formula.—But we are not satisfied with this merely qualitative comparison between the results of formula (3) and the experimental data.

Take, for instance, the polymerization of nitrogen peroxide, represented by the formula

$$2NO_2 = 2N_2O_4.$$

We have here $V_1=4$, $V_1'=2$, so that formula (3) may be written

$$2 \log p_1 - \log p_1' = \frac{M}{2T} + \frac{N}{2} \log T + \frac{Z}{2}.$$

Let D and d be the densities referred to air of the substances N_2O_4 and NO_2, the first being double the second. The equality (27) gives us

$$p_1 d + p_1' D = (p_1 + p_1')\Delta.$$

Furthermore, denoting by π the total pressure, we have

$$p_1 + p_1' = \pi.$$

These two equations of the first degree in p_1 and p_1' give us

$$p_1 = \frac{D-\Delta}{D-d}\pi = \frac{D-\Delta}{d}\pi;$$

$$p_1' = \frac{\Delta - d}{D-d}\pi = \frac{\Delta - d}{d}\pi.$$

Equation (28) then becomes

$$\text{(29)} \qquad \log \frac{(D-\Delta)^2}{d(\Delta - d)} = \frac{M}{2T} + \frac{N}{2} \log T - \log \pi + \frac{Z}{2}.$$

---

[1] Crafts and Meier, *Comptes Rendus*, v. 90, p. 690, 1880.

Gibbs[1] took $N=0$; that is to say, he has admitted that, at constant pressure, the heat of formation of the polymer N_2O_4 from the gas NO_2 was independent of the temperature; he has shown that formula (29) represented in a satisfactory manner the determinations of vapor density of nitrogen peroxide made at different temperatures and under different pressures by Mitscherlich, R. Müller, H Deville Troost, Naumann, Playfair, and Wanklyn.

A similar formula represents the vapor densities of acetic acid observed by Cabours, Bineau, Horstmann, Troost, Naumann, Playfair, and Wanklyn; another the vapor densities of formic acid, determined by Bineau.

After some very exact investigations of the vapor density of nitrogen peroxide E. and L. Natauson have observed that formula (29) did not represent with an entire exactitude the variations of this density; but on the subject of these discrepancies there are two points to note.

1°. For simplicity, Gibbs has attributed the value 0 to the constant N, supposition which is not obligatory;

2°. The preceding theory supposes the two gases NO_2, N_2O_4 are in the state of perfect gases, supposition certainly removed from the truth.

Gibbs's theory does not give the laws of chemical equilibrium in systems containing gases more exactly than the laws of Boyle and Charles make known the compressibility and expansion of a single gas; but it is enough that it renders in chemical mechanics services analogous to those which Boyle's and Charles's laws render in physics, to be extremely valuable.

---

[1] Gibbs, *Transactions of the Connecticut Academy*, v. 3, p. 234, 1876; *American Journal of Arts and Sciences*, v. 18, p. 277, 1879.

[2] E. and L. Natauson, *Wiedemann's Annalen*, v. 24, p. 454, 1885; v. 27, p. 606, 1886.

CHAPTER XVI.

CAPILLARY ACTIONS AND APPARENT FALSE EQUILIBRIA.

265. The preceding theories are often contradicted by experiment.—The comparisons which we have constantly had occasion to make between the results of chemical statics founded on thermodynamics and the experimental data have revealed numerous and precise concordances, but they have likewise shown evidence of too numerous and sharp contradictions for us to pass over them in silence

The decomposition of water absorbs heat; when, therefore, we raise the temperature of a mixture of oxygen hydrogen, and water vapor, the water vapor should dissociate more and more; but if we take a mixture of oxygen and hydrogen and gradually raise the temperature, we shall have at first no chemical reaction; then, all at once, when the temperature attains about 500°, a part of the mixture passes with explosion into the state of water vapor.

The formation of ozone from oxygen absorbs heat; ozone therefore should be the more stable the higher the temperature; now it is sufficient to heat to 200° ozonized oxygen in order to have every trace of ozone disappear.

All the explosive reactions, all rapid combustions are as many exceptions or, better, *objections* to the principle of displacement of equilibrium by variation of the temperature.

The chemical actions are not the only ones which make exception to the rules laid down by thermodynamics; the physical changes of state and allotropic modifications also furnish objections to this theory.

According to this theory, when a liquid is transformed into vapor there exists for every temperature one pressure, and one only, for which there is equilibrium between the liquid and vapor; at a lower pressure the liquid should become vapor; at a higher pressure the vapor must condense. This is not what experiment shows to happen; drops of water, suspended in a liquid of the same density, may, without quitting the liquid state, be carried to a temperature at which the tension of saturated vapor exceeds by a great deal the pressure which they support; very dry and quite pure vapor may, without any condensation being produced, be compressed beyond the tension of the vapor saturated at the temperature of the experiment

When a solid and the liquid resulting from its fusion are submitted to atmospheric pressure, there exists, according to the preceding theory, a temperature, and one only, where the solid is in equilibrium with the liquid; at higher temperatures the solid should melt; at lower temperatures the liquid should freeze. This last prediction is not confirmed by experiment; the temperature of a substance may be lowered far beneath the freezing-point without the substance leaving the liquid state

When a salt is in contact with a solvent there exists, at each temperature and pressure, a concentration for which the system is in equilibrium; in the presence of a less concentrated solution the solid salt should dissolve; from a more concentrated solution the dissolved salt should, in part, be precipitated in the solid state. In this last point the theory is not in accord with observation; a solution may be kept supersaturated without the salt contained in it crystallizing out.

Similarly a gaseous solution may be maintained supersaturated in such conditions of temperature and pressure that, according to thermodynamics, the gas should be liberated

According to the phase rule there should be a temperature where orthorhombic sulphur (octahedric) and clinorhombic sulphur (prismatic) coexist in equilibrium; above this temperature orthorhombic sulphur should be transformed into clinorhombic; below this temperature the clinorhombic form should be changed into orthorhombic sulphur. Actually, below the transformation-point one may have clinorhombic sulphur in the state of crystalline *surfusion;* above the transformation point orthorhombic sulphur may be had in the state of crystalline *superheating.*

266. Rule, stated by J. Moutier, which summarizes these contradictions.—These numberless observations, contradicting the thermodynamic theory, all possess a common character.

Never do we meet, in the domain of experimental facts, a modification declared by thermodynamics to be impossible; we never see two substances combine when the theory declares they will not combine; a compound dissociate, when theory affirms it will not decompose; a liquid change into vapor or freeze when, according to thermodynamics, it should not vaporize or congeal; *without exception*, when thermodynamics states a modification to be impossible, the modification is not produced.

But, on the other hand, when thermodynamics announces that a modification will be produced, the modification is not always produced.

Thermodynamics affirms that at ordinary temperatures oxygen and hydrogen will combine almost entirely, that nitre will decompose; but oxygen and hydrogen rest mixed without combining, and nitre does not decompose. Water, by the retarding of boiling, should become vapor; in surfusion it should freeze; in both cases it remains in the liquid state.

In conclusion, the following proposition, due to Moutier, and already spoken of in Art. 99, may be stated:

Every time that thermodynamics, by aid of the hypotheses and principles mentioned as yet, announces that a certain state will be, for the system studied, a state of equilibrium, observation shows that the system, put in this state, remains there actually in equilibrium; but when thermodynamics states that the system studied, put in a certain state, will undergo a determined modification, it may happen that the system, placed in this state, remains there in equilibrium.

267. True and false equilibria.—In other words, experiment recognizes all the equilibrium states predicted by thermodynamics, states we shall call TRUE EQUILIBRIUM STATES; but besides, it recognizes the existence of a great number of equilibrium states which the principles of thermodynamics deny; to these last we have given the name FALSE EQUILIBRIUM STATES.

268. Internal thermodynamic potential of a homogeneous mass whose various particles are infinitely separated.—A theory permits, in a great number of cases, not only to comprehend the existence of false equilibrium states, but also to predict the circumstances which will assure the maintenance of such states or provoke their rupture. The broad lines of this theory were traced by J. Willard Gibbs [1] in various parts of his admirable memoir on the equilibrium of heterogeneous substances.

Take a certain mass M of water, brought to a definite temperature, say 100°, and having a certain density. Divide this mass into infinitely small particles, identical with each other, and sow them in space at infinite distances from each other.

This mass of water, thus pulverized and disseminated, admits a certain internal thermodynamic potential $\mathfrak{F}$; one may evidently regard this potential as the sum of the internal thermodynamic potentials possessed by the water particles if each of them existed alone in space. Further, as these small parts are supposed identical one with another, all these partial thermodynamic potentials must be equal to each other; to find their sum it will be sufficient to take the value for one of them and multiply this value by the number of parts into which the mass M has been divided. Thus if g denotes the internal thermodynamic potential possessed by one of our small masses of water, in the conditions indicated, absolutely isolated in space, and if n denotes the number of these small masses into which the mass M has been divided, we shall have

$$\mathfrak{F} = ng.$$

Other things being equal, the number n of parts conforms to a given type, which may be considered in the mass M as proportional to the magnitude of this mass; the preceding result may therefore be stated thus:

When the mass M of water is divided into particles infinitely distant from each other, the internal thermodynamic potential of this mass is of the form

$$\mathfrak{F} = M\phi,$$

ϕ being a quantity depending solely upon the temperature and density of water.

---

[1] J. Willard Gibbs, *Transactions of the Connecticut Academy*, v. 3, pp. 120 and 416, 1876.

269. Internal thermodynamic potential of a homogeneous mass when account is taken of the arrangement of its parts.—But what we have the intention to consider is not a mass thus disseminated; it is a coherent mass of water, whose various parts are adjacent to each other, which forms a continuous whole, limited by a certain surface; this mass evidently cannot be considered as being in the same state as the preceding; we may not say whether disseminating the particles crowded together or closing in the scattered parts is an operation which does not modify our mass of water.

Further, if the mass of water divided into infinitely small and infinitely distant parts and the mass of water for which these parts are collected together cannot be regarded as being in the same state; we cannot affirm without hypothesis that these two masses have the same internal potential; to free ourselves from hypotheses, we must think, at least provisionally, that their thermodynamic potentials are different; that if $M\phi$ is the internal thermodynamic potential of the divided mass, the same mass, brought to continuity, will have an internal thermodynamic potential of the form $(M\phi + \Psi)$, Ψ depending not only on the density and temperature of the water, but also on the arrangement of the various parts of the mass M or, in other terms, on the form of this mass.

Concerning this quantity Ψ thermodynamics gives us the following information: The various infinitely small parts into which the mass of water may be divided exercise certain actions on each other which admit a potential, and Ψ is precisely this potential.

We see, therefore, that the value of Ψ will depend on the hypotheses made concerning the actions exerted upon each other by the infinitely small masses into which the total mass may be divided.

270. Hypothesis of molecular attraction.—Concerning the subject of these hypotheses our choice is determined; since Newton, physicists have almost constantly made two hypotheses concerning the actions which two material masses exert on each other; the hypothesis of *universal attraction* and the hypothesis of *molecular attraction;* these two hypotheses have been very fruitful both in celestial mechanics and in physical mechanics;

it is then quite natural to keep them and take them as starting-point in the determination of Ψ.

These hypotheses may be formulated thus:

If two very small masses m, m' are separated by a distance r, each of them exerts on the other an attractive force directed along the line joining them.

The force which the mass m exerts on the mass m' is equal in magnitude to the force exerted by the mass m' on m.

The value of each of these forces is the sum of two terms.

The first term (term of *universal attraction*) has the value

$$K\frac{mm'}{r^2},$$

K being a positive constant coefficient whose value depends neither upon the nature nor the condition of the two masses m and m'.

The second term (term of *molecular attraction*) has the value

$$mm'f.$$

The value of the coefficient f depends not only on the distance r which separates the two masses m, m', but also on the nature and state of these two masses. As to the variation undergone by the coefficient f when the distance r changes in value, the following suppositions are made:

For every sensible value of the distance r, the coefficient f is so small that the term of molecular attraction is negligible compared with the term of universal attraction; on the contrary, when the distance r becomes inferior to a certain limit λ, which is of an extreme smallness and which is called the *radius of molecular activity*, the coefficient f assumes a very great value; it is then the term of universal attraction which is negligible as compared with the term of molecular attraction.

Such are the principles on which are based the determination of Ψ, thence reduced to a problem of mathematical analysis.

The first result reached by the mathematicians is the following:

Given the small magnitude of masses the chemist and physicist have to deal with, we may, in the formation of Ψ, take no account of the universal attraction term and concern ourselves only with molecular attractions.

This first point established, these methods, due to Gauss, allow of demonstrating that Ψ is of the following form:

$$\Psi = M\psi + AS,$$

S being the area of the surface limiting one mass of water and ψ A being two quantities which depend upon the nature of the water and its density.

If we denote by $\mathfrak{G}$ the sum $(\phi + \psi)$ which depends upon the temperature and density of water, we see that the internal potential of our mass of water will be of the form:

$$\mathfrak{F} = M\mathfrak{G} + AS.$$

271. Internal potential of a system divided into a certain number of homogeneous phases.—More generally, let us consider a system formed by a certain number of homogeneous phases 1, 2, . . . , ϕ; let M_1, M_2, . . . , M_ϕ be the masses of these phases and S, S', . . . the areas of the surfaces limiting these various phases or which separate them from each other; the internal potential of the system will be of the following form:

$$(1) \qquad \mathfrak{F} = M_1\mathfrak{F}_1 + M_2\mathfrak{F}_2 + \ldots M_\phi\mathfrak{F}_\phi + AS + A'S' + \ldots$$

$\mathfrak{F}_1$ is a quantity which depends upon the temperature, the nature, the state and density of the substance 1; $\mathfrak{F}_2$, . . . , $\mathfrak{F}_\phi$ depend in an analogous manner upon the substances 2, . . . , ϕ; as to A, it is a quantity depending upon the temperature, nature, state, and density of the substance bounded by the surface S or of the substances which it separates; the quantities A', . . . have analogous properties.

272. Comparison with the form used in the preceding chapters.—The laws developed from the sixth chapter on do not follow from an employment of formula (1), but, as we have seen in Art. 89, from the use of the simpler expression

$$(2) \qquad \mathfrak{F} = M_1\mathfrak{F}_1 + M_2\mathfrak{F}_2 + \ldots + M_\phi\mathfrak{F}_\phi,$$

which is deduced from formula (1) by neglecting the terms AS, $A'S'$, . . .

It is then possible that the laws deduced from this simplified formula may be found inexact for certain cases; a comparison will emphasize the importance of the error which may be committed by neglecting the terms A,S $A'S'$, . . .

Suppose it is desired to find the form assumed by a system composed of one or of several fluid masses under the action of their weight. If the simplified formula (2) is taken as starting-point, propositions are arrived at which are the laws of elementary hydrostatics; compared with observation, these laws are shown to be contradicted by a great number of phenomena called the *capillary phenomena;* to take account of these phenomena, it suffices to take no longer as starting-point the simplified equation (2), but the complete equation (1).

This comparison leads us quite naturally to put the following question:

In what cases is it permissible, in chemical mechanics, to make use of the simplified formula (2)? In what cases, on the contrary, is it necessary to employ the complete formula (1)?

273. **When all the phases are of very great mass, the theories developed in the preceding chapters are exact.**—The reply to this question depends upon an essential remark, namely:

When all the dimensions of a system are multiplied by the same number, the different masses composing the system are multiplied by the cube of this number, while the areas of the surfaces limiting or separating the various bodies of this system are multiplied only by the square of this number; if, therefore, the system is increased in size, the various masses composing it will increase much more rapidly than the surfaces met with; if, on the contrary, the dimensions of the system are reduced, the various masses composing it will diminish much more rapidly than the surfaces of separation.

From this the following consequence: One may always attribute to the various phases which compose a system masses great enough so that the terms AS, $A'S'$ in formula (1) are negligible compared with the term $(M_1\mathfrak{F}_1+M_2\mathfrak{F}_2+ \ldots M_\phi\mathfrak{F}_\phi)$; one may then make use of the simplified formula (2), whence follow all the laws developed in what precedes; in other words, *the laws of chemical mechanics developed in the preceding chapters are exact whenever the various phases into which the system studied is divided have sufficiently great masses.*

When, on the contrary, one or several of these phases have very small masses these laws may be in default.

274. Application to the vaporization of a liquid; case in which the classic theory is exact.—Let us take, for example, the phenomenon of the vaporization of water.

A mass M_1 of water is in contact with a mass M_2 of vapor across a surface Σ; S_1 and S_2 are the surfaces which complete, with the surface Σ, the boundaries of the masses M_1 and M_2; the internal thermodynamic potential of the system is of the form

$$\mathfrak{F}=M_1\mathfrak{F}_1+M_2\mathfrak{F}_2+\alpha\Sigma+A_1S_1+A_2S_2. \tag{3}$$

Suppose, in the first place, the masses of liquid and vapor are both very great; in this case, from the preceding remark, the terms of the potential F which are proportional to these masses are very great compared with the terms proportional to the areas of the limiting surfaces; in the expression for the internal thermodynamic potential one may neglect these last and reduce the expression (3) to the simplified form

$$\mathfrak{F}=M_1\mathfrak{F}_1+M_2\mathfrak{F}_2. \tag{4}$$

Now this simplified form is the same one deduced from the classic theory of evaporation; hence the same laws comprised in this theory are again found:

At every temperature there exists but one pressure for which the liquid remains in equilibrium in contact with the vapor; under a pressure less than this *tension of saturated vapor* the liquid vaporizes; under a higher pressure the vapor condenses.

But these laws, which the classic theory regards as general, appear to us here as subordinated to one condition: it is that the masses of liquid and vapor considered are always great masses. In all cases where this condition is not fulfilled we may, without contradiction, find these laws inexact.

275. Case where the liquid contains a very small vapor bubble. Theory of retardation of ebullition.—Suppose, for instance, a small vapor bubble be surrounded by liquid; we may no longer in equation (3) neglect the term $\alpha\Sigma$ as compared with $M_2\mathfrak{F}_2$; these two terms may be of the same order of greatness, and even, if the bubble is infinitely small, the absolute value of the term $\alpha\Sigma$ will be infinitely great compared with the absolute value of the term $M_2\mathfrak{F}_2$; the presence of the term $\alpha\Sigma$ in the expression for the internal thermodynamic potential will change entirely the conclusions which may be drawn from the study of this potential; so that the laws of equilibrium of a very small bubble of vapor within a liquid may be entirely different from the laws of equilibrium of a great mass of vapor in contact with a great mass of liquid.

These laws of the equilibrium of a small bubble of vapor within a great mass of liquid may be established in detail by means of the principles we have just exposed; they lead to the following consequences:

In order that a bubble of vapor may increase at the expense of the surrounding liquid, it is not sufficient for the pressure at a point near this bubble to be less than the tension of saturated vapor; it is further necessary that the radius of the bubble be greater than a certain limit, a limit depending upon the temperature and the pressure; when the radius of the bubble is less than this limit, not only the bubble cannot grow in size at the expense of the surrounding liquid, but further, the vapor which it encloses necessarily condenses; the bubble collapses.

From this, a bubble of vapor will never be formed in a region where the liquid is continuous; in fact, if such a bubble could begin to form, its radius would be at first infinitely small, less than the limiting radius of which we have spoken; whence, instead of continuing to grow, it would collapse.

We see that boiling can never commence but at those points where gaseous bubbles of a certain size already exist; this is actually the conclusion drawn from numerous and precise observations made on the retardation of ebullition by Donny, Dufour, and Gernez.

276. Generalization of the preceding considerations.—What we have just said on the subject of the transformation of a liquid into vapor may be generalized without difficulty, and we are thus led to the following conclusions:

When a certain substance *a* may be formed at the expense of another substance b, the conditions which permit predicting if the transfer will take place or will not take place are quite different according as a mass a of considerable extent exists beforehand in contact with the substance b or if the substance b exists alone at the beginning of the modification.

It is in the first case only that the consequences habitually deduced from the principles of thermodynamics are legitimate; they are not applicable to the second case; if, for example, to the number of these consequences is added a proposition affirming that, in certain conditions, a considerable mass of the substance a, put in contact with the substance b, will increase at the expense of this substance, one could not conclude that the substance a would be formed within the substance b, originally homogeneous.

277. Various phenomena explained by these consequences.— These considerations do not apply merely to the *retardation of boiling;* they completely explain a great number of phenomena:

The *retarding of condensation* of a vapor compressed beyond the tension of saturated vapor, retardation to which an end is put by the introduction of small liquid drops or of solid dust;

The *supersaturation of gaseous solutions,* which ceases by the introduction of a bubble of gas;

The *retardation of decomposition* of certain endothermic substances (oxygenated water, nitrous acid), retardation ceasing by the introduction of gaseous bubbles or of porous substances containing gas;

The *undercooling of a liquid,* which ceases by the introduction of a particle of the solid to be produced;

The *supersaturation of a salt solution,* to which an end is put by dropping in a crystal of the salt to precipitate or of an isomorphous salt;

The *retarding of transformation* of one crystalline form into another; for instance, the retardation of transformation of clinorhombic sulphur into orthorhombic sulphur at ordinary temperature, retardation stopped by contact with a piece of rhombic sulphur; the retardation of transformation of orthorhombic sulphur into clinorhombic sulphur, at temperatures above 97°.2, retardation ceasing by contact with a clinorhombic particle.

278. These phenomena represent apparent false equilibria.— In no one of the cases we have just cited is there produced, properly speaking, states of *false equilibrium;* all the equilibrium states experiment reveals are predicted by the principles of thermodynamics, provided that, in applying these principles, use is made of the complete equations where account is taken of the terms proportional to the surfaces of contact of the various phases; if there seems to be contradiction in certain cases between observation and theory, it is because the theory has been simplified by means of an unwarranted supposition; in all the cases of which we have just spoken there are produced only *apparent false equilibria.*

CHAPTER XVIII.

GENUINE FALSE EQUILIBRIA.

279. Genuine false equilibria exist. Investigations of H. Pélabon on the formation of sulphuretted hydrogen.—The false equilibria studied in the preceding chapter are apparent false equilibria; they are nowise in disaccord with the principles of thermodynamics; they contradict merely an additional hypothesis which represents, in certain cases, a sufficient approximation and which, in other cases, cannot be conserved.

Must we conclude that all the false equilibria are apparent false equilibria? Does observation never show any case of equilibrium irreconcilable with the principles of thermodynamics? Certain authors seem to have thought so; but we do not think their opinion can be accepted on this point.

Let us analyze the fo lowing observation, which is due to H. Pélabon:[1]

Several glass tubes, containing 0.02 gr. of pure sulphur and of pure hydrogen, were placed in a furnace whose temperature oscillated between 280° and 285°. After six hours' heating the gases in these two tubes were analyzed after sudden cooling; denoting by V the volume, in the normal conditions of temperature and pressure, of the gas contained in the tube, by v the volume after absorption of the sulphuretted hydrogen by potash, and by ρ the ratio of the partial sulphuretted hydrogen pressure in the gaseous mixture to the total pressure of this latter, there resulted;

$$V=8.766 \text{ cm.} \qquad v=8.547 \text{ cm.} \qquad \rho=0.025$$
$$V=10.2 \qquad v=9.95 \qquad \rho=0.0248$$

[1] H. Pélabon, *Mémoires de la Société des Sciences physiques et naturelles de Bordeaux*, 5th S., v. 3, p. 257, 1898.

After 38 hours the gas in the other tube was analyzed and gave

$V = 8.76$ cm. $v = 7.9$ cm. $\rho = 0.098$.

After 162 hours of heating:

$V = 7.135$ cm. $v = 4.75$ cm. $\rho = 0.3356$

After 300 hours:

$V = 9.25$ cm. $v = 6.15$ cm. $\rho = 0.3354$

This shows that the ratio ρ increases at first with the duration of heating; but after 160 hours the ratio ρ attains a value which it afterwards keeps indefinitely, if the temperature does not change; when ρ reaches this value equilibrium is reestablished in the system.

One would expect, from the laws of thermodynamics, that a system containing sulphur, hydrogen, and hydrogen sulphide, where the ratio ρ has a value greater than this limit 0.3355, should be the seat of a partial decomposition of hydrogen sulphide when it is kept at 285°; one would then see the ratio ρ diminish as the time of heating was increased, and approach the same limit 0.3355. This is not so; however rich in sulphuretted oxygen is the gaseous mixture submitted to the temperature of 280°, this sulphuretted hydrogen remains unaltered, and that even if the tube encloses only sulphur and hydrogen sulphide gas without admixture of hydrogen. At the temperature of 280°, in a system containing hydrogen, hydrogen sulphide, and *saturated* sulphur vapor,[1] equilibrium is established every time the value of the ratio ρ equals or surpasses 0.3355.

Does the value $\rho = 0.3355$ correspond to a veritable equilibrium state for the temperature 280°? Sulphuretted hydrogen being a compound strongly exothermic, the value of ρ corresponding to a veritable equilibrium state should diminish as the temperature rises (Art. 174); at the temperature of 440° the system studied is in a state of incontestable true equilibrium, and this state corresponds to a value of ρ included between 97.5 and 98.2; at the

[1] PÉLABON has shown that liquid sulphur absorbs hydrogen sulphide in abundance; this circumstance complicates somewhat the verification of the preceding laws, as may be seen in PÉLABON's memoir, *l. c.*

temperature of 280° the value of ρ which would correspond to a state of veritable equilibrium would differ very slightly from 1.

We may therefore state the following proposition:

At 280°, as long as ρ is included between 0 and 0.3355, there s formed hydrogen sulphide gas, reaction conforming with the predictions of thermodynamics; when ρ is included between 0.3355 and 1 the system is in equilibrium, although, according to the predictions of thermodynamics, there should be hydrogen sulphide gas formed; in this last case the system is in the *state of false equilibrium.*

Is the state we have just defined merely one of apparent false equilibrium? It does not seem that we may, in any manner, apply to it the considerations which allowed us to reduce to thermodynamical laws the retardations of boiling and the analogous phenomena.

Is this a state of illusory equilibrium? May we not admit that the sulphuretted hydrogen continues to be formed in the mixture kept at 280° and in which ρ has a value greater than 0.3355, but formed so slowly that this reaction escapes all control? This is an opinion which may be admitted, that observation evidently cannot prove false, but which it cannot any more prove true.[1] It seems to us simpler and more logical to admit that a system for which the ratio ρ exceeds 0.3355 remains really in equilibrium at the temperature of 280°, that such a state of equilibrium is incompatible with the laws of thermodynamics, and that the latter have need of being modified and extended so as to take into account states of false equilibrium.[2]

---

[1] Thus Van't Hoff thinks that in certain cases even the geological periods are insufficient for the state of veritable equilibrium to be attained (*Archives néerlandaises des Sciences exactes et naturelles*, 2d S., v. 6, p. 489, 1901).

[2] According to Max Bodenstein (*Zeitschrift für physikalische Chemie*, v. 29, pp. 147, 295, 315; 1899) all the effects observed by Pélabon on the formation of sulphydric acid, by Ditte and by Pélabon on the formation of selenhydric acid, by A. Gautier and Hélier on the combination of oxygen and hydrogen would be disputed; there would be produced false equilibria in none of these cases. It seems to me difficult to accept the affirmations of this author, whose researches appear to have been made in a very hasty way, who is in contradiction with all his predecessors, even in the cases where the latter have encountered states of veritable equilibrium,

280. The condition of false equilibrium is not expressed by an equality.—When the laws of false equilibrium are compared with the laws governing the states of true equilibrium, a first difference immediately attracts attention: a law of true equilibrium is expressed by an *equality;* we have seen numerous examples of this in our work; on the contrary, a law of false equilibrium is expressed by an *inequality.* Thus, in the preceding case, at the temperature of 280° the system is in equilibrium if ρ is *at least equal to* 0.3355.

281. Region of false equilibria. Boundary line of false equilibria.—On the two coordinate axes OT, $O\rho$ (Fig. 115) lay off

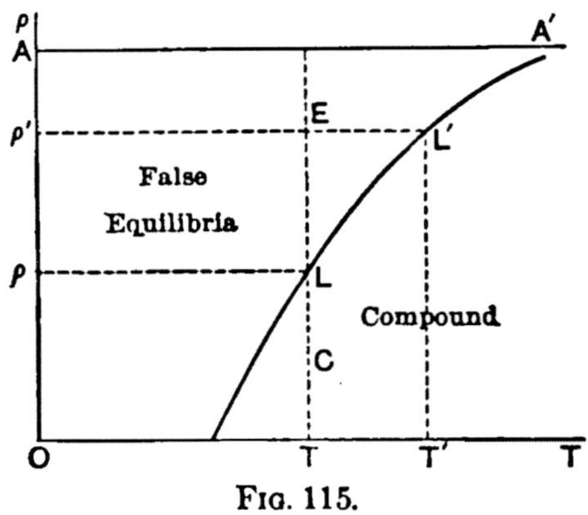

FIG. 115.

the temperatures as abscissæ and the values of ρ as ordinates; let L be the point whose abscissa represents 280° C. and whose ordinate has the value $\rho = 0.3355$; every point C located below L on the straight line TL represents a state of the system where hydrogen combines with sulphur to form hydrogen sulphide; on the contrary, every point E situated above L on the line TL represents a state where the system remains in false equilibrium.

When the temperature T is varied the point L varies; thus, according to the observations of Pélabon, we have, as coordinates of the point L:

who gives no plausible explanation of these contradictions, and finally who has read in a very superficial manner the writings whose conclusions he contests. (See P. DUHEM, *Zeitschrift für physikalische Chemie,* v. 29, p. 711, 1899.)

T	ρ
200° C.	0.0210
235	.0541
255	.13
280–285	.3355
310	.69
350	.972

As the temperature T increases, the point L describes a line LL′ which rises rapidly from left to right.

This line divides the plane into two regions; every point of the region situated below the line LL′ represents a state where the system is the seat of a combination; it is the *region of combination;* every point of the region situated above the line *LL′* represents a state of false equilibrium; it is the *region of false equilibria;* the line *LL′* is the *boundary of the false equilibria.*

When a tube containing 0.02 gr. of sulphur per cubic centimetre in an atmosphere of hydrogen is brought to a temperature T, the proportion ρ in hydrogen sulphide approaches the ordinate corresponding to the boundary of the false equilibria and never goes beyond it; this limiting value of ρ is furthermore attained the more rapidly when the temperature T is higher.

If for each temperature T the number h of hours of heating are laid off as abscissæ and the values of ρ as ordinates, a curve is obtained (Fig. 116) which rises at first from left to right, then

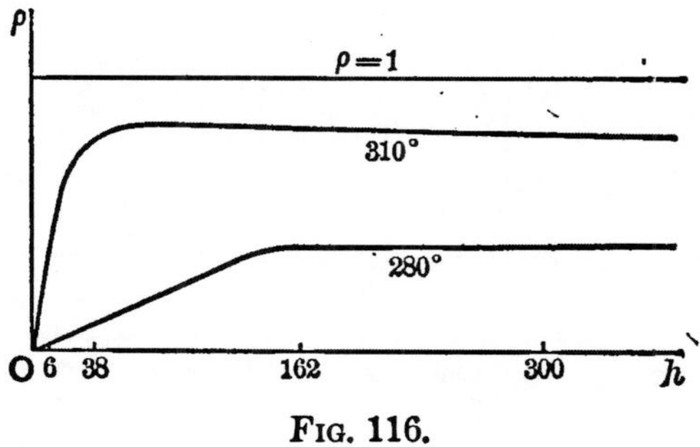

Fig. 116.

becomes parallel to Oh; for the first part of this curve the rise is the more rapid as the temperature rises; at temperatures above 350° equilibrium is reached in a few minutes.

282. Case in which the region of false equilibria separates two regions corresponding to two reactions the inverse of each other. Work of Jouniaux on the reduction of silver chloride by hydrogen.—In the case just spoken of, the boundary of the false equilibria is single; it separates the region of false equilibria from a region where a reaction of well-determined direction is produced, a combination.

In other cases the condition in order that the system be in the state of false equilibrium is expressed by a double inequality; the region of false equilibria is comprised between two boundaries; one of them separates this region from one where a certain reaction is produced; the other, from a region where the inverse reaction is produced.

The investigations of A. Jouniaux [1] on the two reactions, the inverse of each other,

$$AgCl + H = Ag + HCl,$$
$$Ag + HCl = AgCl + H,$$

will give us an example.

The composition of a definite volume of a mixture of hydrogen and of hydrochloric acid gas is represented, in Jouniaux's researches, by the ratio ρ of the volume of hydrochloric acid gas to the total volume of the mixture, these volumes being read in the same pressure and temperature conditions.

At ordinary temperature and under a pressure of 380 millimetres of mercury a tube of Jena glass containing silver chloride is filled with pure, dry hydrogen; the tube is brought to 448°; after being heated a time h it is suddenly cooled and the gaseous mixture it contains is analyzed, with the following results:

h	ρ	h	ρ
7 hours	0.7109	70 hours	0.8866
24 "	0.8157	408 "	0.8888
36 "	0.8246	504 "	0.8842

[1] A. Jouniaux, *Comptes Rendus*, v. 129, p. 883, 1889; *Actions des hydracides halogènes sur l'argent et réactions inverses*, Lille Thesis, 1901.

If the values of h are laid off as abscissæ and of ρ as ordinates, a curve aa' is obtained which rises at first from left to right, then becomes parallel to Oh (Fig. 117). After about 60 hours' heating

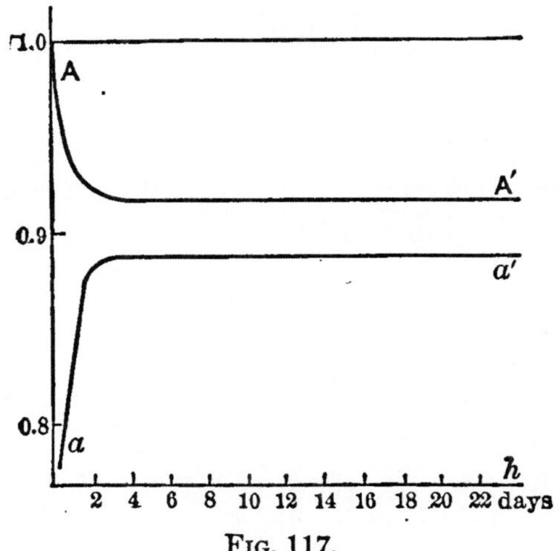

FIG. 117.

the reduction of silver chloride by hydrogen stops; the value of ρ is then close to

$$r = 0.8888.$$

The system is in equilibrium.

If the hydrogen introduced into the tube were employed entirely to reduce silver chloride, the hydrochloric acid set free would exert a pressure exactly double that exerted by the hydrogen absorbed; the inverse system of this one we have taken as starting-point in the preceding experiment is therefore formed of silver in the presence of hydrochloric acid, this latter exercising, at the ordinary temperature, a pressure measured by 760 millimetres of mercury. Let us take such a system; heat it to 448° for a time h, which we lay off as abscissæ, and lay off the value ρ as ordinates; the following results are obtained:

h	ρ	h	ρ
8 hours	0.9598	70 hours	0.9158
24 "	0.9392	408 "	0.9167
36 "	0.9292	504 "	0.9155

These results are represented by the curve AA', which descends at first from left to right, then becomes parallel to Oh. After 60 or 70 hours the formation of silver chloride stops; the value of ρ is sensibly

$$R=0.9155,$$

and the system is then in equilibrium.

Thus at 448° a system of same percentage composition as one of the two inverse systems studied in what precedes will be in a state of false equi ibrium whenever

$$r \leqq \rho \leqq R.$$

It is here a double inequality which defines the condition of equilibrium.

The two values of r, R, always determined by taking at ordinary temperature hydrogen under a pressure of 380 millimetres of mercury and hydrochloric acid gas at 760 millimetres, vary with the temperature T to which the system is brought. The values obtained by Jouniaux are the following:

T	r	R
200°	hardly appreciable	1
250	0.05	1
350	0.7588	0.95
448	0.8888	0.9155
490	0.9036	0.9094

If the values of T are laid off as abscissæ and of r and R as ordinates, *two boundaries ll'* and LL′ are obtained (Fig. 119, p. 380). Between these two lines extends the *region of false equilibria;* the line ll' separates this region from that where *hydrogen reduces silver chloride;* the line LL′ separates this region from that where *hydrochloric acid attacks silver.*

283. Another example: Carbonate of magnesium and bicarbonate of potassium. Engel's studies.—Here is another example [1]

[1] Engel, *Comptes Rendus*, v. 101, p. 749, 1885.

where the region of false equilibria separates two regions which correspond respectively to two reactions the inverse of each other.

Carbonate of magnesium combines with bicarbonate of potassium, forming the double salt of formula

$$CO_3Mg,CO_3HK+4H_2O.$$

Put in the presence of water, this salt decomposes; the carbonate of magnesium, almost insoluble, is deposited while the potassium bicarbonate is dissolved; when the solution of potassium bicarbonate is sufficiently concentrated, the decomposition stops and equilibrium is established.

At this moment the system is divided into three phases: the solution, the double salt, the solid carbonate of magnesium; it is besides formed of three independent components, bicarbonate of potassium, magnesium carbonate, water; it is therefore a bivariant system; under atmospheric pressure there should correspond to every temperature a state of true equilibrium defined by a given composition of the solution. The solution containing almost exclusively potassium bicarbonate and water, its composition may be fixed by its concentration s. The equilibrium would then correspond, for each temperature T, to a value S of the concentration; at this temperature, in presence of a solution of concentration less than $S(s<S)$, the double salt would decompose; on the contrary, in presence of a solution of concentration greater than $S(s>S)$, the bicarbonate of potassium would combine with the magnesium carbonate.

This is not at all how things actually happen.

At a given temperature T the solution decomposes the double salt while the concentration s is less than a certain given limit $\sigma(s<\sigma)$; the concentration being comprised between two limits σ, Σ, the second higher than the first,

$$\sigma \leqq s \leqq \Sigma,$$

the system remains in equilibrium; finally, when the concentration exceeds $\Sigma(s>\Sigma)$ the potassium bicarbonate combines with the magnesium carbonate.

On the two axes of rectangular coordinates OT, Os (Fig. 118) lay off the temperatures T as abscissæ, and the concentrations as ordinates; for the same temperature T let l and L be the points having the ordinates σ and Σ respectively.

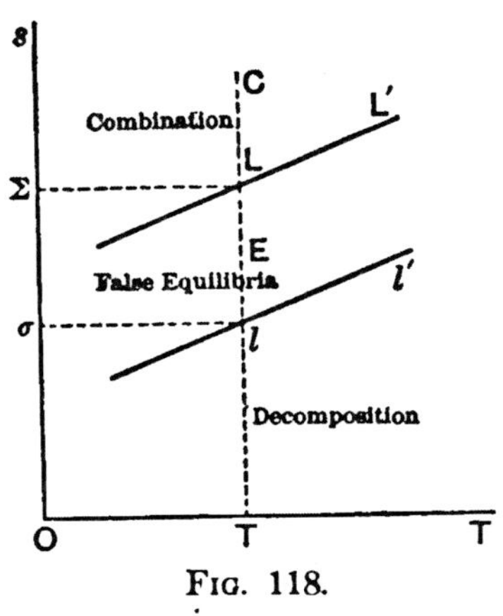

FIG. 118.

Every point D located below the point l on the straight line Tl represents a state of the system within which the double salt decomposes; every point C situated above L represents a state in which potassium bicarbonate combines with magnesium carbonate; any point E included between l and L represents a state of false equilibrium.

When the temperature T varies, the concentrations σ, Σ vary likewise and the two points l, L describe two lines ll', LL'; according to Engel's observations, at temperatures included between 14° and 40° these lines both rise from left to right. They divide the plane into three regions; the region situated below the line ll' is the *region of decomposition;* the region above the line LL' is the *region of combination;* the one between the lines ll' and LL' is the *region of false equilibria.*

284. Return to the idea of reversible modification.—The chemical equilibria studied in the preceding chapters all had a common property; each of these equilibria was the common limit of two reactions the inverse of each other (Arts. **46, 47, 53, 54, 55, 56**). From this property a consequence of capital importance followed; a continuous series of equilibrium states constituted a reversible transformation (Art. **59, 60, 61**). Now, it is because a continuous series of equilibrium states constituted a reversible transformation that it was possible to apply to these states of equilibrium the various theorems of thermodynamics.

All these theorems, all the corollaries which may be deduced from them, become useless when the chemical equilibrium is no longer the common limit of two reactions the inverse of each other. We must not be astonished, therefore, to find the states of false equilibrium in contradiction with propositions such as the phase rule or the laws of the displacement of equilibrium.

These contradictions are actually met with at every instant; let us cite one as example:

The combination of hydrogen with sulphur is exothermic. According to the law of the displacement of equilibrium by variation of temperature, the mass of hydrogen sulphide gas formed within a system where hydrogen and sulphur are heated at constant volume should be, at the instant of equilibrium, the feebler as the temperature were the higher. In reality (Art. 281) the proportion of hydrogen sulphide gas in the system, at the moment the reaction ceases, increases indefinitely with the temperature when this latter is raised up to 448°.

285. Relation between the states of veritable equilibrium and the states of false equilibrium. Action of hydrogen on silver chloride and the inverse action.—Often a chemical system, susceptible of possessing states of false equilibrium at certain temperatures, may have states of veritable equilibrium at other temperatures, in general higher than the first. In certain cases it is possible to follow the continuous passage of one of the forms of equilibrium to the other.

Take, for instance, the action of hydrogen on silver chloride and the inverse action of hydrochloric acid on silver (Art. 282). At a temperature such as 350° or 448° the values of ρ for which the system can be in equilibrium are included between two limits r and R which are notably different; but as the temperature rises these two limits approach each other; at 490°, where r is hardly less than R, we had

$$r = 0.9036, \qquad R = 0.9094.$$

At higher temperatures[1] the two limits r and R are identical; the same equilibrium state is reached in starting from the system hydrogen–silver chloride, or in starting from the system hydrochloric acid–silver. This common value of r and R, which we shall denote by $\mathcal{R}$, is the following at various temperatures:

---

[1] A. Jouniaux, *Comptes Rendus*, v. 132, p. 1270, 1901; *Actions des hydracides . . . ,* Lille Thesis, 1901.

T	$\mathcal{R}$	T	$\mathcal{R}$
540°	0.9155	650°	0.938
600	0.928	700	0.95

In place of the two boundaries of false equilibria, we have at temperatures above 500° but a single line of veritable equilibria, VV' (Fig. 119), whose every point has for abscissa a value of T

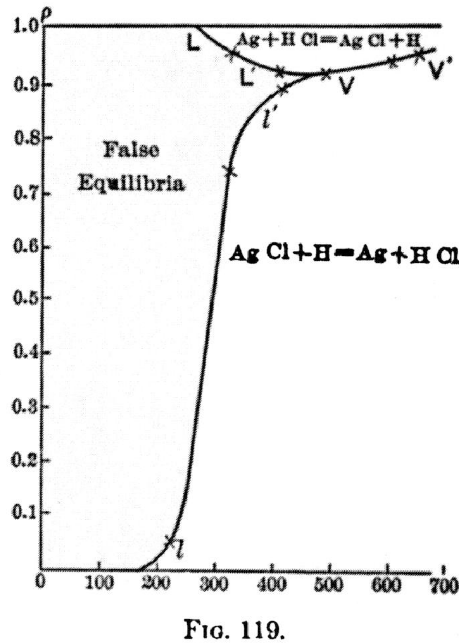

Fig. 119.

and for ordinate the corresponding value of $\mathcal{R}$. This line rises from left to right, as required by the law of the displacement of equilibrium by variation of the temperature, for the reaction

$$AgCl + H = Ag + HCl$$

is endothermic.

286. Action of hydrogen on selenium and the inverse action. Pélabon's investigations.—The relation between the states of false equilibrium and the states of veritable equilibrium is more sharply and completely brought out in the example given us by the dissociation of selenhydric acid and the inverse action of selenium on hydrogen. First studied by Ditte,[1] this example has been the object of researches which are among the most important of physical chemistry. These researches are due to Pélabon.[2]

The system studied encloses, at constant volume, liquid selenium, vapors of selenium, hydrogen, and selenhydric acid.

We have studied (Chap. XVI, Art. 258) the condition which controls the states of veritable equilibrium of such a system. We have seen that if p denote the partial hydrogen pressure in the gaseous mixture and p' the partial pressure of selenhydric acid, we should have, within a system in veritable equilibrium at the *absolute* temperature T,

$$(1) \qquad \log \frac{p}{p_1} = \frac{m}{T} + n \log T + z,$$

m, n, and z being three constants suitably chosen.

At temperatures above 350° the system possesses quite characteristic states of veritable equilibrium; we may therefore, by means of three experiments properly chosen and performed at temperatures above 350°, determine the values which should be attributed, in the preceding equation, to m, n, z.

These values once determined, we may, by means of equation (1), calculate for each absolute temperature T, and consequently for each centigrade temperature t, the value that $\frac{p}{p'}$ should have in order that the system be in a state of veritable equilibrium; take two rectangular axes and, following Pélabon, plot the centigrade temperatures as abscissæ (Fig. 120), and as ordinates the values of the ratio $\rho = \frac{p'}{p+p'}$, deduced from the preceding computation; we shall obtain the curve V_1VV' which will represent the states of veritable equilibrium of the system.

At a temperature higher than 350° the system will be in equilibrium only if the representative point is on the curve VV'; if

[1] Ditte, *Annales de l'Ecole normale supérieure*, 2d S., v. 1, p. 293, 1872.

[2] Pélabon, *Sur la dissociation de l'acide selenhydrique* (*Mém. de la Soc. d. Sciences Phys. et Nat. d. Bordeaux*, 5th S., v. 3, p. 141, and Paris, A. Hermann, 1898).

it is below this curve, selenhydric acid will be formed in the system; if the representative point is above this curve, the selenhydric acid contained in the system will be in part destroyed.

It will be quite otherwise for temperatures below 350°.

Let us operate, for example, at 270°.

Take tubes which contain at the start only hydrogen and selenium; the initial value of ρ is equal to 0; heat them a long time

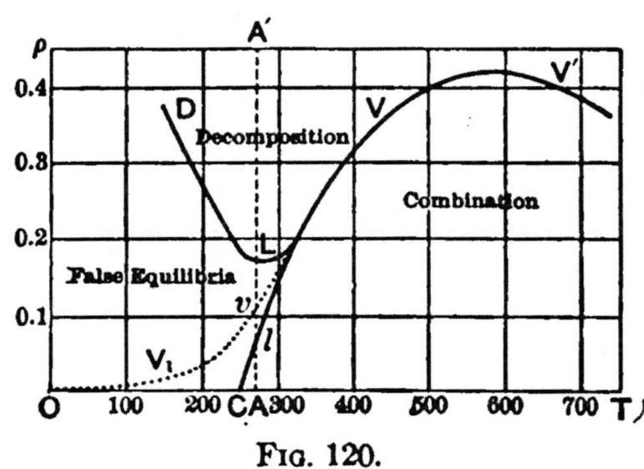

FIG. 120.

at 270°; ρ increases at first on account of the formation of selenhydric acid; then after a sufficiently long time equilibrium is established and ρ keeps a constant value which is nearly

$$r=0.048.$$

Thus in an experiment where the tube was heated for 490 hours ρ had the value 0.0491; in another in which the tube had been heated for a month ρ had the value 0.0478.

Suppose, on the contrary, we took, at the beginning of the experiment, mixtures containing a large proportion of selenhydric acid; imagine, for instance, that the initial value of ρ is about 0.40; maintain these mixtures at 270°; the selenhydric acid they contain will partly decompose, the value of ρ will diminish; after a sufficient time of heating equilibrium will be established and ρ will then keep a constant value near to

$$R=0.16.$$

Thus in four experiments during which the times of heating were respectively

192 hours, 288 hours, 480 hours, 490 hours,

the limiting values of ρ were respectively equal to

0.171, 0.165, 0.1605, 0.163.

Therefore every time ρ verifies the inequality

$$\rho<r$$

at 270°, hydrogen combines with selenium; whenever

$$\rho>R$$

at 270°, selenhydric acid decomposes; finally, whenever ρ is comprised between r and R,

$$r \leqq \rho \leqq R \tag{2}$$

at 270°, the system is in equilibrium.

If one computes, from formula (1), the value of ρ which, at the temperature of 270°, would put the system in a state of veritable equilibrium, we find that this value of ρ is in the neighborhood of

$$\mathcal{R}=0.10.$$

It is therefore included between r and R and verifies the condition (2).

Draw a parallel AA' (Fig. 120) to the straight line $O\rho$, having the constant abscissa 270°; running up the length of this line from A to A', we shall meet successively a point l of ordinate r, a point v of ordinate $\mathcal{R}$, and a point L of ordinate R. The points of the straight line AA' located below the point l represent systems within which selenhydric acid is formed; the points situated between l and L, among which is the point v, represent systems in false equilibrium; the points situated above L represent systems in which selenhydric acid is decomposed.

When the temperature, which we have supposed to equal 270°, assumes other values, the points l and L vary and describe respectively the lines Cl and DL (Fig. 120). These lines divide the plane into three regions; from the properties possessed by a system according as the representative point is in one or another of these three regions, we may give to these regions the following denominations: *region of combination,* located below the line Cl; *region of false equilibria,* situated between the lines Cl and DL; *region of decomposition,* above the line DL.

The line V_1v, theoretical *prolongation of the line of veritable equilibria,* determined by means of equation (1), is located wholly within the region of false equilibria.

The line *Cl* leaves the temperature axis at a point whose abscissa is about 250°; it rises continuously from left to right.

The line DL has been followed by Pélabon from the temperature 150°, to which corresponds a value of *R* closely equal to 0.3824; this curve descends at first from left to right to about the temperature of 270°, where it possesses a minimum ordinate nearly equal to 0.16; then it rises again from left to right.

How is the passage made from the law of formation and decomposition of selenhydric acid, such as we have just treated it, to the law which controls these same phenomena at temperatures above 350°, where we meet no more false equilibria and where a *simple line of veritable equilibria VV′ separates the region of combination* from the *region of decomposition?* When the temperature exceeds 300°, the three lines *Cl*, DL, *Vv* approach one another; they meet at about 325°, and remain coincident above 325°. Below are some observations due to Pélabon which show this effect:

Temperatures.	Duration of Heating.	*r*	$\mathcal{R}$	*R*
300°	212 hours	0.124	0.15	0.172
300	322 "	0.127	0.15	0.170
315	196 "	0.164	0.174	0.185
315	320 "	0.1625	0.174	0.180
325	175 "	0.187	0.192	0.193
325	213 "	0.1882	0.192	0.192

287. The region of false equilibria separated from that of veritable equilibria by a region of unlimited reaction. Action of hydrogen on sulphur and the inverse action.—In the two cases we have just analyzed there is passage by a gradual transition from low temperatures where the system admits of a region of false equilibrium which, by two lines of limiting false equilibria, confines inverse reactions to two regions, to high temperatures where the system admits none other than states of veritable equilibrium.

This is not always so.

Let us, for example, reconsider the case studied in Art. 279.

Up to about 350° the hydrogen sulphide gas is not decomposable by heat; on the contrary, above 200° hydrogen and sulphur combine; the reaction ceases when the gaseous mixture reaches a certain content of hydrogen sulphide; this content is the greater as the temperature is the higher.

Therefore from 200° to 358° a limiting line of false equilibria *ll'* (Fig. 121) rises from left to right; the points located below this

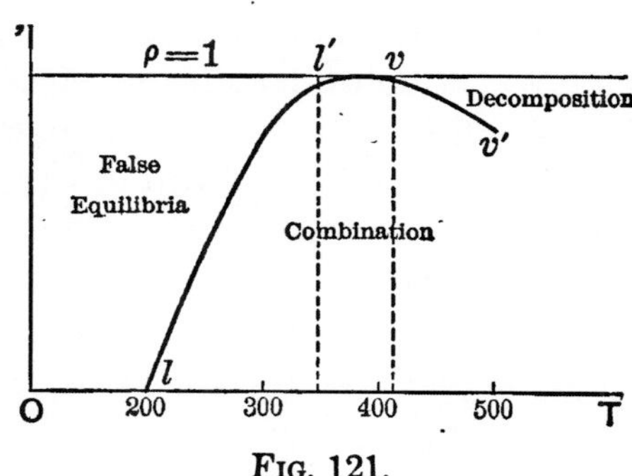

Fig. 121.

line represent states of the system such that hydrogen and sulphur combine; the points located above this line represent states of false equilibria.

At temperatures included between 350° and 400° the hydrogen sulphide remains undecomposable by heat; in return hydrogen combines *in totality* with sulphur. The combination is *unlimited.* The equilibrium states of the system are represented by the part $l'v$ of the line $\rho=1$.

When the temperature assumes a value greater than 400° the system possesses a state of veritable equilibrium, common limit of these two inverse reactions: combination of hydrogen and sulphur, decomposition of hydrogen sulphide. At 440°, for instance, the limit obtained, starting either from the pure compound or from the components, is the same, or, at least, the difference is of the order of experimental error.

Two tubes, enclosing each 0.02 gr. of sulphur per cubic centimetre, were kept for 6 hours at 440°.

The first, filled with hydrogen sulphide, gave for ρ the number 0.975; the second, which contained only hydrogen and sulphur, gave for ρ the value 0.982, practically equal to the preceding.

At temperatures above 400° the equilibrium states of the system, which are veritable equilibrium states, are represented by the line vv'; the region of combination is below this line, the region of decomposition above.

The formation of sulphuretted hydrogen being exothermic in the conditions indicated, the law of the displacement of equilibrium by variation of the temperature requires the line vv' to descend from left to right.

It is clear, from these important observations of Pélabon, that we may describe in the following way the influence exerted by the temperature on the formation or destruction of hydrogen sulphide:

At temperatures below $t=200°$ hydrogen sulphide is not decomposed; hydrogen does not act upon sulphur.

Between the temperature $t=200°$ and the temperature $\tau=350°$ hydrogen sulphide does not decompose; hydrogen combines with sulphur and the combination is limited; the gaseous mixture obtained is the richer in sulphuretted hydrogen as the temperature is higher.

At temperatures included between $\tau=350°$ and $\theta=400°$ sulphuretted hydrogen is decomposable by heat; hydrogen combines entirely with sulphur.

Above $\theta=400°$ hydrogen sulphide dissociates; this dissociation is limited by the inverse action, and it is the more marked as the temperature is higher.

288. Action of oxygen on hydrogen. Work of A. Gautier and H. Hélier.[1]—The history of a great number of exothermic combinations appears to be the following:

At temperatures less than t the compound substance is indestructible; the elements of this substance cannot combine.

At temperatures comprised between t and τ the compound is indestructible; the elements of this substance may combine and this combination is limited; the limit corresponds to a degree of combination higher as the temperature increases.

Between the temperatures τ and θ the compound is indestructible; the elements combine; this reaction stops only when the combination is complete.

---

[1] Armand Gautier and H. Hélier, *Comptes Rendus*, v. 122, p. 566, 1896; H. Hélier, *Annales de Chimie et de Physique*, 7th S., v. 10, p. 521, 1897.

Above the temperature θ the compound decomposes; the elements combine; these two reactions are limited; at a given temperature the same state of equilibrium limits the states of the system within which a decomposition is produced and the states within which a combination is produced; this state of equilibrium corresponds to a decomposition the more complete as the temperature is higher; temperatures above θ form properly the domain of DISSOCIATION.

Let us take water as an example.

It has been known since Lavoisier's time that at low temperatures water does not decompose, that oxygen and hydrogen do not combine; it is known also that at sufficiently high temperatures where water is decomposable oxygen and hydrogen combine in totality. It may therefore be said that since the origin of chemistry the existence of temperatures less than t has been recognized, as well as temperatures between τ and θ.

In demonstrating that water was dissociable at a very high temperature (Arts. 49 and 50), H. Sainte-Claire Deville showed the existence of temperatures, above θ, where veritable equilibria are established.

Finally, A. Gautier and Hélier have explored recently the zone of limited combination comprised between t and τ.

At atmospheric pressure, heat a mixture containing 16 grammes of oxygen to 1 gramme of hydrogen.

At 180° the oxygen and hydrogen begin to combine; at 200° the combination becomes measurable; by employing an artifice of which we shall say a word in the last chapter (Art. 320), Gautier and Hélier have been able to follow the phenomena up to 825° without obtaining explosion. In all this temperature interval the combination of hydrogen and oxygen is limited; the value of the ratio x of the mass of water formed to the possible mass of water, which limits the combination, increases with the temperature as indicated by the following table.

At these temperatures the water vapor, either alone or mixed with a certain quantity of explosive gas, is undecomposable; the combination of oxygen and hydrogen is therefore not limited by the inverse reaction, but by a region of false equilibria. The temperature interval within which were made Gautier and Hélier's experiments is wholly below the temperature we have called τ;

Tempera-tures.	x	Tempera-tures.	x
180° C.	0.0004	416° C.	0.3570
200	0.0012	433	0.3981
239	0.0130	498	0.5638
260	0.0160	620	0.8452
331	0.0978	637	0.8565
376	0.02511	875	0.9610

the temperature τ is therefore higher than 875°; still higher on the temperature scale, and probably above 1000°, is located the temperature we have called θ, at which is entered the region of the dissociation of water vapor, object of the researches of H. Sainte-Claire Deville.

From Hélier's researches, the value of x which, at a given temperature less than τ, limits the combination of hydrogen with oxygen at atmospheric pressure, changes when the mixture, instead of containing 16 grammes of oxygen to 2 grammes of hydrogen, contains an excess of one of these component gases; it changes also if an inert gas, as nitrogen, for example, is added to the mixture.

289. Action of oxygen on carbon dioxide.—What we have just said on the subject of the formation of water vapor and its dissociation may be almost textually repeated for what concerns carbonic anhydride.

At temperatures below a certain limit t the carbonic gas is undecomposable, the oxide of carbon does not combine with oxygen; at temperatures high enough to be included between two certain limits τ and θ the carbonic gas is undecomposable, but combines integrally with oxygen; these facts have long been known. The observations of H. Sainte-Deville Claire have shown us (Art. 51) that beyond the temperature θ the carbonic gas may dissociate; veritable equilibria are then established. Hélier has made known the region of limited combination, comprised between t and τ.

He has found that when a mixture containing two molecules of oxide of carbon and one molecule of oxygen are heated with sufficient precautions to avoid any explosion, the formation of carbonic acid ceased when the ratio x of the mass of this acid formed to the possible mass had attained a certain value, variable with the temperature and increasing with it, as the following table shows:

Temperatures.	x	Temperatures.	x
195° C.	0.0013	504° C.	0.0730
302	0.0044	566	0.1443
365	0.0101	575	0.1727
408	0.0303	600	0.2114
418	0.0341	689	0.4636
468	0.0464	788	0.6030
500	0.0620	855	0.6500

Throughout this temperature interval the carbonic gas is undecomposable, so that the formation of this substance is limited not by the inverse action, but by the establishment of a false equilibrium. It is only at very much higher temperatures that we penetrate, as was demonstrated by H. Sainte-Claire Deville, into the region of dissociation for carbonic acid gas.

290. Analogous phenomena shown by endothermic combinations.—A compound formed from its elements with absorption of heat may very well present phenomena analogous to those we have just described for an exothermic compound; for Fig. 121 we should then substitute a representation such as Fig. 122. Here also x continues to denote the ratio of the mass of the compound existing in the system to the possible mass.

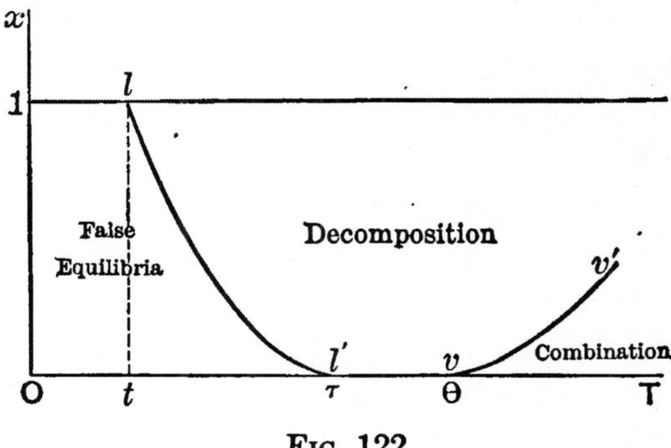

FIG. 122.

There are four regions to distinguish:

1°. *At temperatures below t no reaction is produced in a system containing the compound and the elements capable of forming it whatever the value of x;*

2°. *At temperatures included between t and τ the compound cannot be formed at the expense of its elements; on the contrary, if the initial value of x is sufficiently great, the compound is partly destroyed; the composition is limited; the value of x which limits the decomposition is the less as the temperature is higher; the reaction is limited not by the inverse action, but by the production of false equilibria;*

3°. *At temperatures between τ and θ the compound cannot be formed; it decomposes; this reaction is unlimited;*

4°. *At temperatures above θ, according to the value possessed by the ratio x in the system, the latter may be the seat either of a decomposition or of a combination; at a given temperature the two reactions, the inverse of each other, are limited by the same value of x; this limiting value of x increases with the temperature.*

These distinctions enable us to classify the properties of a great number of endothermic compounds.

291. Ozone.—For ozonized oxygen ordinary temperature is already higher than the temperature τ; at this temperature, more rapidly at 100°, and still more so at 200°, ozone undergoes decomposition which may be regarded as complete; we know, further, that Troost and Hautefeuille (Art. 176) have shown the direct, but partial, transformation of oxygen into ozone at temperatures of about 1200°; these temperatures are therefore higher than θ.

292. Silicon trichloride. Investigations of Troost and Hautefeuille.—Troost and Hautefeuille were able to completely explore the various parts of the field represented by Fig. 122 for certain silicon compounds,[1] and in particular for silicon trichloride, Si_2Cl_6.

At 250° this substance is not formed by the action of tetrachloride vapor, $SiCl_4$, on silicon; in return, silicon trichloride vapors are undecomposable at this temperature.

At 350° the trichloride vapor undergoes decomposition very slowly and the reaction is limited; the deposit of silicon on the walls of the vase is hardly sensible after 24 hours; the tetrachloride vapors are still without action on silicon.

When the temperature is raised, the decomposition of the trichloride of silicon into the tetrachloride and silicon becomes more

[1] Troost and Hautefeuille, *Annales de Chimie et de Physique*, 5th S., v. 7, 1876.

and more marked; it destroys at 440° a notable fraction of the compound substance; at 800°, if the experiment is sufficiently prolonged, the decomposition is complete.

On the contrary, at 1000° the decomposition is but partial; on the other hand, the tetrachloride combines at this temperature partially with silicon, giving the trichloride.

293. Systems with unlimited reaction and the principle of maximum work.—A considerable number of chemical reactions are classed in the category of which the formation of sulphuretted hydrogen and the decomposition of silicon trichloride are types. All these reactions give rise to an important observation: *At temperatures included between t and τ, when the only possible reaction is limited by states of false equilibrium, and at temperatures between τ and θ, where this reaction is unlimited, it is exothermic, so that the principle of maximum work is verified; to find the principle of maximum work in default it is necessary to attain temperatures, above θ, where states of veritable equilibrium may be established.*

294. Systems with unlimited reaction are not essentially distinct from systems with limited reaction.—It seems at first sight that a radical difference separates the systems incapable of unlimited reaction which we have studied in Arts. 285 and 286, from those which may, between certain temperatures, give rise to unlimited reactions; such are the systems studied in Arts. 287 to 293. In reality, as we shall show, it may be very well admitted that such a difference is a difference not in nature, but in degree.

Take a system which contains nearly perfect gases and where an exothermic compound may be produced or dissociated; for instance, a system containing sulphur in liquid and vapor states, hydrogen and hydrogen sulphide; let x be the ratio between the mass of the compound the system contains and the mass of this same compound it would contain if the combination of its elements were pushed as far as possible; heat the system either at constant pressure or at constant volume. The line VV' (Fig. 123) of veritable equilibria has, in the plane TOx, a form we have already traced in Fig. 111; as far as point B, of abscissa θ, it remains practically identical with the line AA', parallel to OT, and having as constant ordinate $x=1$; it then detaches itself and descends from left to right.

Let CC' be the line separating the region of false equilibria from the region of combination, and DD' the line which separates the region of false equilibria from that of decomposition; if things hap-

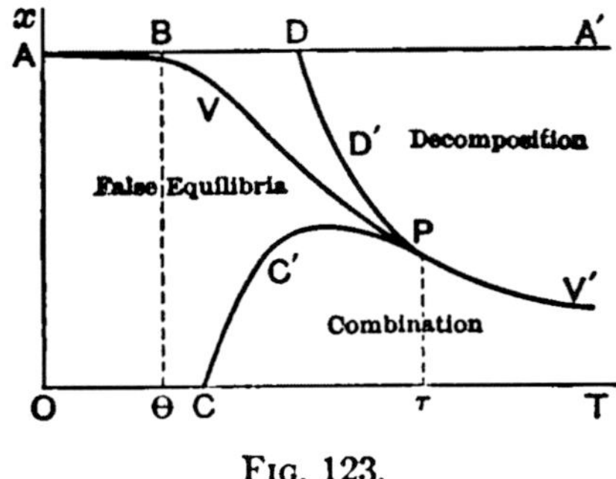

FIG. 123.

pen in accordance with what we have seen in Art. 286, these two lines should coincide with the line of veritable equilibria VV' beyond a certain point P of abscissa τ.

The temperature τ may be considerably above θ; we obtain then an arrangement analogous to that we studied in the system hydrogen, selenium, selenhydric acid; the new arrangement is simply symmetrical with the other with respect to an axis parallel to OT.

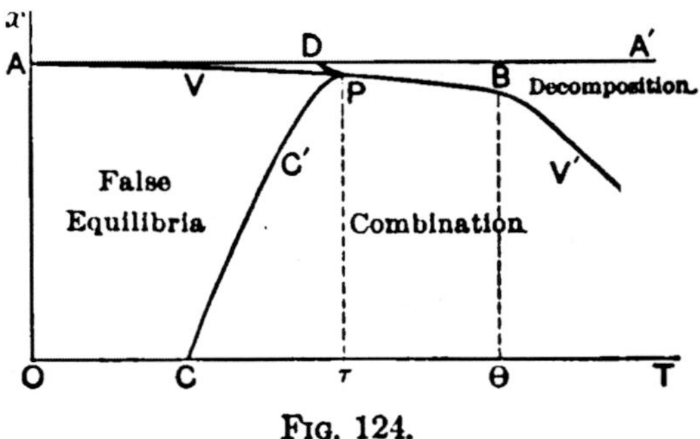

FIG. 124.

The temperature τ may, on the contrary, be considerably below θ, the point P being well to the left of the point B; in this case, shown in Fig. 124, the line CC' alone is discernible; the line DD' is reduced to a negligible segment DP. *Practically* the chemical statics of our system is resumed in the propositions stated at the close of Art. 287.

295. One may always cool a chemical system sufficiently for it to exist in the state of false equilibrium.—We have therefore only three kinds of temperature to distinguish in the study of a chemical system:

1°. *High temperatures*, where the system is susceptible of two reactions, the inverse of each other, having as common limit a series of states of veritable equilibrium;

2°. *Moderate temperatures*, where the system can undergo two reactions, limited no longer by the inverse reaction but by states of false equilibria.

3°. *Low temperatures*, where the system is not susceptible of any reaction.

All these observations seem to accord with this principle: *Given a chemical system, the temperature may always be sufficiently lowered to cause the system to remain in a state of false equilibrium.*

Thus below 250° a mixture containing only selenium and hydrogen, without trace of selenhydric acid, remains in a state of false equilibrium; no reaction is produced in it; below 215° a system containing sulphur and hydrogen is in the state of false chemical equilibrium.

296. False equilibria at very low temperatures. Pictet's researches.—For certain systems the state of false chemical equilibrium cannot be obtained except by greatly lowering the temperature; this was shown by R. Pictet.[1] At −125° a mixture of frozen sulphuric acid and caustic soda may be compressed without any reaction being produced; as long as the temperature is below −80° C. no combination takes place; it is produced abruptly at this temperature of −80° C., liberating such a quantity of heat that the eprouvette containing the mixture is broken.

Sulphuric acid and potash remain in equilibrium at temperatures less than −90° C.; sulphuric acid and a concentrated ammoniacal solution at temperatures less than −65° C.; at −120° C. sulphuric and hydrochloric acids leave litmus its blue color; litmus turns red at −110° C. with hydrochloric acid, and at −105° C. with sulphuric acid.

It is to be remarked that certain of the systems of which we have just spoken are not, perhaps, at the temperatures realized

[1] R. Pictet, *Comptes Rendus*, v. 115, p. 814, 1892.

by Pictet, systems in equilibrium, but merely systems where an excessively slow reaction is produced; according to Besson[1] hydrochloric acid which has remained at a very low temperature, at $-80°$ for instance, in contact with sodium, encloses small quantities of sodium chloride.

297. **The reaction-point.**—Take, at a very low temperature, a system in the state of false equilibrium and gradually raise the temperature; at a certain moment the system will cease to be in false equilibrium and a reaction will be produced. The temperature at which a given system, under a given pressure or maintained at a given volume, ceases to be in the state of false equilibrium and becomes the seat of a chemical modification, is called the *reaction-point* of this system. Thus the reaction-point of a system which contains hydrogen and selenium, without trace of selenhydric acid, and which is heated at constant volume, is close to 250°; at this temperature selenhydric acid begins to be formed.

For certain systems the reaction-point may correspond to a very low temperature; we have seen that the litmus reaction with hydrochloric acid was about $-110°$ C.

In other cases, on the contrary, this reaction-point corresponds to an extremely elevated temperature; one of these cases is given us by a mixture of hydrogen and nitrogen.

Ammonia gas would be formed, starting with hydrogen and nitrogen, with a great liberation of heat; if, therefore, a mixture of these three gases kept, either at constant pressure or at constant volume, was in a state of veritable equilibrium, the combination in it would be almost complete at low temperature; it is only at a high temperature that ammonia gas would show appreciable dissociation.

In fact a mixture of hydrogen and nitrogen, whether or not containing ammonia gas, may be kept in the state of false equilibrium at almost any of the temperatures produced by our furnaces; it is only at very high temperatures, generated by very hot electric sparks, that the combination begins to take place, as was shown by Morren;[2] his observation was confirmed by means of the hot and cold tube apparatus of H. Sainte-Claire Deville.[3]

---

[1] Besson, *Comptes Rendus*, v. 124, p. 763, 1897.

[2] Morren, *Comptes Rendus*, v. 48, p. 342, 1859.

[3] H. Sainte-Claire Deville, *Leçons sur la dissociation*, 1864.

The reaction-point of a system may depend upon a number of circumstances: on the initial pressure supported by the system if heated at constant volume; on the volume if heated at constant pressure; on the initial composition of the system and the foreign substances which may be mixed with it.

Finally, in certain cases these various circumstances may influence not only the temperature of the reaction, but also the nature of the reaction which begins to be produced the instant this temperature is reached.

Take, for example, a system formed of hydrogen, selenium, and selenhydric acid. Trace (Fig. 125) the line CC', which sepa-

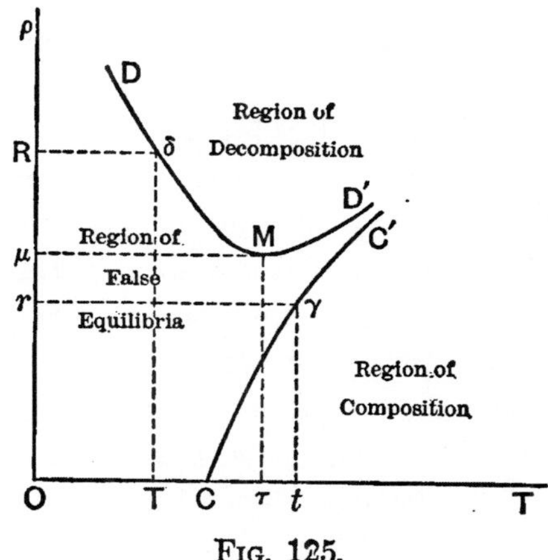

FIG. 125.

rates the region of combination from the region of false equilibria, and the line DD', which separates the region of decomposition from the region of false equilibria; this last line has a point M lower than all the others; let $\mu = O\mu$ be the ordinate of this point.

Consider a system in which the initial value r of the ratio ρ is less than μ; suppose the temperature low enough for the system to be in the state of false equilibrium and gradually raise this temperature; the representative point describes the straight line $r\gamma$ which meets the line CC' at γ; the abscissa t of the point γ is the reaction-point of the system; at the instant the temperature reaches, then exceeds this value t, the system becomes the seat of a *combination.*

Take, on the contrary, a system in which the initial value of R of the ratio ρ is greater than μ and whose temperature is low enough for equilibrium to be reached; when the temperature increases, the representative point will describe the line $R\delta$, parallel to OT, meeting DD' in δ; the abscissa T of the point δ will be the reaction-point of the system; the system, when it attains this point, becomes the seat of a *decomposition*.

298. Reaction-point for the phosphorescence of phosphorus. Joubert's studies.—For the majority of cases the complexity is less; at the moment when the syst m reaches the reaction-point there is produced a reaction whose nature does not depend upon the initial composition of the system.

Thus, whatever the initial composition of a system containing sulphur, hydrogen, sulphuretted hydrogen, the reaction-point always corresponds to the beginning of combination.

The same is true in the case of the combination of oxygen and phosphorus, studied in detail by Joubert.[1]

Consider a space which contains oxygen and the saturated vapor of phosphorus, in the presence of an excess of phosphorus. The oxygen and phosphorus may combine either rapidly, which constitutes the phenomenon of the combustion of phosphorus, or slowly, which produces phsophorescence.

In such a system there exists a *reaction-point;* below this temperature no combination is produced in the system; above, phosphorescence is produced and then combustion.

This reaction-point is not fixed; it depends upon the pressure supported by the system; it is the higher as the pressure is higher.

Take as abscissæ the pressures π (Fig. 126), as coordinates the temperatures T. For every pressure π there corresponds a reaction-point T; the point M, of coordinates π, T, has a certain curve CC' for locus.

This curve divides the plane into two regions.

Take a point a of abscissa π, of ordinate θ, less than T, reaction-point at the pressure π; this point represents a system in which no reaction is produced; the point a is therefore in *the region of false equilibria, which coincides with the region situated below the curve CC'*.

Take, on the contrary, a point A, of abscissa π and ordinate θ,

[1] JOUBERT, *Annales de l'Ecole normale supérieure*, 2d Series, v 3, p. 209, 1874.

greater than T; this point represents a system in which oxygen and phosphorus combine; the point A is therefore in the *region of combination, which coincides with the region located above the curve CC′.*

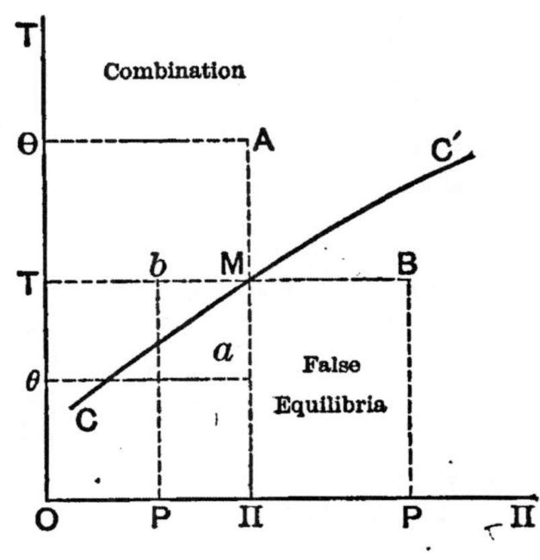

FIG. 126.

The curve CC′ rises from left to right; the region of combination is therefore to the left of the curve CC′, and the region of false equilibria to the right of the same curve.

Whence, if we take a point b, of ordinate T and of abscissa p, less than π, this point represents a system where the oxygen and the phosphorus combine; a point B, of the same ordinate T, but of abscissa P, greater than π, represents a system where no reaction takes place. Therefore, *at every temperature* T *corresponds a certain limiting pressure π; under a pressure less than π oxygen combines with phosphorus; at a pressure greater than π a system containing oxygen and phosphorus is in equilibrium; the pressure π is the higher as the temperature is higher.*

This is the law stated and verified by Joubert, who gives the following values for π:

Temperatures.	π	Temperatures.	π
1°.4 C.	355 mm.	9°.3 C.	538 mm.
3 .0	387 "	11 .5	580 "
4 .4	408 "	14 .2	650 "
5 .0	428 "	18 .0	730 "
6 .0	460 "	19 .2	760 "
8 .9	519 "		

The line CC' is sensibly straight.

The form and position of this line vary greatly when certain inert gases are mixed with the oxygen. Joubert has made a very complete study of this variation.

The combustion of phosphoretted hydrogen in oxygen has given rise to analogous observations on the part of Van de Stadt.[1]

299. Analogy of the states of false equilibria with the mechanical equilibria due to friction.—The several examples we have just studied give us a clear idea of the principal characteristics of false equilibria; in particular they show us readily that, in a system capable of false equilibria, the condition for equilibrium is not expressed by an equality, but by an inequality or by a double inequality.

Is this characteristic incompatible with the analogy between chemical statics and statics properly so called, analogy which we regard as one of the guiding ideas of the science? Quite the contrary, and this characteristic establishes a close resemblance between the chemical systems capable of false equilibria and the mechanical systems possessing *friction*.

Let us take the following example, ingeniously imagined by Pélabon:

Consider a cylinder full of air with axis vertical, closed at the base; within this cylinder moves a piston upon which weights may be put; to simplify, suppose the area of the cross-section of the cylinder to be unity.

Denote by H the pressure of the gaseous atmosphere above the piston, by ω the weight of the latter, by p the additional weight it carries.

If the friction on the walls of the tube is neglected, there is for each weight p one equilibrium position for the piston and only one; for example, if V represent the distance from the base of the piston to the bottom of the cylinder when $p=0$, and x the value of the same length when the additional weight has the value p, we shall have, applying Boyle's law to the gaseous mass enclosed in the cylinder, the relation

$$(3) \qquad (H+\omega)V=(H+\omega+p)x,$$

which will determine the position of equilibrium in question.

[1] Van de Stadt, *Zeitschrift für physikalische Chemie*, v. 12, p. 322, 1893.

Take two rectangular coordinate axes (Fig. 127); along the abscissa axis Op lay off the values of the additional weight p; as ordinate take the corresponding values of x calculated by equation (3); the point v, of coordinates p, x will represent an equilibrium state of the piston supposed *without friction;* as p varies, the point v describes a line VV_1, which we shall call the line of *veritable equilibrium* of the piston.

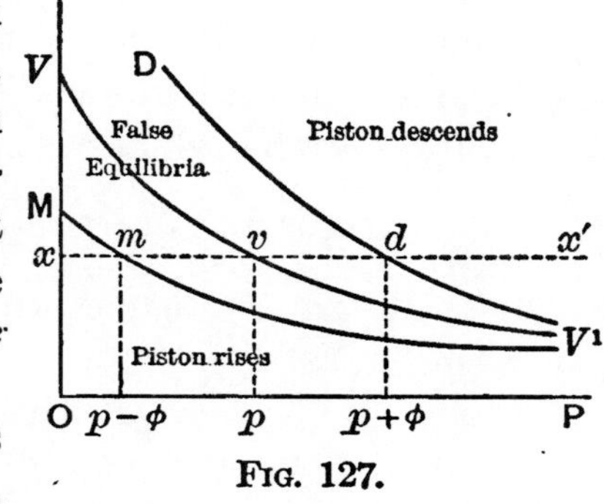

Fig. 127.

Suppose now the piston rubs along the interior of the cylinder, and denote by P the additional weight; in order for equilibrium to exist it will be no longer necessary for the pressure $(H+\omega+P)$ exerted by the cylinder to be equal to the pressure $(H+\omega)\frac{V}{x}$ of the gaseous mass; it will be sufficient for the absolute value of the difference of these two pressures not to exceed a magnitude ϕ, dependent upon the nature of the contiguous surfaces of the cylinder and piston. The condition for equilibrium of the piston is then, *taking friction into account,*

$$(4) \qquad -\phi \leqq H+\omega+P-(H+\omega)\frac{V}{x} \leqq \phi.$$

This may be written otherwise. Denote by

$$(5) \qquad p=(H+\omega)\left(\frac{V}{x}-1\right)$$

the value of P which would correspond to the state of veritable equilibrium for which the bottom of the cylinder and base of the piston are at a distance x; the point v, of coordinates (p, x), will be evidently the point of ordinate x on the line of veritable equilibria, for equation (5) is but equation (4) solved for p. The double inequality (4) may be written

$$(6) \qquad p-\phi \leqq P \leqq p+\phi.$$

Through the point v draw a parallel to the axis OP; on this line mark off two points m, d having respectively as abscissæ $(p-\phi)$ and $(p+\phi)$; the distances dv, vm are both equal to ϕ; every point of the segment md represents a state of equilibrium of the piston.

A point on the line xm, situated to the left of the point m, represents a system where the distance from the base of the piston to the bottom of the cylinder is x, but where the pressure of the gas $(\mathrm{H}+\omega)\frac{V}{x}$ exceeds the pressure $(\mathrm{H}+P+\omega)$ exerted by the piston by a quantity greater than ϕ; in these conditions the gas expands and the piston rises.

A point on the line dx', situated to the right of the point d, represents a system where the distance from the base of the piston to the bottom of the cylinder is still x, but where the pressure $(H+P+\omega)$ exerted by the piston exceeds the pressure of the gas $(\mathrm{H}+\omega)\frac{V}{x}$ by a quantity greater than ϕ; in these conditions the gas is compressed and the piston descends.

For each value of x analogous considerations may be repeated.

When x is varied, the point m describes a line Mm, and the point d a line Dd; these two lines divide the plane into three regions; every point of the region situated to the left of the line Mm represents a system where, without initial speed, *the piston rises;* every point of the region situated to the right of the line Dd represents a system where, without initial speed, *the piston descends;* finally, every point in the region situated between these two lines, including the points on the lines, represents a system where, without initial speed, the piston rests stationary; this is the *region of false equilibria.*

The line of veritable equilibria is in its entirety drawn in the region of false equilibria.

This example renders tangible the analogy which exists between mechanical systems with friction and chemical systems with false equilibria.

300. The existence of false equilibria in chemical systems is not exceptional, but regular.—The existence of friction in a mechanism must not be regarded as an exception, but the rule; in a great number of cases this friction is feeble enough to be neglected, and freed from this complication the laws of mechanics assume the simple form in which they are ordinarily exposed; but it would be dangerous to forget that these forms are incomplete and constitute, in the most favorable cases, but an approximation; one would be led otherwise to seek for a perpetual motion.

Analogy leads us to suppose that false chemical equilibria are not exceptional facts, but are the rule; every chemical system is capable of possessing such states of equilibrium; only, in a great number of cases the states of false equilibrium are all so near to the state of veritable equilibrium that one may not practically distinguish them from this last, which alone seems realizable.

Thus Pélabon's experiments do not give evidence of a region of false equilibrium, within the system formed of hydrogen, selenium, and selenhydric acid, at temperatures above 325°. But these results may very well be interpreted in the following way: The two curves CC', DD' (Fig. 128) which limit the region of

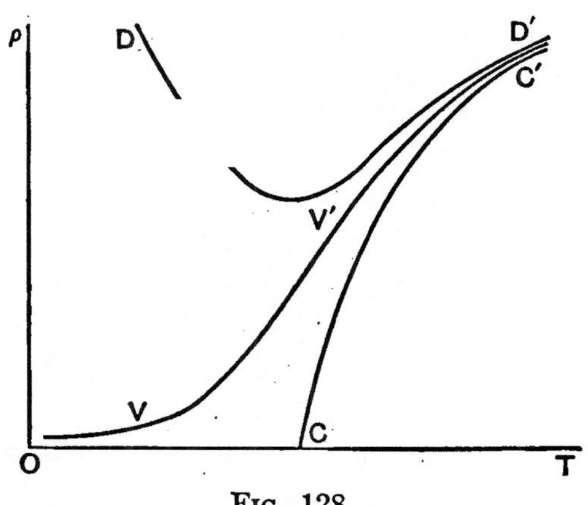

Fig. 128.

false equilibria remain, at every temperature, distinct from each other and from the line VV' of veritable equilibrium; but at temperatures above 325° the two lines CC', DD' are too close together and the region of false equilibria is reduced to a too narrow band to permit distinguishing by experiment the false from the veritable equilibria.

This manner of looking at the matter carries with it a profound change in the ideas we have admitted so far concerning chemical equilibrium. The state of chemical equilibrium has appeared (Chap. IV, Art. 61) as the common frontier between the states where the system undergoes a modification in a definite direction, and the states where the system undergoes a modification in the opposite direction; it was the state into which two reactions of opposite dir ctions limit each other; its essential property is expressed by this proposition: A series of equilibrium states is a *reversible transformation.*

These ideas grouped about the notion of reversibility appear to us now as notions incapable of representing exactly the facts; chemical statics constructed by means of these notions is a too simple statics; it gives only laws for an ideal case, for a limiting case to which certain systems approach more or less. Similarly, mechanics where abstraction is made of friction is a mechanics too simplified; its laws are limiting laws, to which, in certain cases, the real laws of motion approach more or less.

A last similarity is to be noticed between the evolution of mechanics and the development of chemical mechanics.

The mechanical systems which ordinarily surround us are rendered extremely complex by the continual presence of friction; thus, unless one attacks, with Kepler and Galileo, celestial mechanics free from friction, one is compelled to consider groups of bodies susceptible of rubbing against each other and it becomes impossible to discover simple laws, such as the law of inertia which must serve as the basis for dynamics—for a dynamics no doubt too abstract and ideal, but whose creation had necessarily to precede the theory of friction.

Similarly, chemical actions which are produced at ordinary temperatures give rise to continued false equilibria. As long as they alone were considered, chemical mechanics could not be placed upon a secure foundation. The principles of this science were not clearly appreciated until after H. Sainte-Claire Deville, by creating the chemistry of high temperatures, had eliminated chemical friction.

CHAPTER XIX.

UNEQUALLY HEATED SPACES.

301. Formation and dissociation of selenhydric acid in an unequally heated space. Three cases to distinguish.—The principles we have just been stating lend themselves to the discussion of phenomena produced in unequally heated spaces.

We shall take as example the formation of selenhydric acid from selenium and hydrogen, and we shall neglect the volatility of selenium; this volatility affects the laws we are going to state by slight perturbations easy to allow for; but in neglecting it we shall have the advantage of obtaining expressions applicable to non-volatile or very slightly volatile substances.

Take the temperatures T as abscissæ, and as ordinates the ratio x of the mass of the compound existing in the system to the mass which would be found there if the combination were pushed as far as possible (Fig. 129).

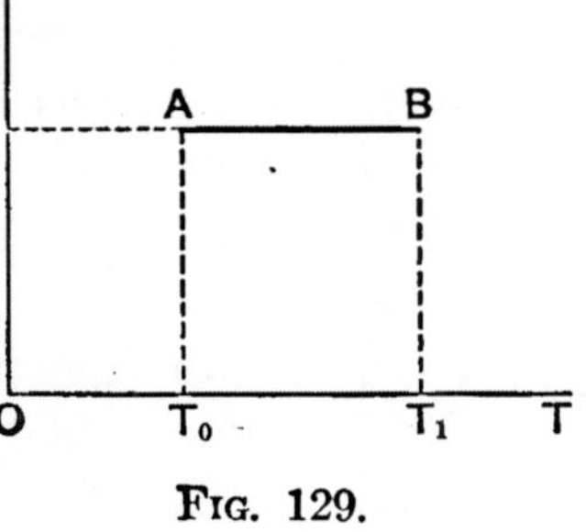

FIG. 129.

We shall suppose the selenium, hydrogen and selenhydric acid enclosed in a tube where the temperature varies between a lower limit T_0 (*temperature of the cold extremity*) and a higher limit T (*temperature of the hot extremity*).

Due to the diffusion of the gases, the ratio x, here equal to the ratio $\rho=\dfrac{p'}{p+p'}$ considered by Pélabon, has sensibly the same value in all the parts of the gaseous mass at a given instant; the various parts of the tube correspond, therefore, at the same instant, to different values of T, but to the same value of x; they are represented by the various points of a straight line *AB*, parallel to *C*T, whose extremities *A* and *B* have respectively for abscissæ T_0 and T_1.

Suppose any portion whatever of the line *AB* to be traced in the region of decomposition; in the portions of the heated tube represented by the points of the straight line *AB* situated in the region of decomposition, selenhydric acid is certainly split into selenium and hydrogen; whence this first proposition: *The system contained in the unequally heated tube cannot be in equilibrium as long as some point of the representative straight line AB exists in the region of decomposition.*

Suppose, in the second place, certain points of the representative straight line *AB* are in the region of combination; if in the portions of the tube represented by these points there is solid or liquid selenium, selenhydric acid will be formed in these portions: *the system contained in the uneqally heated tube cannot be in equilibrium, therefore, so long as there exists solid or liquid selenium in a portion of the tube represented by a point in the region of combination.*

These principles stated, let us examine some problems which have been studied experimentally by Pélabon.[1]

Let us turn back to the sketch (Fig. 130) which gives, for the systems studied, the disposition of the line of veritable equilibria, of the region of false equilibria, of the region of decomposition, and of the region of combination. Several temperatures merit remark.

The two limiting lines of the false equilibria coalesce practically with the line of veritable equilibria at a point *P* whose abscissa does not exceed 350°.

The line *PV′* of the veritable equilibria has a point *M*, of maximum ordinate, whose abscissa is close to 575°.

The line *DP* has a point *m*, of minimum ordinate; the abscissa of this point is sensibly 270°; if, through the point *m*, the tangent to the line *DP* is drawn, tangent which is parallel to *OT*, this tangent cuts the line *CP* in a point μ whose abscissa is 310°.

Finally, the point *C* of the axis *OT*, where the line *CP* starts, corresponds to a temperature less than 250°.

---

[1] H. Pélabon, *Sur la dissociation de l'acide sélenhydrique* (Mém. de la Soc. d. Sciences physiques et naturelles de Bordeaux, 5th S., v. 3, p. 232).

The experiments made by Pélabon are concerned with the three following cases:

FIRST CASE.—*All portions of the tube are at temperatures included between* 350° *and* 575°.

The straight line representing the state of the system in equilibrium cannot in any way encroach on the region of decomposition; neither can it be entirely traced in the region of combination, for

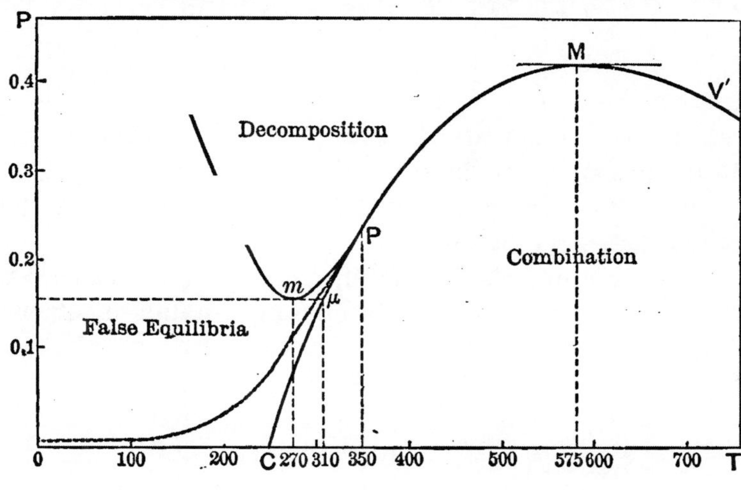

FIG. 130.

the excess of liquid selenium would be found in a portion of the tube represented by a point in the region of combination; therefore, in order to have equilibrium, it is necessary for the representative straight line AB of the system in equilibrium to have a point in common with the line of veritable equilibria and all the other points in the region of combination; besides, selenium must be all collected in the portion of the tube represented by the first point; it is clear, furthermore, that this disposition assures equilibrium.

Whence the final position of the straight line AB is that represented by Fig. 131; *at the end of the experiment, the excess of selenium is entirely collected in the coldest region of the heated tube; the composition of the system is the same as if it had been maintained wholly at the temperature* T_0 *of the cold extremity;* we may say that *the principle of Watt or of the cold boundary is here applicable.*

Pélabon's observations justifying this statement are:

$$T_0=425°, \quad T_1=658°, \quad \rho=0.3678.$$

A tube kept at the same time *entirely* at $T_0=425°$ would give for ρ the value $\rho=0.342$; kept at $T_1=660°$, it would give $\rho=0.395$.

$$T_0=440°, \quad T_1=640°, \quad \rho=0.3628.$$

A tube kept at the temperature $T_0=440°$ would give $\rho=0.352$; heated wholly at the temperature $T_1=640°$ it would give $\rho=0.401$.

$$T_0=350° \quad T_1=510°, \quad \rho=0.245.$$

A tube maintained entirely at 350° would have given $\rho=0.234$; at 510° the value of ρ would have been $\rho=0.398$.

In all these experiments the selenium had been placed, at the start, in the extremity of the tube which was to be brought to the higher temperature; at the end of the experiment it was found collected at the colder end.

In spite of the uncertainty which hangs over the exact values of the extreme temperatures, the preceding results are sufficiently conclusive.

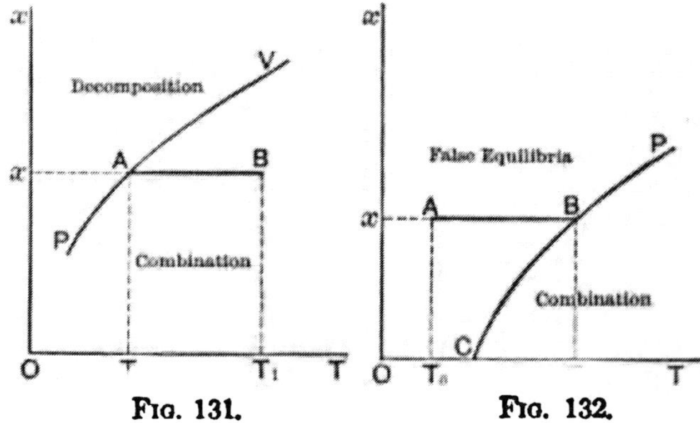

Fig. 131. Fig. 132.

Second Case.—*The temperatures of the various parts of the tube are all less than* 310°; certain of these parts are nevertheless brought to temperatures higher than 250°; initially the system does not contain hydroselenic acid; the selenium was placed in the part of the tube which is to be heated the strongest.

At the start, the representative points of the various parts of the system are on the axis OT between the point T_0 and T_1; those points (Fig. 132) between C and T are in the region of combination; among these points are those representing the parts of the tube where selenium is collected in excess; hydroselenic acid is therefore formed, x increases and the representative straight line rises until the position AB is reached, at the moment this position is attained the equilibrium is evidently established in the whole system. *The composition of the system in equilibrium is the same as if it were entirely brought to the temperature* T_1 *of its hottest point; the excess of selenium remains at the hot extremity of the tube, where it was initially put.*

One may say that in the case at hand use is being made of *Watt's principle reversed.*

Here are two experiments confirming this law:

$$T_0 = \text{laboratory temperature}, \quad T_1 = 260°, \quad \rho = 0.0312.$$

A tube kept during the same time *wholly* at the temperature $T_1 = 260°$ gave also $\rho = 0.0312$.

$$T_0 = \text{laboratory temperature}, \quad T_1 = 285°, \quad \rho = 0.084.$$

A tube heated throughout at the temperature $T_1 = 285°$ gave $\rho = 0.085$.

THIRD CASE.—*The cold extremity of the tube is at a temperature less than* 270°; *the hot end is at a temperature higher than* 310°. Initially the tube does not contain hydroselenic acid; the selenium is collected at the hot end of the tube. At the beginning of the experiment the state of the system is represented by the straight line T_0T_1 (Fig. 133), of which one part CT_1 is in the region of one combination; certain points of this portion represent precisely the parts of the tube where the selenium is; in these parts

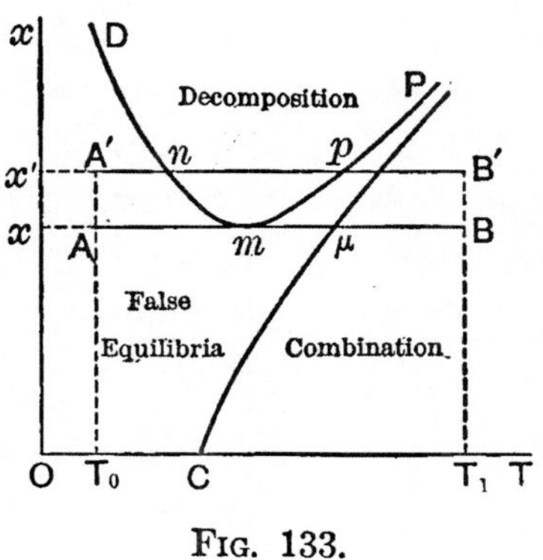

FIG. 133.

selenhydric acid is produced and diffuses into the part of the tube where it may be destroyed, no points of the representative line being in the region of decomposition; x increases, therefore, and the representative line rises to the position AB, where it touches the line DP at m.

Will it stop at this position?

The warm portions of the tube are in the region of combination and they still contain selenium; there will be formed, therefore, in these portions selenhydric acid, and x will increase, which will bring the representative straight line to $A'B'$; but a small part of this line np, composed of points whose temperature is about 270°, will be in the region of decomposition; in the corresponding portions of the system hydroselenic acid will decompose; selenium will be deposited, having been transported by *apparent volatilization* from the warm regions of the tube. During this time selenhydric acid will continue to be formed at the expense of the selenium contained in the warm parts of the tube. It is clear that, *in order for equilibrium to be established, it will be necessary and sufficient*—

1°. *That all the selenium be transported by apparent volatilization from the warm portion of the tube whose temperature is about* 270°.

2°. *That the repres ntative line occupy the position AB, that is to say, that the omposition of the gaseous mixture, independent of the extreme temperatures T_0 and T_1, is the same as in a system kept wholly at the temperature* 270°.

The carrying of selenium by apparent volatilization into the part of the tube whose temperature is about 270° may be masked by the real volatalization which carries selenium from the warm parts to the cold. Ditte and Pélabon devised ingenious experiments, but too long to describe here, to separate these two phenomena and to establish the first beyond dispute.

As to the second law, it is confirmed by the following experiments, due to Pélabon. The temperature T_0 was, in all the observations, the laboratory temperature; the temperature T_1 varied greatly without ρ changing sensibly:

$T_1=592°$,	$\rho=0.1986$ after 160 hours	
$T_1=680°$,	$\rho=0.2002$ " 162 "	
$T_1=700°$,	$\rho=0.1977$ " 69 "	

Furthermore, in tubes kept entirely at 270°, after heating periods of

192, 288, 480, and 490 hours,

the following values were found for ρ:

0.171 0.165 0.1605 0.163

302. Phenomena of apparent volatilization.—The formation, at a high temperature, of selenhydric acid from selenium and hydrogen and its destruction at a lower temperature produce a transportation of selenium by *apparent volatilization.*

This phenomenon is not isolated; in conditions similar in all points to the preceding, save the absolute values of the temperatures, which are here higher, Ditte has obtained a transportation of tellurium by apparent volatilization.

Troost and Hautefeuille [1] likewise discovered certain remarkable facts concerning apparent volatilization.

If a very slow current of silicon tetrachloride, $SiCl_4$, is passed over silicon at 1200° in a reverberating furnace, kept quite fixed at this temperature, it is noticed that after a certain time silicon has been carried by apparent volatilization into the moderately warm region of the tube, whose temperature is comprised between 500° and 800°. In reality, by the action of the silicon tetrachloride on silicon, there is formed silicon trichloride, Si_2Cl_6, volatile at these temperatures, which a sudden cooling would allow to be collected, but which, as we have seen, completely decomposes at temperatures included between 700° and 800°.

Silicon fluoride, $SiFl_4$, passing over silicon at white heat, transports the latter likewise, by apparent volatilization, into the parts of the tube at red heat; a sudden cooling allows collecting the subfluoride to which this carrying is due.

Chlorine, passing over platinum at 1400°, carries it, by apparent volatilization, into a cooler region of the tube; there is formed a protochloride of platinum which a sudden cooling allows to collect.

Hydrogen,[2] passing at red heat over zinc oxide, kept at this temperature, may transform it, with absorption of heat, into water vapor and zinc vapor; inversely, this last mixture passing into the cooler regions of the tube, is transformed anew, with liberation of heat, into hydrogen and zinc oxide which is deposited in the crystallized state.

---

[1] TROOST and HAUTEFEUILLE, *Comptes Rendus*, v. 73, pp. 443 and 563, 1871.

[2] H. SAINTE-CLAIRE DEVILLE, *Annales de Chimie et de Physique*, 3d S., v. 43, p. 477, 1855.

By an analogous phenomenon of *mineralization,*[1] hydrogen passing over amorphous zinc sulphide, displaces it and transforms it into hexagonal crystals of *würtzite* (hexagonal blende).

303. Vaporization presents phenomena analogous to those just studied.—The study of simple phenomena of vaporization furnishes examples of each of the three cases realized by Pélabon by means of selenhydric acid.

At the temperatures where ordinarily the vaporization of liquids, such as water and alcohol, is observed we may admit that the states of false equilibria are so close to the states of veritable equilibria that they are practically indistinguishable; whence, whatever are the temperatures of an enclosure containing a liquid and its vapor, the conditions in which the enclosure exists are those of the first case; the final tension of the vapor in the enclosure will be the same as if the enclosure were wholly brought to the temperaature of its coldest point; it is in the cold region that the liquid will be found wholly condensed; this proposition constitutes one of the forms of *Watt's principle.*

The condensation of phosphorus vapor to the state of white phosphorus conforms to this law; it is not the same with the condensation of phosphorus vapor to the state of red phosphorus; at the ordinary temperature saturated white phosphorus vapor has a tension greatly exceeding the tension of saturated red phosphorus vapor, the latter being practically zero; nevertheless the saturated vapor of white phosphorus remains indefinitely, *at least in the dark*, without changing into red phosphorus. This phenomenon of false equilibrium makes it possible to predict the realization of an enclosure unequally heated, filled with phosphorus vapor at a high pressure, and in which phosphorus would condense, in the state of red phosphorus, otherwhere than on the coldest walls of the enclosure; this experiment was realized by Troost and Hautefeuille;[2] it is the more convincing as the condensation of vapors to the state of white phosphorus, modification free from

---

[1] H. SAINTE-CLAIRE DEVILLE and TROOST, *Annales de Chimie et de Physique*, 4th S., v. 5, p. 118, 1865.

[2] TROOST and HAUTEFEUILLE, *Annales de l'Ecole normale supérieure*, 2d S., v. 2, p. 253, 1873.

false equilibrium, takes place, at the same time, on the cold walls of the tube.

This experiment is as follows:

White phosphorus is heated to about 500° in the middle portion of a glass tube whose two ends are kept the one at 350° (vapor of boiling mercury) and the other at 324° (vapor of mercury bromide). After an hour and a half, the part of the tube at 350° showed an orange-red coating, uniform and translucent, while the other end, at 324°, showed not the least trace of this, but instead a few drops of liquid white phosphorus.

In another series of observations Troost and Hautefeuille carried one of the ends to 445° (vapor of sulphur, boiling at atmospheric pressure) and the other end to 425° (vapor of sulphur boiling under 0.470 m.) of mercury; after fifteen or twenty minutes a good red coating was seen in the end at 445°, and at the most a yellow layer extremely thin in the end at 425°.

The transformation of the vapors of cyanic acid into solid cyamelide gives rise to analogous observations, likewise due to Troost and Hautefeuille.

While cyanic acid vapor is transformed into cyamelide after several hours at 250°, and after a few minutes at 350°, it resists for several days at the ordinary temperature.

If cyanic acid vapor be introduced into an enclosure a portion of which is at 350°, while the rest is maintained at 100°, this vapor is condensed into the state of cyamelide on the walls heated to 350°, and the tension of the vapor of cyanic acid has the value of 1200 mm., which is that of saturated vapor of cyamelide at 350°; the equilibrium which is established is the same as if the enclosure were wholly brought to the temperature of its hottest part.

CHAPTER XX.

CHEMICAL DYNAMICS AND EXPLOSIONS.

304. Chemical dynamics.—Up to this point we have been especially occupied with the conditions in which a chemical system exists in the state of equilibrium, whether it be a question of veritable equilibrium or of false equilibrium; we have treated of *chemical statics*.

When a system is not in chemical equilibrium, it is transformed; its true condition varies from one instant to the next; what laws govern these variations? To establish these laws is the object of *chemical dynamics*, the part of chemical mechanics much less advanced than statics.

We intend to indicate here, in a concise manner, some of the main ideas of this science; in order not to enter into complications of little use, we shall suppose in general, and except as noted to the contrary, that we are dealing with a *homogeneous system.*

305. Velocity of a reaction.—To be definite, let us suppose that the reaction produced in the system studied is a combination; at a certain instant t the system contains a mass m of the compound formed by this combination; this mass increases with the time, so that at an instant t', later than t, this mass has a value m', greater than m.

The ratio $\frac{m'-m}{t'-t}$ is what is called the *mean velocity of the combination* between the instants t and t'. If we suppose that a time t' nearer and nearer to t is taken, $(t'-t)$ approaches zero and similarly for $(m'-m)$; but the ratio $\frac{m'-m}{t'-t}$ approaches a limit which we shall denote by v:

$$\text{Lim}\,\frac{m'-m}{t'-t}=v,$$

which we shall call the *velocity of combination at the instant t.*

306. Fundamental principle of chemical dynamics.—The fundamental principle of chemical dynamics is the following:

The velocity of the combination which is produced at a given instant within a homogeneous system is determined when for this instant the nature and state of the substances forming the system considered, the temperature to which the system is brought, and the pressure it supports are known.

307. Acceleration of a reaction.—The above principle is widely different from the one which underlies the dynamics of local motions or dynamics properly so called; in order to indicate clearly this difference, we shall introduce a notion which will also be useful in what follows; it is the idea of *acceleration of a reaction.*

Let v be the velocity of reaction at the instant t, and v' the velocity at the instant t', later than t; the ratio $\frac{v'-v}{t'-t}$ is the *mean acceleration of the combination* between the instants t and t'; if we take as the instant t' a time nearer and nearer to t, $(t'-t)$ approaches 0; it is the same with $(v'-v)$; but the ratio $\frac{v'-v}{t'-t}$ approaches a positive or negative limit, which we shall denote by γ,

$$\text{Lim}\,\frac{v'-v}{t'-t}=\gamma,$$

and which we shall call the *acceleration of combination at the instant t.*

308. Comparison of the fundamental principle of chemical dynamics and the fundamental principle of dynamics properly so called.—Let us take now a material point M which moves along a straight line; let l be the distance which, at the instant t, this point has moved from a certain origin O; l' be the distance at the time t', later than t; we have that the *velocity of this point at the instant t* is the limit approached by the ratio $\frac{l'-l}{t'-t}$ as t' becomes nearer and nearer to t; once the velocity of this moving point is defined at the instant t, the *acceleration of the point at the instant t* is defined in exactly the same way as we have defined the acceleration of combination.

For the particular case of rectilinear motion to which we shall limit ourselves for greater simplicity, the fundamental principle of dynamics is the following:

If m is the mass of the moving point, and F is the component, taken along a straight line, of the force acting on this point, the acceleration is at every instant equal to the quotient $\frac{F}{m}$:

$$\gamma=\frac{F}{m}.$$

Now, in general, the force F is known when the nature of the mass m, its position, the nature and position of the substances which act on this mass, are known; we may therefore say the *acceleration is determined when the state and circumstances of the moving body are known.*

But if the acceleration is determined by the conditions which determine the force, it does not follow that the velocity is so determined; the same point, movable over the same straight line and undergoing the influence of the same bodies, may pass through the same position, in different circumstances, with different velocities.

Suppose, for example, that there is, on the straight line considered, a position of equilibrium for the point undergoing the action of the forces studied, that is to say, a position where the point, acted upon by these forces, would remain indefinitely in equilibrium if it were placed there with zero velocity.

It may very well happen that the movable point arrive in this position with a velocity different from zero; then it will not remain there; it will go beyond *by virtue of the velocity acquired.*

Nothing similar to this can occur in chemical dynamics; the system being placed in a given state and undergoing *definite actions*, the reaction of which it is the seat has a definite velocity; in particular, if the system is placed in a state which fulfils the equilibrium conditions, the velocity of reaction is necessarily equal to zero; if the system is brought in any way whatever into this state, it remains there in equilibrium; here *nothing can be observed analogous to an acquired velocity.*

We have supposed that we are concerned with a combination; but all we have just said may be applied to a decomposition, on the condition of denoting by m the mass of the compound which has been destroyed or, what amounts to the same thing, the mass of the elements this destruction has set free; all that has been said may likewise be applied to a double decomposition or to any other complicated chemical reaction, on the condition of denoting by m the mass of the substance or substances generated by this reaction.

309. Influence of the composition of the system on the velocity of reaction.—Is it possible to further specify the fundamental principle of chemical dynamics and formulate the law which, for a system, joins the velocity of combination to the conditions in which this system is placed? There may be stated, in a general and certain manner only, some very simple propositions.

Imagine, in the first place, the temperature T of the system to be kept constant; further, that either the pressure it supports or the volume in which it is contained is kept constant. In these conditions we may state the following proposition:

The velocity of reaction diminishes as the mass m of the substance or substances generated by this reaction increases.

Another proposition may be joined to this one:

If the value μ of the mass m corresponds to a state of equilibrium of the system, the velocity of combination approaches zero when the mass m approaches μ.

310. Every isothermal reaction is a moderated reaction.—This first law has a consequence which it is important to bring out.

Consider a reaction which is produced in the conditions supposed: on the one hand, the temperature is kept constant; on the other hand, either the volume or the pressure is kept constant. At the instant t, m is the mass of the compound formed, and v the velocity of reaction; at the instant t', later than t, these same quantities have the respective values m' and v'. The system having been the seat of a reaction in the direction considered between the instants t and t', the mass m' is necessarily greater than m, so that the velocity v' is less than v.

When the temperature T of the system is kept constant (isothermal reaction), and moreover if the volume it occupies or the pressure it supports is constant, the velocity of the reaction of which it is the seat decreases from one instant to the next.

We shall call *moderated reaction* one whose velocity diminishes from one instant to the next; we may then say that *every isothermal reaction accomplished either at constant pressure or at constant volume is a moderated reaction.*

311. The acceleration of a moderated reaction is negative.—Let v be the velocity of a reaction at the instant t; let v' be the velocity of the same reaction at an instant t', later than t; if the reaction is moderated, v' is less than v and the ratio $\frac{v'-v}{t'-t}$, mean acceleration of the reaction between the instants t and t', is negative; this is true however near t' is to t, which fact allows to say:

A reaction which is moderated at the instant t is a reaction whose acceleration is negative at this instant.

312. Influence of temperature on the velocity of reaction.—Let us consider a combination, which at the temperature of the experiment is unlimited and does not cease until the elements contained in the system are entirely combined; or else a limited combination, but whose limit, in the conditions of operating (either at constant volume or at constant pressure), corresponds to a value μ of m, value independent of the temperature.

In these conditions the following law may be stated:

Other things being equal, and in particular the composition of the system corresponding to the same value of m, the velocity v of combination is the greater as the temperature is higher.

This law is verified *a fortiori* if the value μ of m limiting the combination rises with the temperature. It may, on the contrary, become inexact if the value of μ decreases while the temperature T increases.

Thus, imagine that at temperatures T and T′, the latter being the higher, there correspond two limiting values μ and μ' of m, and that μ' is less than μ. At the temperature T a system for which m equals μ' is the seat of a certain reaction whose velocity is not zero; on the contrary, at the temperature T′ a system where m has the value μ' is in equilibrium and the reaction there has zero velocity.

313. Example: Phenomena of etherification.—A great number of examples illustrating the above law may be cited; the numerous investigations, qualitative and quantitative, which have been made on the velocities of reaction all confirm this law; we shall mention but one particularly typical example.

The etherification of an alcohol by an acid, in a closed vessel, hence at constant volume, reaches a limit which is a veritable equilibrium state; this limit corresponds (Art. 179) to a proportion of ether formed which is independent of the temperature, so that the example considered is within the conditions for which our law was stated; according to Berthelot[1] the velocity of this reaction is 22,000 times greater at + 200° C. than in the neighborhood of + 7° C.

314. Variation of the velocity due to a small change of composition and temperature.—Let v be the velocity of the reaction considered, when the mass of the substance or of the collection of substances to which reaction gives rise has the value m and when the temperature has the value T. If the mass m underwent alone a small variation $(m'-m)$, the velocity would undergo an increase $A\ (m'-m)$, A being a coefficient whose value depends on the mass m, the temperature T, and the other conditions in which the system is; an increase of the mass m always produces, other things being equal, a decrease in the velocity and conversely; it follows that *the coefficient A is always negative.*

If the temperature T underwent alone a small variation $(T'-\mathrm{T})$, the velocity would undergo an increase $B\ (T'-T)$, where B is a coefficient depending on the mass m, the temperature T, and the other conditions of the system; we know that, if the reaction is unlimited, or if the value μ of m which limits the reaction does not decrease as the temperature increases, a rise of temperature produces, other things being equal, an increase in the velocity of reaction; in these conditions *the coefficient B is positive.*

Suppose in particular that the reaction studied obeys, as it so often happens, those laws (Arts. 287 to 292) whose existence was illustrated by the reaction

$$H_2+S=H_2S$$

[1] BERTHELOT, *Essai de Mécanique chimique fondée sur la Thermochimie,* v. 2, p. 93.

or by the reaction

$$3SiCl_4 + Si = 2Si_2Cl_6,$$

as was shown in the preceding chapter; as long as the temperature is less than that we have denoted by θ, and starting from which begins the *region of veritable equilibria,* the reaction which may be produced in the system is either limited by a value of μ which increases with the temperature (temperatures less than τ) or else practically unlimited (temperatures included between r and θ). Therefore, *for reactions which follow the laws of which the formation of hydrogen sulphide is the type, the coefficient B is surely positive at temperatures below the point θ.*

If simultaneously an increase of $(m'-m)$ is imposed upon the mass m and $(T'-T)$ upon the temperature T, the velocity v undergoes the increase

$$(1) \qquad v'-v=A(m'-m)+B(T'-T),$$

the sum of the two partial increments of which we have just spoken.

315. Return to isothermal reactions.—Suppose first the temperature kept constant or the *reaction isothermal;* suppose that T, m, v refer to the instant t, and T′, m', v' to the temperature t', later than t and very close to t; we shall have

$$T'-T=0.$$

Then by definition

$$m'-m=v(t'-t)$$

and

$$v'-v=\gamma(t'-t).$$

Equation (1) then becomes

$$(2) \qquad \gamma=Av.$$

We have seen that A is always negative; it is the same with γ, as we have already found.

316. Adiabatic reactions.—Suppose now that while the reaction lasts there is no interchange of heat between the system and surrounding bodies; *the reaction is then adiabatic;* this supposition is approximately verified for very rapid reactions; the temperature T varies from one instant to another in the same time as the mass m; it is easy to determine the law of this variation.

Between neighboring instants t and t' there is formed in the system a mass $(m'-m)=v(t'-t)$ of the substance or collection of substances produced by the reaction; if this formation takes place at constant temperature, it will be accompanied by a *liberation* of heat.

$$L(m'-m)=Lv(t'-t),$$

L being the heat called out by the reaction which, for the condition in which the system is placed, would transform one gramme of matter.

Furthermore, the temperature of the system changes from T to T′; if this modification were produced alone it would absorb a quantity of heat $C(T'-T)$, where C is the *total heat capacity of the system for the conditions in which it is placed.*

The total heat set free by the system has the value

$$Lv(t'-t)-C(T'-T).$$

This liberation of heat being zero for an adiabatic transformation, we have

$$T'-T=\frac{L}{C}v(t'-t). \tag{3}$$

Substituting this value of $T'-T$ in equation (1), and noting also that by definition

$$m'-m=v(t'-t), \qquad v'-v=\gamma(t'-t),$$

we find

$$\gamma=\left(A+\frac{BL}{C}\right)v. \tag{4}$$

The ratio of the acceleration to the velocity has not the same value in an adiabatic reaction as in an isothermal reaction.

317. An adiabatic reaction may have a positive acceleration.—It may even happen that the acceleration of the reaction, necessarily negative for an isothermal reaction, is positive for an adiabatic reaction.

Let us consider in particular a reaction following the law of which the formation of hydrogen sulphide, studied in the last chapter, is the type, and let us suppose the temperature is less than the temperature θ, above which veritable equilibrium states may be observed.

From what has been said previously, B is then positive; as we have already indicated in the preceding chapter (Art. 293), the only reaction which can be observed is exothermic, so that L is positive; C is also positive, hence the ratio $\frac{BL}{C}$ is assuredly positive. The ratio $\frac{\gamma}{v}$ has, for an adiabatic modification, a value $\left(A+\frac{BL}{C}\right)$, which is surely greater than the value A of the same ratio for an isothermal transformation; this latter is necessarily negative, but it may happen that the first is positive.

318. Reactions with positive acceleration and explosive reactions.—What are the properties of this reaction if $\left(A+\frac{BL}{C}\right)$ is positive?

The acceleration being positive, the velocity of the reaction will increase with the time; at the same time, as shown in equation (3), where L and C are positive, the temperature will rise from one instant to the next.

It may be said that these are the two characteristics by which *explosive reactions* are recognized; or rather, chemists generally consider explosive a reaction which has these two characteristics in a very high degree; but to give the expression *explosive reaction* a clear definition, we shall henceforth agree that it denotes a *reaction whose acceleration is positive and which is accompanied by an elevation of temperature in the system.*

We may then state the following proposition:

While an isothermal reaction is forcibly a moderated reaction, an adiabatic reaction may be explosive.

319. Condition in order that an adiabatic reaction be explosive.—In particular, *a reaction of the type*

$$H_2+S=H_2S,$$

adiabatic and produced at a temperature lower than the point θ, where states of veritable equilibria begin to manifest themselves, is explosive if the essentially positive ratio $\frac{BI}{C'}$ is greater than the absolute value of the negative coefficient A.

320. Indetermination of the temperature which renders a reaction explosive.—It is evident from the above that, in order to decide if a reaction is or is not explosive, it does not suffice to indicate the composition of the system, the temperature and the pressure (or volume); all these conditions remaining the same, it may happen that the same reaction is or is not explosive according to the law which rules the variations of temperature; a reaction, explosive if the system is enclosed in a vessel impermeable to heat, becomes moderated if it is rendered isothermal.

This explains why various observers, operating by different methods, may give very different indications concerning the conditions in which a reaction becomes explosive; Mitscherlich found that the combination of oxygen and hydrogen became explosive at 674°; Mallard and Le Chatelier have indicated for this phenomenon the temperature of about 550°; A. Gautier and H. Hélier, by heating the mixture of oxygen and hydrogen in a porcelain vessel filled with porcelain chips which increased the heating surface and rendered the reaction almost isothermal, were able to retard until 845° the temperature at which the formation of water becomes explosive.

In the future, when we speak of the conditions in which a reaction becomes explosive, we shall always suppose the system placed in an enclosure impermeable to heat, so that the reaction is adiabatic.

321. Stability and instability of limiting false equilibria.—Related to the questions we have just examined is the study of the *stability of limiting false equilibria*, as we shall see. This question may be treated in a very general manner; but in view of the applications we wish to make of it, it will be sufficient to discuss a reaction of the type

$$H_2+S=H_2S.$$

To be definite, we shall suppose there is question of a *combination*. The system will be, for example, treated at constant pressure.

Take the temperature T as abscissæ (Fig. 134), and for ordinates the ratio x between the mass m of the compound contained by the system and the mass m of the same compound which it would contain if the compound were pushed as far as possible. From this definition

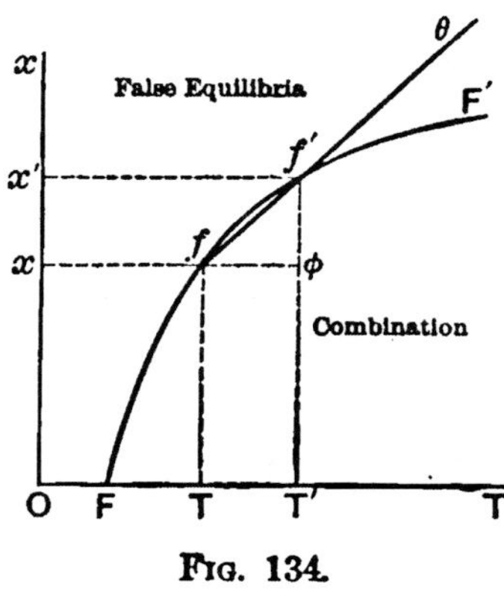

FIG. 134.

$$m=Mx.$$

The region of false equilibria is separated from that of combination by a line FF', which rises from left to right; the various points of this line represent the limiting states of false equilibrium of the system.

Let f and f' be two points close together taken on the line FF'; T and x are the coordinates of the first, T' and x' of the second. The velocity v of the combination approaches zero while the system approaches the state represented by the point f; it also approaches zero when the system approaches f'; the difference $(v'-v)$ relative to these two states is therefore equal to zero, which may be written, in accordance with equation (1),

$$A(m'-m)+B(T'-T)=0,$$

or, since

$$m=Mx, \quad m'=Mx',$$

$$AM(x'-x)+B(T'-T)=0.$$

Now in the right triangle $f\phi f'$ we have

$$f\phi=T'-T, \qquad \phi f'=x'-x;$$

consequently

$$\tan f'f\phi=\frac{x'-x}{T'-T}.$$

By the preceding equation this becomes

$$\tan f'f\phi = -\frac{B}{AM}. \tag{5}$$

But the point f' being very near to f, the line ff' becomes identical with the straight line $f\theta$ which touches at f the line FF', and the tangent $f'f\phi$ is what we have called (Art. 146) *trigonometrical tangent at f of the line* FF'; we see that the *trigonometrical tangent has the value* $-\frac{B}{AM}$. This result will be useful to us shortly.

322. Every state of false equilibrium which is not limiting is indifferent.—Let us take a case of false equilibrium and ask ourselves if this state of false equilibrium is stable, indifferent, or unstable.

Suppose in the first place that the state of false equilibrium considered is not a case of limiting false equilibrium; the point representing it is in the interior of the region of false equilibria and not on the limiting line; give the system a slight change, corresponding to small variations of T and x; we may always take these variations small enough so that the state of the deranged system is still represented by a point in the region of false equilibria, case in which the deranged system will be still in equilibrium; we may therefore state the proposition:

Every state of false equilibrium which is not a state of limiting false equilibrium is a state of indifferent equilibrium.

323. If the temperature is constant, every false equilibrium is stable.—Take now a limiting state of false equlibrium, represented by a point f on the line FF', and impose a slight change upon the system; it may happen that this change causes the point representing the state of the system to penetrate into the interior of the region of false equilibria; for such a derangement the system is surely in indifferent equilibrium and we have no need to concern ourselves further with such a change; it may happen, on the contrary, that this change brings the point representing the system to g (Fig. 135) in the region of combination, and this is the case we are going to discuss.

Suppose, in the first place, that after having slightly deranged the system, it is placed in such conditions that it can no longer undergo other than isothermal changes; the point g representing the state of the system being in the region of combination, x will increase without T varying; the representative point g will rise along a parallel to Ox through the point g, and, as this last point is assuredly beneath the line FF', the point representing the state of the system will approach this line; the reaction will tend to bring back the system to the equilibrium state. *A state of limiting false equilibrium is a state of stable equilibrium for a system, which once deranged can no longer undergo other than isothermal transformations.*

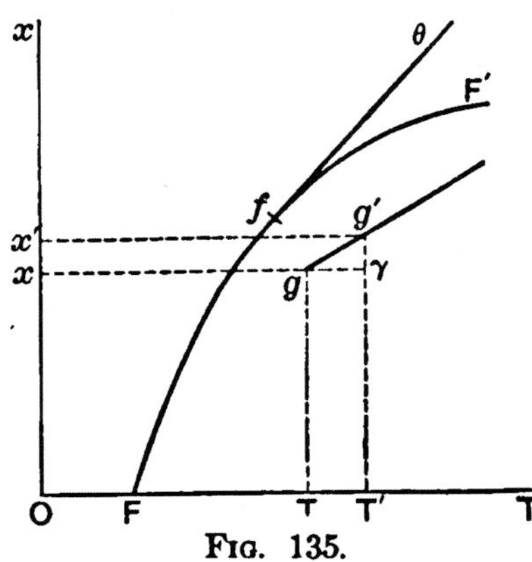

Fig. 135.

324. **If the reactions are all adiabatic, the limiting false equilibria may be stable or unstable.**—Suppose, now, that the system once deranged can undergo only adiabatic transformations. The representative point is at g at the instant t; let g' be the position it occupies at the instant t', near to t but later than t: let T and x be the coordinates of the point g, and T', x' the coordinates of the point g'.

We shall have $$m'-m=M(x'-x),$$

and as $$m'-m=v(t'-t),$$

we shall have $$x'-x=\frac{v}{M}(t'-t).$$

Further, the transformation being adiabatic, $(T'-T)$ is given by equation (3):

$$T'-T=\frac{L}{C}v(t'-t). \qquad (3)$$

In the triangle $g\gamma g'$ we have

$$g\gamma=T'-T, \quad \gamma g'=x'-x,$$

consequently

$$\tan g'g\gamma = \tan\,(gg', OT) = \frac{x'-x}{T'-T}.$$

We may therefore write the following equality:

$$\tan\,(gg', OT) = \frac{C}{LM}.$$

Let $f\theta$ be the line tangent at f to the curve ff'; we have seen that

(5) $$\tan\,(f\theta, OT) = -\frac{B}{AM}.$$

These two results allow us to discuss the stability or instability of one state of equilibrium.

According to equation (3), where L, C, v, $(t'-t)$ are positive quantities, $(T'-T)$ is positive; the point g' is to the right of the point g on the line whose tangent is given by equation (5); there are then two principal cases to distinguish:

1°. We have the equation

(7) $$\frac{C}{LM} > -\frac{B}{AM}.$$

The line gg' rises from left to right more rapidly than the line $f\theta$; the representative point of the system which, on account of the reaction, moves from left to right on the line gg', approaches the line $f\theta$, tangent to the line FF'; and since over a short distance a line may be considered identical with its tangent, the point representing the system approaches the limiting line of the false equilibria; the system tends to resume a state of equilibrium.

A system compelled, after being deranged, to undergo only adiabatic modifications is in stable equilibrium in a state of limiting false equilibrium if condition (7) *is verified.*

2°. We have the equation

(8) $$\frac{C}{LM} < -\frac{B}{AM}.$$

By similar reasoning, it is found that the adiabatic reaction has the same effect of separating farther the representative point from the line FF'.

A system compelled, after being deranged, to undergo only adiabatic changes is in unstable equilibrium in a state of limiting false equilibrium if condition (8) *is verified.*

The line FF' rises from left to right (Fig. 136), rapidly at first,

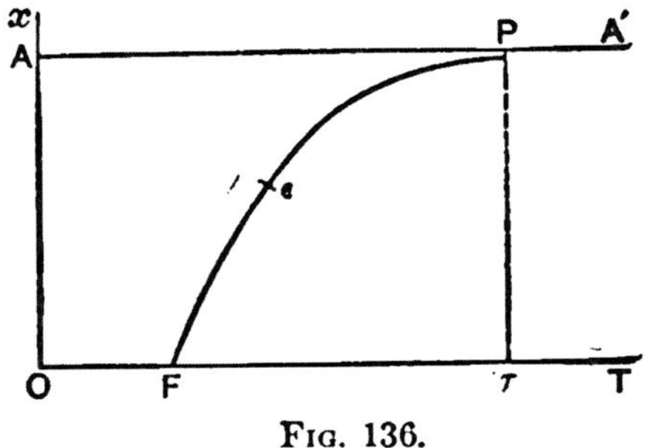

Fig. 136.

then more and more gradually; at the temperature which in the preceding chapter (Art. 288) we have denoted by τ it becomes sensibly tangent at P to the straight line AA', parallel to OT, and whose constant ordinate is equal to 1. At this moment its tangent is very nearly 0; therefore for limiting false equilibria relative to temperatures close to τ but less than τ, $-\frac{B}{AM}$ has very small positive values.

On the other hand, there is no reason why the positive ratio $\frac{C}{LM}$ should assume, in the neighborhood of the states of false equilibrium in question, a very small value; the inequality (7) will therefore be verified for these states of false equilibrium; *the states of limiting false equilibrium which are sufficiently near to the point P are assuredly stable even for a system enclosed in a covering impermeable to heat.*

Is it the same for all the limiting states of false equilibrium, represented by the various points of the line FF'? It may be that this is so; it may be, on the contrary, that there exists on the line FF' a point ϵ where we would have the equation

$$\frac{C}{LM} = -\frac{B}{AM}. \tag{9}$$

In this case the limiting false equilibria represented by the various points of the line εP would be stable in a system maintained in an enclosure impermeable to heat, while, in the same circumstances, the limiting false equilibria represented by the various points of the line $F\varepsilon$ would be unstable.

325. Relation between the limiting false equilibria which are unstable and the explosive reactions.—The letter A which figures in equations (7) and (8) represents a negative quantity; B, on the contrary, for the reactions we studied, is positive; as to the letters M, C, L, they represent quantities essentially positive; it is then evident that equation (7) may be written

$$A+\frac{BL}{C}<0, \tag{7'}$$

while equation (8) may be expressed as

$$A+\frac{BL}{C}>0. \tag{8'}$$

We are therefore led to the following conclusion:

A system is kept in an enclosure impermeable to heat; every point situated in the region of combination and near to a part of the line FF′ which represents stable limiting states figures a state where the system is the seat of a moderated combination; every point situated in the region of combination and near to a point of the line FF′ which represents unstable states figures a state where the system is the seat of an explosive combination.

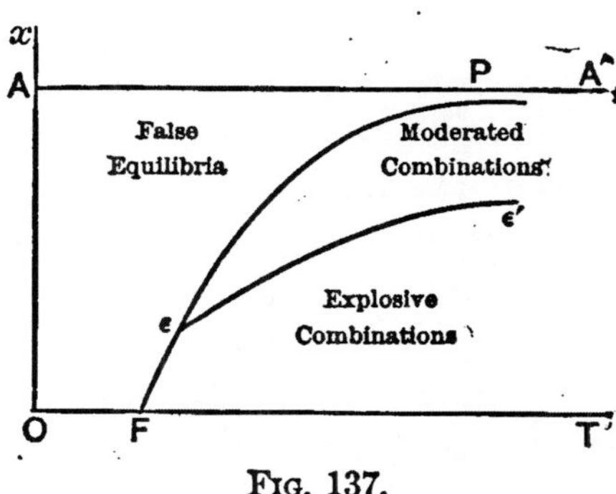

Fig. 137.

326. Three cases to distinguish.—Consider in the first place the case where, on the line FF' (Fig. 137), there exists a point ε separating the limiting stable states, represented by the various points of ϵP, from the limiting unstable states represented by the various points of $F\epsilon$. In this case there exists surely a line $\epsilon\epsilon'$, starting from the point ϵ, and dividing the region of combination into two subregions. Every point of the subregion located above $\epsilon\epsilon'$ represents a state where the system, enclosed in an envelope impermeable to heat, is the seat of a moderated combination; every point of the subregion situated below $\epsilon\epsilon'$ represents a state where, in the same circumstances, the system is the seat of an explosive combination.

If, on the contrary, the line FF' has no point such as ϵ, all the states of this line represent states of limiting equilibrium which are stable if the system is in an enclosure impermeable to heat; this line is confined, therefore, throughout its length to the region of moderated combination.

There are then two cases to distinguish:

In the first case (Fig. 138) there exists a line $\eta\eta'$ which divides

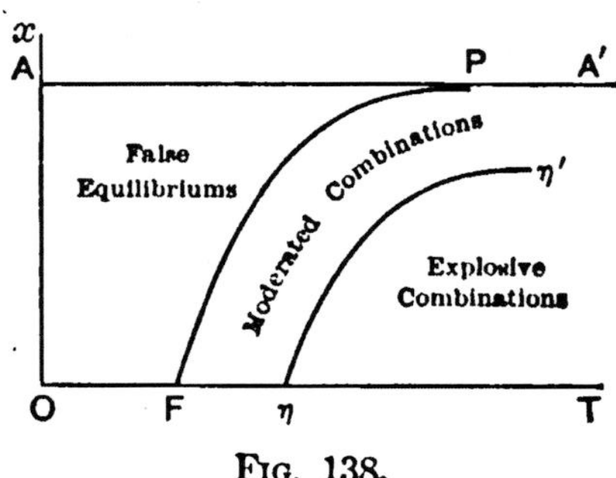

Fig. 138.

the region of combination into two subregions: a subregion of moderated combination situated between $\eta\eta'$ and FP, and a subregion of explosive combination situated below $\eta\eta'$.

In the second case every combination produced within the systems enclosed in a vessel impermeable to heat is a moderated combination.

Return to the case represented in Fig. 137.

Take in this case a system which contains the elements suitable to form a compound, but which contain no trace of the compound; x is equal to 0, and the representative point of the state of the system lies on the straight line OT.

As long as the temperature is less than OF the system is in the state of false equilibrium; at the moment the temperature attains the value OF, which is the *reaction-point* of the system, a combination is produced; *if the system is in an enclosure impermeable to heat, this combination is explosive.*

327. The reaction-point of a mixture is in general below the explosion-point.—We cannot cite with certainty any homogeneous system having these properties; for all, the temperature of the reaction-point is well below that for which the reaction may become explosive; thus, according to A. Gautier and Hélier, the reaction-point of a mixture of oxygen and hydrogen does not exceed 180°, while the explosive formation of water vapor has never been observed at less than 500°. Further, it has been known a long time[1] that the mixtures of formene and oxygen, carbon bisulphide and oxygen, chlorine and hydrogen, at temperatures included between 350° and 500°, unite slowly without explosion; they are therefore as many mixtures for which the point ε does not exist.

These various mixtures therefore form systems for which it is proper to employ Fig. 138 or, what amounts to the same thing, Fig. 139.

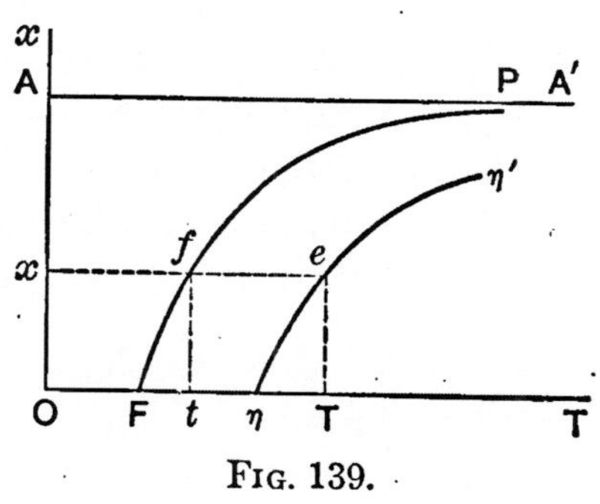

Fig. 139.

Given the composition x of such a system; draw a parallel to OT whose constant ordinate is $Ox=x$; this straight line intersects the line FP in a point f, of abscissa t, and the line $\eta\eta'$ in a point e of abscissa T.

At temperatures less than t the mixture of composition x is in the state of false equilibrium.

At temperatures above t, *reaction-point* of the mixture, a combination is produced. If the temperature is comprised between t and T, the combination is moderated, even if the system is enclosed in a covering impermeable to heat. At temperatures above T the combination is explosive, provided the system is enclosed in a

[1] A. Gautier, *Bull. de la Soc. chimique de Paris*, v. 13, p. 1, 1869; J. H. Van't Hoff and V. Meyer later confirmed these observations.

non-conducting envelope; the temperature T may be called the *explosive temperature* of the mixture which contains a proportion x of the compound substance; the line $\eta\eta'$ is the *line of temperatures of explosion* of the system taken in the given conditions, either volume or pressure constant.

328. The interval between these two points and safety explosives.—The explosive temperature T of a mixture exceeds the reaction-point t of the same mixture; the interval between these two temperatures may be very great; this is true, according to Mallard and Le Chatelier,[1] for mixtures of oxygen and methane; also it is possible in coal-mines to employ explosives which produce a temperature higher, it is true, than the reaction-point of these mixtures, but below their explosive temperature; these explosives cannot cause fire-damp mixtures to detonate.

329. Mixtures which are never detonating.—It may happen that the system studied does not admit a line of temperatures of explosion, and that the combination in it is always moderated, even in an envelope impermeable to heat; this is, for instance, the case for the combination of hydrogen and sulphur, of which we have spoken at length in the preceding chapter (Art. 287).

330. Explosive combinations.—All we have just said on the subject of a system in which an exothermic compound is formed, according to the laws of which the reaction

$$H_2+S=H_2S$$

has given us an example, may be repeated *mutatis mutandis* regarding a system in which an endothermic system is destroyed, according to the laws of which the reaction

$$2Si_2Cl_6=3SiCl_4+Si$$

has furnished us the type (Chap. XVIII, Art. 292).

Here again all the cases studied experimentally seem to range themselves in two classes.

In certain systems the destruction of the compound substance is always a moderated reaction, even when this reaction is adiabatic.

---

[1] *Commission des Substances explosives: Sous-commission spéciale* (E. Mallard, rapporteur) (*Annales des Mines*, 8th S., v. 14, p. 197, 1888).

In others, on the contrary, there exists a line $\eta\eta'$ (Fig. 140) of explosive temperatures which divides the region of decomposition into two regions: .

1°. A region of moderated decomposition having the form of a band more or less wide between the lines $\eta\eta'$ aad FP;

2°. A region of explosive decomposition, situated above $\eta\eta'$.

That such a disposition is indeed that proper to the greater number of explosive substances is what chemists accustomed to their manipulation have found. "*B*elow the temperature at which they become explosive," says *B*erthelot,[1] "and during an interval

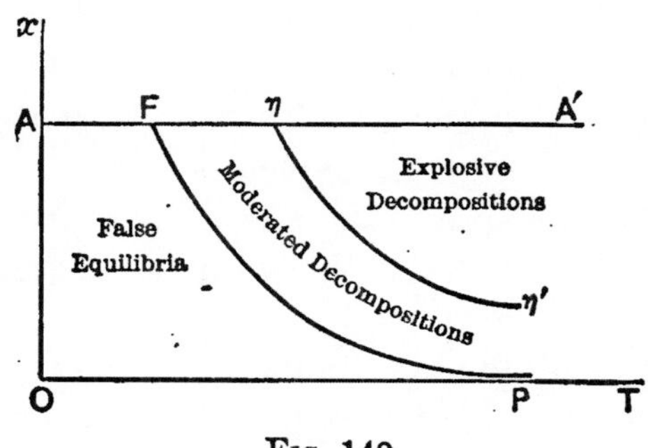

FIG. 140.

of temperature more or less extended, all the exothermic decompositions must be produced in a progressive manner." "*C*ertain explosive matters are sometimes decomposed very slowly, starting from room temperature, and produce detonations only if the temperature is raised intentionally or by accident." [2]

331. Influence of pressure on the point of explosion.—All we have said, all the figures we have drawn, suppose the system heated either at constant pressure or at constant volume; to fix our attention, suppose it is question of a system heated at constant pressure; the form and position of the lines $\eta\eta'$, FP depend on the value of this constant pressure and change with this value; in particular it may happen that the line $\eta\eta'$ exists when the value of the pressure is taken within a certain interval, and no longer exists when it is taken within another interval; under all the pres-

[1] BERTHELOT, *Essai de mécanique chimique, fondée sur la Thermodynamique*, v. 2, p. 66.

[2] BERTHELOT, *Sur la force des matières explosives*, v. 2, p. 71.

sures of the first interval, the system may be the seat of an explosive reaction; on the contrary, under pressures belonging to the second interval there will never occur other than a moderated reaction.

Within ozonized oxygen, under ordinary pressure, the ozone undergoes moderated decomposition from the temperature of the laboratory; but at no temperature does this decomposition become explosive; when, on the contrary, ozonized oxygen is strongly compressed, the ozone may decompose with explosion.[1]

Hydrogen arsenide is an endothermic compound which is slowly destroyed at room temperature, and which under atmospheric pressure does not become detonating at any temperature, not even that developed by the electric spark; if one detonates in hydrogen arsenide a fulminate-of-mercury cap, the gas undergoes not only a great rise in temperature, but also an energetic compression, and it decomposes with explosion.[2]

This influence of pressure on the possibility of explosion has been clearly shown by Berthelot and Vielle[3] in their studies on acetylene.

Acetylene is an endothermic compound which, according to the principle of the displacement of equilibrium by variation of temperature, is formed directly at the very high temperature of the electric arc (Art. 177).

Liquid acetylene is an explosive compound whose effects are of the order of those produced by guncotton.

The same is not true of acetylene taken in the gaseous state under atmospheric pressure; in these conditions this gas does not detonate either by the action of a red-hot platinum wire or by an electric spark, not even by the explosion of a fulminating mercury cartridge.

Under a pressure of two atmospheres, on the contrary, acetylene gas behaves like an explosive compound; it decomposes with

---

[1] HAUTEFEUILLE and J. CHAPPUIS, *Comptes Rendus*, v. 91, p. 522, 1880.

[2] BERTHELOT, *Sur la force des matières explosives*, v. 1, p. 114.

[3] BERTHELOT and VIELLE, *Annales de chimie et de physique*, 7th S., v. 11, 1897. There is a good résumé of the explosive properties of acetylene by L. MARCHIS, *Leçons sur les machines thermiques: Moteurs à gaz et à pétrole*, given at Bordeaux University in 1899–1900.

detonation under the action of the various agencies we have just enumerated.

Berthelot and Vielle have shown that, in order for acetylene to explode in contact with a platinum wire brought to incandescence, it was necessary to submit the gas to an initial pressure measured by 137 centimetres of mercury. But acetylene detonates by the explosion of a cartridge containing 0.1 gr. of mercury fulminate, as soon as the initial pressure is measured by 100 centimetres of mercury.

We shall cease here these observations on chemical dynamics and explosions.

While the chemical statics of veritable equilibria is already so well advanced that it furnishes a great number of exact theorems, and finds numerous, varied and exact, experimental confirmations, the chemical statics of false equilibria and chemical dynamics are still in an undeveloped state. Already, however, they allow of classifying the greater number of the reactions observed when the temperature and pressure in a chemical system are varied; until recently, this study had remained a veritable chaos.

LIST OF AUTHORS CITED IN THIS WOR

INDEX OF CHEMICAL SUBSTANCES

STUDIED IN THIS WORK

Numbers refer to pages.

GENERAL INDEX.

Numbers refer to pages.